Reservoir Engineering

The Fundamentals, Simulation, and Management of Conventional and Unconventional Recoveries

Reservoir Engineering

The Fundamentals, Simulation, and Management of Conventional and Unconventional Recoveries

Abdus Satter

Ghulam M. Iqbal

AMSTERDAM • BOSTON • HEIDELBERG • LONDON
NEW YORK • OXFORD • PARIS • SAN DIEGO
SAN FRANCISCO • SINGAPORE • SYDNEY • TOKYO
Gulf Professional Publishing is an imprint of Elsevier

G | P
P | ⍦

Gulf Professional Publishing is an imprint of Elsevier
225 Wyman Street, Waltham, MA 02451, USA
The Boulevard, Langford Lane, Kidlington, Oxford, OX5 1GB, UK

Notices
Knowledge and best practice in this field are constantly changing. As new research and experience broaden
our understanding, changes in research methods, professional practices, or medical treatment may become
necessary.

Practitioners and researchers must always rely on their own experience and knowledge in evaluating and
using any information, methods, compounds, or experiments described herein. In using such information
or methods they should be mindful of their own safety and the safety of others, including parties for whom
they have a professional responsibility.

To the fullest extent of the law, neither the Publisher nor the authors, contributors, or editors, assume any
liability for any injury and/or damage to persons or property as a matter of products liability, negligence or
otherwise, or from any use or operation of any methods, products, instructions, or ideas contained in the
material herein.

Library of Congress Cataloging-in-Publication Data
A catalog record for this book is available from the Library of Congress

British Library Cataloguing-in-Publication Data
A catalogue record for this book is available from the British Library

ISBN: 978-0-12-800219-3

For information on all Gulf Professional Publishing publications
visit our website at http://store.elsevier.com/

Working together
to grow libraries in
developing countries

www.elsevier.com • www.bookaid.org

Dedication

The authors would like to dedicate this book to their parents, who motivated them when they were young, and continue to motivate them unto this day after they are long gone.

Contents

Acknowledgment

The authors would like to acknowledge the valuable contributions made by Barclay Macaul, Reyaz Siddiqui, Kiran Venepalli, and Raya Iqbal in making this book a reality.

An introduction to reservoir engineering: Advances in conventional and unconventional recoveries

1

Introduction

Reservoir engineering, a core discipline of petroleum engineering, involves the efficient management of oil and gas reservoirs in a technical and economic sense. It evolved as a separate discipline in the first part of the twentieth century in order to maximize the production of oil and gas. Reservoir engineering teams set up a comprehensive plan to produce oil and gas based on reservoir modeling and economic analysis, which implements a development plan, conducts reservoir surveillance on a continuous basis, evaluates reservoir performance, and implements corrective actions as necessary. Reservoir engineering is dynamic and poses unique challenges, as new frontiers and resources in oil and gas are discovered across the world. Reservoir engineers are expected to come up with innovative technologies and novel strategies to extract oil and gas in the most efficient, safe, and economic way possible.

Modern reservoir engineering studies, projects, and practices are based on teamwork and an integrated approach. Geology, geophysics, geochemistry, petrophysics, drilling, production, computer-based simulation, and other areas of science and engineering come together to make it all happen. Regulatory, economic, and environmental aspects are included as well. Reservoir-related studies and efforts come to fruition in the form of reservoir engineering projects that optimize oil and gas production and maximize the economic value of the reservoir.

This book focuses on the fundamental concepts of reservoir engineering and how these concepts are applied in the oil and gas industry to meet technical challenges. Field case studies, highlighting the applications of reservoir engineering and simulation in both conventional and unconventional reservoirs, are presented. In essence, the book strives to prepare students for the job from day one, and provides professionals with valuable information regarding present-day tools, techniques, and technologies.

Advances in reservoir technologies

In the early twentieth century, production of petroleum was mostly based on onshore fields that were relatively easy to manage. Nevertheless, the ultimate recovery from the fields was less than satisfactory, with large portions of oil left in the ground. Reservoir engineering advanced rapidly in recent decades to meet the challenges posed by

Reservoir Engineering. http://dx.doi.org/10.1016/B978-0-12-800219-3.00001-2

the new discoveries of oil and gas. Some of the state-of-the-art tools and technologies include the following:

- Horizontal drilling up to several miles underground, having one or more lateral branches
- Multistage hydraulic fracturing that facilitates production from shale – until recently this was thought to be impossible
- Fluid injection into reservoirs with complex geology to recover oil efficiently
- Thermal treatment of immobile oil sands
- Seismic monitoring of fine fractures and fluid fronts
- Simulation of robust reservoir models that are utilized to optimize the recovery of oil and gas

Wells are being drilled to produce oil economically in many geologic settings that were not accessible before, including deep-sea reservoirs, ultratight formations, and matured fields where large amounts of oil were previously left behind. As technology forges ahead, oil and gas are recovered in significant quantities from reservoirs that were not considered to be reservoirs at all only a few decades ago.

Some of the recent advances in reservoir engineering and related technologies are outlined in the following:

- *Horizontal wells*: Horizontal drilling is a game-changing technology that enables the effective development of many reservoirs in adverse geologic settings, onshore and offshore. Some horizontal wells are drilled as long as 7 miles in the lateral direction. The wells drill through oil and gas-bearing formations across various heterogeneities such as faults and compartments, which was not possible with vertical or deviated wells. Due to the large exposure in the formation, commercial production from very tight formations is possible. This holds the key to the development of certain unconventional reservoirs. As a horizontal well is drilled, detailed rock properties are obtained over the entire length of the drilled portion of the formation by employing measurement while drilling techniques. The wells have a smaller footprint on the ground as one horizontal well may replace the need to drill several vertical wells to produce the same amount of oil or gas.
- *Multistage fracturing*: Hydraulic fracturing technology, sometimes referred to as fracking, has revolutionized shale gas production. Unconventional shale gas and oil reservoirs are continuous over hundreds of miles. The volume of petroleum in place is substantial and the probability of finding the deposits are much higher than that of conventional drilling. However, the reservoirs are ultratight and were thought to be nonproducible in commercial quantities only a decade ago. Multistage fracturing of horizontal wells drilled in the ultratight organic-rich shale changed all that. A horizontal well is hydraulically fractured every few hundred feet to create a fracture network that combines with any natural fractures present and facilitates production from the semipermeable formation. The technology has changed the energy landscape in the United States, and the reverberations of multistage fracturing are felt across the world. In a related development, microseismic studies have enabled the visualization and characterization of the fine fractures created by multistage fracturing.
- *Extraction of oil sands*: Heavy and extra heavy oil were considered to be hardly producible in large quantities only a few decades ago. Drilling of horizontal wells along with steam injection ushered in a new era of extraction of oil sands, also referred to as tar sands or bitumen. A widely recognized technique comprises drilling dual horizontal wells in the formation that are vertically apart by a short distance, injecting steam through the upper well, and producing relatively light hydrocarbons from the lower well. The technology is referred to as steam-assisted gravity drive, as heated oil with reduced viscosity is moved toward the

producer by the force of gravity. Advancements in oil refining technology have enabled the upgrading of the produced hydrocarbons to marketable standards.

- *Reservoir simulation and integrated studies*: Reservoir development projects generally require substantial capital investment. With the advent of the digital age, virtually all major decisions in reservoir development are based on reservoir simulation. It utilizes mathematical models to replicate the real-world processes and events that take place in the petroleum reservoir. Robust models can be built upon more than a million cells and multiple realizations of the reservoir. What-if scenarios are generated within relatively short periods of time, projecting the range of performance that can be expected from a reservoir under various development schemes and options. Integrated reservoir studies are based on information obtained from various disciplines of earth sciences and engineering, which brings oil and gas industry professionals together to work as a team.

Classification of petroleum reservoirs

Reservoir engineering deals with petroleum reservoirs that may be classified in different ways. The categorization goes a long way in determining how the development and management of a reservoir can be strategized. The major classification of reservoirs include in the following.

Type of petroleum fluid:

- Oil (light, intermediate, heavy, and ultraheavy, including bitumen)
- Dry gas (gas remains dry throughout production without any dropout of hydrocarbon components)
- Gas condensate (gas containing relatively heavier hydrocarbons that may condense out as reservoir pressure declines below the dew point)

Technology:

- Conventional – reservoirs that are developed and produced by traditional tools and techniques; rock and fluid characteristics are favorable for production on a commercial scale
- Unconventional – reservoirs that require innovative approaches and emerging technologies to develop economically due to unfavorable conditions; unconventional reservoirs are characterized by ultratight formation, extra heavy oil, or location of the reservoir at great depths, among others

As the technology to produce an unconventional resource matures over the years, unconventional may be regarded as conventional.

Lithology of petroleum-bearing rock:

- Sandstone
- Carbonate
- Shale, silt, clay
- Coalbed
- Salt dome
- Combinations of the above

Nature of rock:

- Source rock (petroleum is produced from where it was generated)
- Reservoir rock (oil and gas migrated to a separate location from the source rock)

Rock characteristics:

- Unconsolidated
- Consolidated
- Tight

Geologic complexity:

- Single layered
- Multilayered or stratified (communicating, partially communicating, noncommunicating)
- Fractured
- Faulted (sealing, partially sealing, nonsealing)
- Compartmental
- Tight (poor oil and gas conductivity characteristics)
- Highly heterogeneous (rock properties vary significantly)

Location:

- Onshore
- Offshore, including deep-sea reservoirs
- Shallow, including oil sands
- Deep, including basin-centered reservoirs

Reservoir pressure:

- Overpressured
- Underpressured

Reservoir drive energy:

- Depletion
- Gas cap
- Fluid and rock expansion
- Gravity
- Aquifer
- Rock compaction
- External fluid injection, including water and chemical flooding
- Thermal

Reservoir boundary:

- Closed
- Edge-water drive
- Bottom-water drive

Reservoir dip:

- Steep inclination – dictates location of wells

Mode of production:

- Primary (production by natural reservoir energy)
- Secondary (production augmented by water flooding)
- Tertiary (production enhanced by injecting chemical, foam, and thermal treatment)

Production characteristics:

- Single-phase flow (oil or gas)
- Multiphase flow (oil and gas, oil and water, oil, gas and water, gas and water)
- High water cut
- High gas/oil ratio

Reservoir life:

- Early stage in production
- Peak production
- Declining production
- Matured reservoir

Reservoir engineering functions

No two petroleum reservoirs have the same characteristics. Each type of reservoir requires a unique approach to develop and produce optimally, often involving the validation, interpretation, and integration of vast amounts of reservoir data, characterization of geologic complexities, visualization of fluid flow processes, and utilization of analytic or computer-based fluid flow models. Typical reservoir engineering tasks include, but are not limited to, the following:

- Detailed understanding of the reservoir, including the conceptualization and visualization of rock and fluid flow characteristics, and the mechanisms by which a reservoir is produced; unconventional reservoirs pose new challenges
- Integration of reservoir engineering data with geophysical, geological, petrophysical, and production information, among others, to develop a conceptual model of the reservoir
- Estimation of oil and gas in place based on various methodologies, including volumetric calculations, study of declining production trends, material balance of fluids involved in production and injection, and simulation of a reservoir model
- Estimation of petroleum reserves of oil and gas fields with various degrees of probability
- Design, placement, and completion of producers and injectors in order to optimize production
- Plan, design, execution, and monitoring of water flood and enhanced oil recovery operations
- Implementation of a strategy for incremental oil recovery from matured fields
- Meeting challenges posed by declining well productivity, premature breakthrough of water and gas, unexpected reservoir heterogeneities, operational issues, economic aspects, environmental concerns, statutory regulations, and others
- Development and simulation of computer-based models that predict reservoir performance
- Reservoir surveillance that enhances the knowledge of the reservoir and charts future courses of action
- Working closely with a multidisciplinary team of engineers and earth scientists in order to manage the reservoir effectively
- Adhering to the best practices in reservoir engineering and management

Two workflows are presented. The first workflow presents an overview of the responsibilities of reservoir engineering team in managing conventional oil reservoirs, and second workflow is little more specific, highlighting the development of unconventional shale gas reservoirs (Figures 1.1 and 1.2).

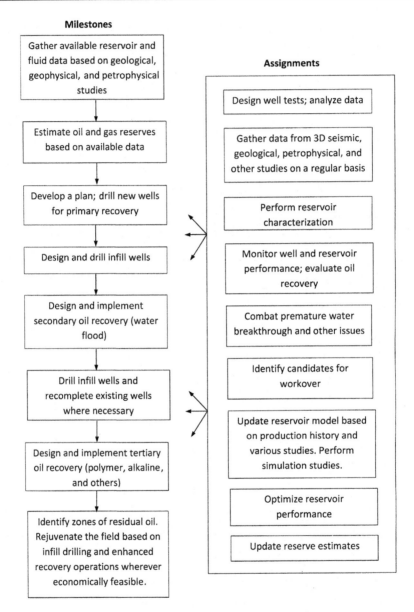

Figure 1.1 Reservoir engineering workflow. Milestones are depicted at left, while the ongoing reservoir engineering activities are shown at right.

Walkthrough

The workflows presented above suggest the breadth and depth of the wide-ranging skills required to effectively manage conventional and unconventional petroleum reservoirs. The following is a quick walkthrough highlighting the contents of various chapters presented in the book.

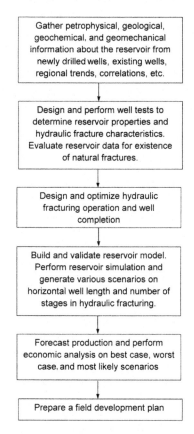

Figure 1.2 Workflow highlighting the development of an unconventional shale gas reservoir.

Chapter 2: Origin of Petroleum Reservoirs

In order to evaluate reservoir characteristics including geologic complexities, knowledge of how petroleum reservoirs were formed in ancient times is necessary. This chapter provides an overview of depositional environments that ultimately influence reservoir performance in producing oil and gas. In recent times, the topic has gained significance for reservoir engineers as certain unconventional reservoirs produce from source rock, i.e., from the rock where petroleum was generated.

Chapters 3, 4, and 5: Rock and Fluid Properties, and Phase Behavior of Petroleum Fluids

Fundamental to reservoir engineering are reservoir rock and fluid properties, including fluid phase behavior. These determine how the reservoir will be developed and managed, including the location and spacing of wells, design of water flood and enhanced recovery operations, range of oil and gas recoveries that can be expected, and overall management of the reservoir. In unconventional reservoirs such as shale gas, geochemical and geomechanical properties play important roles. Petrophysical properties are traditionally determined to help develop these reservoirs.

Chapter 6: Reservoir Characterization

Any reservoir development begins with three words: "Know your reservoir." A reservoir must be characterized in terms of geologic complexities and rock properties in micro- as well as macroscale in order to determine their effects on fluid flow and reservoir performance. Various disciplines of science and engineering contribute to reservoir characterization studies.

Chapter 7: Reservoir Life Cycle

All reservoirs go through a life cycle, from exploration to discovery, and finally to abandonment. Included in the cycle is the delineation of the extent of the reservoir, development based on drilling of wells, and production in various phases, namely, primary, secondary, and tertiary. As a reservoir moves through the cycle, the role of engineers and earth scientists changes according to the skills that are required to manage the reservoir.

Chapter 8: Reservoir Management Process

Efficient management of a reservoir requires a well-laid-out process that must be planned, implemented, monitored, and reviewed for lessons learned. Corrective measures are implemented as and when necessary. The management process is demonstrated by a case study. The field has been produced commercially over many decades by applying various innovative technologies throughout the life of the reservoir.

Chapter 9: Fluid Flow Characteristics in Porous Media

Understanding the fluid flow behavior in porous media serves as the backbone of conceptualizing reservoir dynamics. Analytic equations and models predict the flow rate, pressure and saturation of various fluid phases under various flow regimes, and reservoir boundary conditions.

Chapter 10: Well Transient Pressure Testing

One of the most valuable tools in evaluating a reservoir, including the wells, is transient pressure, or well testing. A pressure pulse or transient is created at the well, and the response is monitored for a period of time. Based on well condition, rock characteristics, and fluid properties, the response creates distinct signatures that are analyzed to obtain valuable information.

Chapter 11: Primary Drive Mechanisms of Reservoirs

Most reservoirs have the help of natural energy for production, up to a set point. The sources of energy include, but are not limited to, high pressure, expansion of fluids, water influx from adjacent aquifers, and gravity. Based on the mechanism or mechanisms at work, the range of primary recovery is determined.

Chapters 12, 13, and 14: Volumetric Analysis, Decline Curves, and Material Balance Method

Estimation of oil and gas in place, and petroleum reserves, is a core task of the reservoir engineers. Various techniques are available to accomplish this. Volumetric estimates are based on geological and geophysical studies, which depend on static data. On the other hand, decline curve analysis and material balance requires dynamic data, including production rates and fluid volumes.

Chapter 15: Reservoir Simulation

Major reservoir engineering decisions rely heavily on reservoir model simulations. Integrated reservoir models are built, simulated, and updated to predict reservoir performance in the future under various scenarios, including the number and location of wells, water flooding, and enhanced oil recovery operations.

Chapters 16 and 17: Improved Oil Recovery Methods

Improved recovery operations are planned and implemented for most conventional oil reservoirs to augment recovery. Once the natural energy to produce oil is depleted, additional energy is provided by water and chemical injection. Thermal methods are applied to heavy oil to increase mobility.

Chapter 18: Horizontal Wells

Horizontal drilling is a success story. In recent decades, it brought vast improvements in oil and gas recovery not envisioned before. Horizontal wells contact a large reservoir area, and are particularly suitable in producing from ultratight formations such as shale, compartmental reservoirs, and others.

Chapter 19: Oil and Gas Recovery Methods

Recovery of petroleum is engineered in various ways in difficult settings, including highly heterogeneous formations and low to ultralow permeability reservoirs. Methods include infill drilling once the relatively largely spaced wells decline in production. In tight reservoirs, horizontal drilling is a major practice to produce commercially.

Chapter 20: Rejuvenation of Matured Reservoirs

Reservoir performance inevitably declines with time; however, reservoir engineers attempt to rejuvenate a reservoir by targeting the areas and geologic layers where a significant portion of oil is left behind. Various tools and techniques, including 3D seismic studies and reservoir simulation, are utilized to accomplish this.

Chapters 21 and 22: Unconventional Oil and Gas

With the advent of technology, unconventional resources of petroleum are rapidly becoming a major player in meeting the demands for oil and gas in the world. Most notable are the production of shale gas and tight oil based on horizontal drilling and multistage fracturing, referred to as fracking. Extraction of oil sands is another important technology where innovative thermal methods are used.

Chapter 23: Estimation of Petroleum Reserves

As indicated earlier, reservoir engineers are required to provide estimates of oil and gas reserves. Apart from evaluating the assets of a company, reporting of reserves to the authorities is a law in most petroleum producing countries. Due to the inherent uncertainties associated with petroleum accumulations, reserves are categorized as proved, probable, and possible, depending on the probability that can be associated with each category.

Chapter 24: Reservoir Management Economics

Each reservoir project needs to be justified in an economic sense. In addition to technical expertise, reservoir engineers are required to perform economic analysis of the reservoir on a regular basis. Frequently, the merit of the project depends on various economic criteria such as net present value, payout period, and rate of internal return.

Elements of conventional and unconventional petroleum reservoirs ▪2▪

Introduction

It is important for reservoir engineering professionals to have a clear understanding of the basic elements and events of nature that influence petroleum reservoirs from inception until the present day. A detailed knowledge of the origin, migration, and entrapment of hydrocarbons in geologic formations aids in evaluating the characteristics, behavior, and potential of the reservoir. The petroleum industry utilizes the valuable information in the exploration of the new frontiers of oil and gas; a case study demonstrating the above is presented in this chapter. Furthermore, the knowledge aids in the interpretation of geologic events that shaped the petroleum basins, regional geologic trends, extent of the reservoirs, estimates of hydrocarbon volume, and the analysis of subsurface pressure anomalies, among others. There is a new focus on the origin of petroleum due to the fact that the source rock of petroleum plays a direct role in the exploration of unconventional reservoirs. Wells are drilled in the source rock to produce oil and gas wherever geologic and other conditions are favorable.

Study of the reservoir elements leads to the following queries:

- How are petroleum reservoirs formed?
- How, when, and where did oil and gas originate?
- What are the types of the reservoir rocks?
- How are the fluids accumulated and trapped in a reservoir?
- What are the essential rock properties to store and produce petroleum?
- Did petroleum fluids originate at the same location as discovered today?
- What is a petroleum system? What are its elements?
- Is there any distinction between the elements of conventional and unconventional reservoirs?
- How do computer models aid in petroleum exploration and production?

The answers to these queries can be found in the results of wide-ranging studies pertaining to the petroleum basin, the reservoir, and the rocks. The studies include, but not limited to, geological, geochemical, petrophysical, geophysical, hydrodynamic, and geothermal. The organic matter found in the rocks is also the subject of intense scrutiny. Tools and methodologies involved in the studies range from very basic, such as field observation, to the most sophisticated, including simulation of robust computer models.

Reservoir rock types and production of petroleum

Shale is the most abundant rock type in sedimentary basins, comprising about 80% or more of the total rock volume in many instances. However, conventional oil and gas reservoirs are mostly composed of sandstone and carbonate formations, often

Reservoir Engineering. http://dx.doi.org/10.1016/B978-0-12-800219-3.00002-4

interbedded with shale. Carbonate reservoirs are highly prolific producers, about 60% of the world's production of petroleum is based on these reservoirs. Sandstone reservoirs account for over 30% of production. In recent times, however, production potential from shale and other unconventional resources is rapidly gaining intense industry interest since the early years of this century. A sizeable portion of natural gas in the United States is currently produced from unconventional shale gas reservoirs. Certain metamorphic or igneous rocks are known to be producers of petroleum. However, the source of petroleum is believed to be sedimentary rock, mostly shale, from which oil moved to the other rock types mentioned above.

Sandstones are widely composed of feldspar and quartz grains with their origin rooted in desert, stream, or coastal environments in prehistoric ages. The grains range from micrometers to millimeters and are typically cemented by silica. Carbonate rocks (limestone or dolomite) are based on the skeletal remains and shells of organisms that chiefly lived in shallow marine environments. Carbonates may have inorganic origin too, where calcite is precipitated in water. Certain limestones transformed into dolomites following postdepositional processes involving the evaporation of marine water, transformation of calcium carbonate to magnesium carbonate, and recrystallization. Shale, the most abundant of reservoir rock types, is composed of clay and silt particles. It is not uncommon to encounter petroleum reservoirs having a combination of the various rock types mentioned above. For example, a sandstone reservoir with appreciable shale content is referred to have a shaley sandstone lithology.

Origin of petroleum

Over decades, scientists have proposed several theories regarding the origin of petroleum, including organic, abiogenic, and cosmic. Based on field evidence, laboratory investigations, mathematical modeling, and analyses, the organic origin of petroleum has been largely accepted by the petroleum industry. In the following, the elements of petroleum reservoirs are discussed in brief.

Deposition of sediments and organic matters: the process begins

The origin of petroleum is rooted in the transportation and deposition of sediments in marine, shallow marine, deltaic, lagoon, swamps, mud, desert, and various other environments by the natural forces of wind, water, ice, and gravity over long periods in ancient times. Pertaining details for various rock types related to deposition of sediments are presented in Table 2.1. A typical depositional process involving mountains, land, and sea shelf is depicted in Figure 2.1.

The depositional process continued through prehistoric ages. Deposited along with the sediments was organic matter such as marine organisms and remnants of woody plant material, among others. These organic resources ultimately led to the origination of oil and gas found in present day reservoirs in a span of tens to hundreds of millions of years.

Table 2.1 **Origin of sedimentary rocks [1]**

Rock type	Sediment	Transport and accumulation	Notes
Sandstone	Sand	Desert dunes – windblown sands (eolian), river channels (fluvial), low gradient stream valleys (alluvial), deltas, shorelines, and shallow seas	Light beige to tan in color; sometimes dark brown to rusty red. Composed of grains of quartz, feldspar, etc. and cemented by silica
Conglomerate	Gravel	River channels, alluvial fans, and wind-swept coastlines	Grains of sandstone and conglomerate originate from pre-existing rocks and minerals
Limestone (calcium carbonate) and dolomite (calcium–magnesium carbonate)	Shells, algae, and coral; precipitation of calcite	Warm shallow seas	Usually light to dark gray in color; exhibits fossil molds and casts; void spaces largely due to dissolution and vugs
Chalk (calcium carbonate)	Produced by marine plankton	Deep seas	Fine textured
Shale	Clay, silt	Lakes (lacustrine), tidal flats, river flood plains, deltas, and deep seas	Dark brown to black in color; sometimes dark green; composed of fine grains of clay and silt. Exhibits lamination in the horizontal direction
Coal	Woody plant matter, peat	Swamps	
Chert (silicon dioxide)	Produced by marine plankton	Deep seas	
Rock salt	Salt	Lagoons or marginal seas	

Types of sediments

Sediments are of clastic, biochemical, and chemical origin as in the following:

- Clastic (detrital) rocks such as sandstone and siltstone are formed by the particles or grains of pre-existing rocks, which in turn were created by the effects of weathering.
- Limestone and dolomite, referred to as carbonates, have a biochemical origin as these rocks are based on the skeletal remains and shells of organisms that chiefly lived in shallow marine environments. Certain limestones transform into dolomites following postdepositional

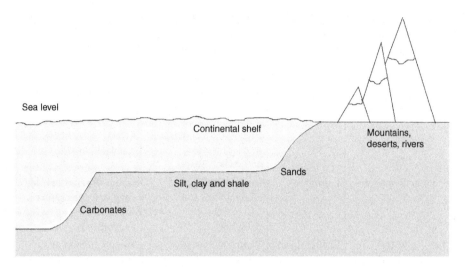

Figure 2.1 Typical depositional environment of sediments and organic matter in shallow and deep marine. The accumulation of sand, shale, silt, clay, and carbonates depends on the location, available energy, and other natural processes.

processes involving the evaporation of marine water, transformation of calcium carbonate to magnesium carbonate, and recrystallization.
- Chemical sediments originate from minerals that precipitate from water. Examples of chemical sediments are gypsum and calcite.

Geologic basins and occurrences of petroleum: an overview

Deposition, burial, and subsequent compaction of sediments that continued for very long periods in a geologic time scale resulted in the creation of sedimentary basins. The geologic time scale is presented in Table 2.2. Many petroleum basins extend over a large area and are thousands of feet thick. Some basins have a depression or concavity toward the center and rifts at the periphery, as depicted in Figure 2.2. Some other basins are gently sloping.

There are about 600 basins known to exist worldwide, of which 26 are significant producers of oil and gas [4]. It is estimated that about 65% of the world's petroleum is concentrated in the giant oil fields located in a relatively small number of sedimentary basins.

The geologic time scale

All numbers shown in Table 2.2 are approximate, and vary somewhat from source to source. According to a 1991 study, over 50% of the world's petroleum reservoirs date back to the Jurassic and Cretaceous periods in the geologic time scale.

The formation of basins is associated with the geologic events related to plate tectonics, which deals with the movement of the earth's crustal plates. Interestingly, the

Table 2.2 Geologic time scale [2,3]

Eon	Era	Period	Epoch	Millions of years old
		Quarterary	Holocene	0.01
			Pleistocene	2.6
			Pliocene	5.3
		Neogene	Miocene	23.7
			Oligocene	36.6
			Eocene	57.8
	Cenozoic	Paleogene	Paleocene	66
		Cretaceous	Late	100
			Early	145
		Jurassic	Late	164
			Middle	174
			Early	201
	Mesozoic	Triassic	Late	237
			Middle	247
			Early	252
		Permian		299
		Pennsylvanian		323
		Mississippian		359
		Devonian		419
		Silurian		444
		Ordovician		485
Phanerozoic	Paleozoic	Cambrian		541
	Proterozoic			2500
Precambrian	Archean			3800

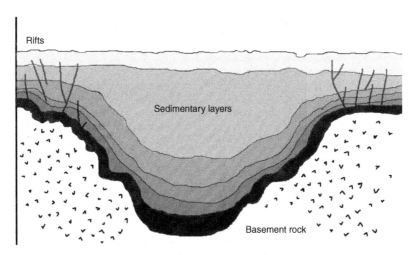

Figure 2.2 Cross-sectional view of a typical petroleum basin showing multiple depositional sequences and rifts due to regional stresses. Large numbers of oil and gas accumulations are found in multiple geologic strata of the basin trapped by various mechanisms.

depositional as well as other geologic processes related to the origin of petroleum continue to this day in the giant laboratory of the earth.

Stratigraphic sequence

A typical sedimentary basin is composed of alternating layers of sedimentary rocks. The stratigraphic sequence of sand, shale, and carbonate rocks is presented in Figure 2.3

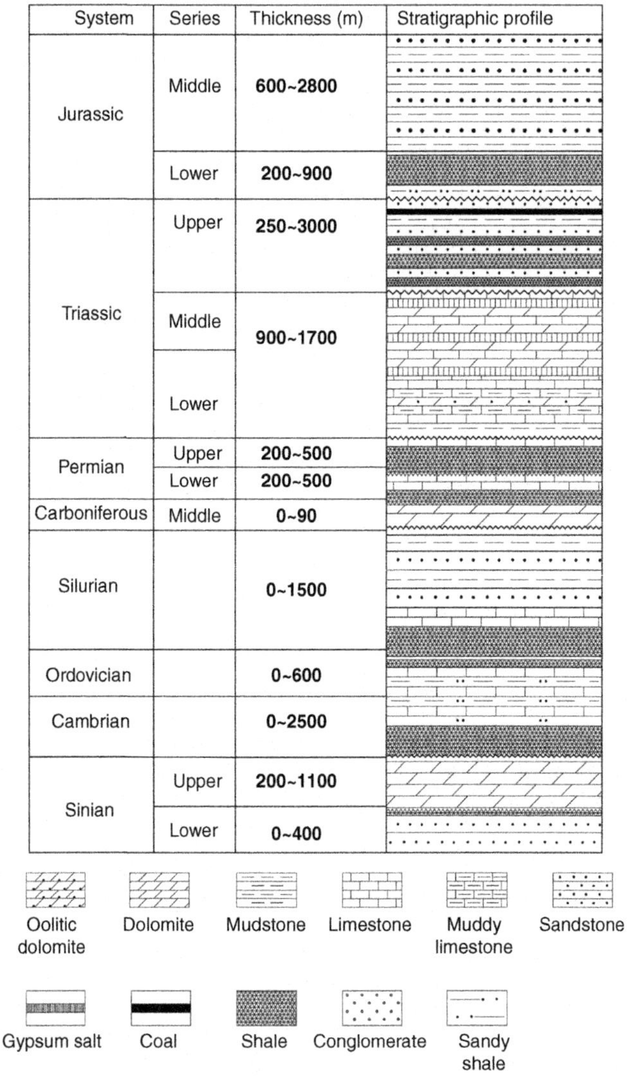

System	Series	Thickness (m)	Stratigraphic profile
Jurassic	Middle	600~2800	
	Lower	200~900	
Triassic	Upper	250~3000	
	Middle	900~1700	
	Lower		
Permian	Upper	200~500	
	Lower	200~500	
Carboniferous	Middle	0~90	
Silurian		0~1500	
Ordovician		0~600	
Cambrian		0~2500	
Sinian	Upper	200~1100	
	Lower	0~400	

Oolitic dolomite   Dolomite   Mudstone   Limestone   Muddy limestone   Sandstone

Gypsum salt   Coal   Shale   Conglomerate   Sandy shale

Figure 2.3 Stratigraphic sequence showing alternating beds of sand, shale, and carbonates formed over long geologic periods.

as an example. Some of these geologic formations can store and produce significant quantities of petroleum. It is important to note that the formations are usually subjected to major geologic events throughout the postdepositional periods, including folding, faulting, fracturing, uplifting, and erosion, to name a few. The above events profoundly affect reservoir geometry and heterogeneity, requiring various reservoir engineering strategies to recover oil and gas efficiently.

Rock geochemistry: formation of kerogen

As the sediments are deposited, the following processes take place leading to the formation of a dark and waxy substance called kerogen, which is the precursor to oil and gas:

- The sediments are buried to increasing depths with the continued discharge and overloading of sedimentary particles in large quantities by the streams and rivers over long periods of time.
- The unconsolidated sediments undergo a process called lithification, which involves compaction and cementation of the sediments. Compaction occurs due to overloading by massive amounts of sediment over time, which creates an enormous confining pressure.
- Cementation occurs due to the work of certain minerals, such as silica and calcite, which precipitate from water, form around the sediments, and finally create bonding between the grains by cementation. The cementation process results in the formation of consolidated rocks.
- Oil and gas are hydrocarbon compounds, generally believed to originate from the organic matter that was buried along with the sediments. Due to the high pressure and temperature in an oxygen deficient environment, the organic matter contained in rock transforms into kerogen. It is insoluble in common solvents.

The types of kerogen, including various characteristics and associated depositional environments, are listed in Table 2.3.

Additionally, there is a Type IV kerogen where the hydrogen/carbon ratio is insignificant. It does not produce any oil or gas (Figure 2.4).

Table 2.3 Types and characteristics of kerogen

Kerogen	Type I – sapropelic	Type II – planktonic	Type III – humic
Origin	Algal material reworked by bacteria and microorganisms	Planktonic remains reworked by bacteria	Woody plant matter
Converts to	Oil	Oil as well as gas	Gas and coal
H/C ratio	>1.25	<1.25	<1.0
O/C ratio	<0.15	0.03–0.18	0.03–0.3
Depositional environment	Lake deposits (lucustrine) and marine	Moderately deep marine (reducing environment)	Nonmarine and shallow to deep marine

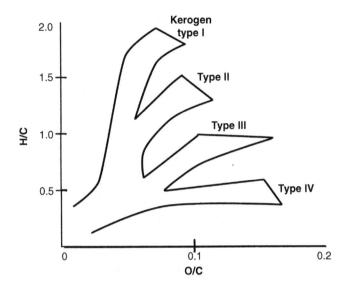

Figure 2.4 Types of kerogen depending upon the elements present. Ranges of kerogen types are plotted as H/C versus O/C ratios.

Generation of hydrocarbons

Under subsurface conditions, the organic matter present in rock is subjected to intense heat. As a result, kerogen is produced initially. Bitumen can also be produced to a lesser degree. With increasing depth of burial, kerogen is exposed to further heat. As a result, it is thermally cracked or degraded to produce oil and gas. The hydrocarbon compounds that are produced have relatively less and less molecular weight and complexity as the heat intensifies and the rock thermally "matures."

The thermal maturity of rock, indicated by vitrinite reflectance, is an important parameter for source rock evaluation in unconventional reservoirs. Vitrinite reflectance is described in Chapter 3. The stages associated with the thermal maturity of rock, namely, diagenesis, catagenesis, and metagenesis, are described in Table 2.4.

Table 2.4 Stages of thermal maturity of rock

Process	Diagenesis	Catagenesis	Metagenesis
Temperature range (°F)	<125	125–275	225–400
Product	Kerogen, bitumen	Oil and gas	Dry gas
Vitrinite reflectance (R_o)		0.5–1.5	1.5–3.0
Notes	Biogenic gas may be produced due to bacterial action at low temperature range	Optimum temperature range for oil	Kerogen finally reduces to graphite beyond the temperature range

Note: All values of temperature and R_o cited in the table are approximate.

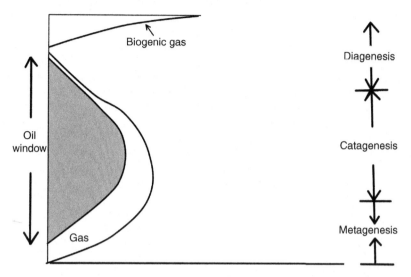

Figure 2.5 Oil and gas windows as a function of subsurface temperature in the y-axis.
The intensity of generation is plotted in the x-axis. (Figure is not to scale.)

Oil and gas generation depth

Since the depth of burial is correlated to subsurface temperature, oil and gas are produced at particular depths where the temperature is conducive to petroleum generation. In petroleum basins, heavy oil is typically found in shallower depths where
the subsurface temperature is relatively low. Light oil is found at further depths as the
temperature increases. "Oil window" refers to a depth interval, approximately ranging
from few thousand feet to about 10,000 ft., where subsurface temperature supports oil
generation by catagenesis (Figure 2.5). At further depths, the temperature is higher,
only gas is generated as a result. Hardly any hydrocarbon is generated below 15,000 ft.
due to the intensity of heat. It is noteworthy that the gas produced in rocks can also be
biogenic, which results from the work of bacteria present in the rock in relatively low
temperatures and at much shallower depths (Table 2.5).

Table 2.5 Generation of oil and gas

Type	Temperature (°F)	Typical depth (ft.)	Notes
Oil window	125–275	5,000–10,000+	Heavy oil is formed near the top of oil window. Oil is lighter toward the bottom of the window.
Gas window (thermogenic)	225–400	7,000–15,000+	Gas condensate and wet gas are formed near the top of gas window.
Biogenic gas	<125	Near surface	

Note: All values of temperature and depth are approximate.

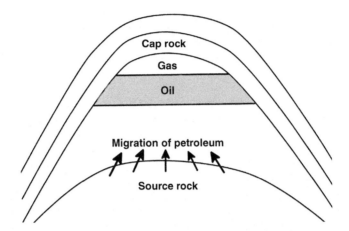

Figure 2.6 Vertical migration and accumulation of petroleum in conventional reservoirs.
Lateral migration is also commonplace. For certain unconventional reservoirs such as shale oil
and gas, the source rock acts as a reservoir rock.

Source rock, reservoir rock, and migration of petroleum

Rock containing kerogen is referred to as the source rock for petroleum. These rocks
are composed of fine-grained shale and mudstone and enriched in clay having a dark
gray to black color. Certain carbonates are also known to be source rock.

Petroleum formed in the source rock is eventually expelled under pressure to migrate to the reservoir rock where it undergoes accumulation under a suitable sealing
and trapping mechanism (Figure 2.6). The seal can be provided by impervious or
semipervious caprock, among others. The above is a key element for conventional reservoirs. Continuous pathways such as pore channels, microfractures, faults, and joints
must exist in rock for the movement of oil and gas to take place. Geologic studies have
indicated that migration of petroleum can occur in a horizontal or vertical direction,
and continue over hundreds of kilometers in certain cases.

Migration of petroleum can be either primary or secondary. The migration of oil
and gas from the source rock to the edges of the petroleum reservoir is referred to as
primary migration. The driving force is the compaction of source rock under overburden pressure. The above action results in the expulsion of pore fluids. The mechanism of primary migration also includes diffusion and solution. Diffusion is a process
by which oil moves from areas of relatively high concentration to adjacent areas of
low concentration. Lighter components of petroleum, including methane and ethane,
may also be transported in a dissolved state in formation water. Since petroleum in
source rocks is generated when the rock pores are significantly reduced in size due to
compaction, the mechanism of primary migration is a subject of debate in the scientific community.

Secondary migration takes place within the petroleum reservoir where oil moves
updip by buoyant forces. Buoyancy of oil is created as it is lighter than formation

water. However, oil needs to overcome capillary pressure to displace water from the rock pores. Capillary pressure arises due to the fact that oil and water are not soluble in each other, and oil must exert a pressure to displace water present in rock pores. In essence, gravity and capillary forces counteract during the migration of oil where water is displaced by oil. The mechanism of secondary migration is better understood than that of primary migration. It has also been observed that oil and gas can seep to the earth's surface in the absence of an effective seal. The phenomenon is referred to as tertiary migration in literature. According to some estimates, only 10% of petroleum generated in source rocks is trapped in the reservoirs.

An important distinction between conventional and unconventional reservoirs is based on the role played by the source rock of petroleum. Unconventional reservoirs, which are capable of producing oil and gas as a result of modern technology, are also the source rock where hydrocarbon is generated in the first place. Migration of oil or gas plays little or no role in most unconventional reservoirs.

Traps associated with conventional reservoirs

Conventional resources of oil and gas accumulate under a suitable trapping mechanism following migration from the source rock. Traps can be classified as structural, stratigraphic, or a combination of both. Structural traps are formed by folding and faulting of geologic strata as a result of tectonic forces. A common example of a structural trap is a dome-shaped structure or anticline (Figure 2.7). Trapping of oil and gas may also occur due to the presence of an impermeable fault. Stratigraphic traps

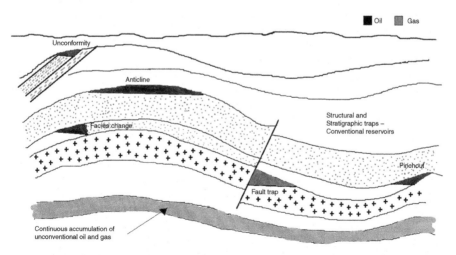

Figure 2.7 Depiction of structural and stratigraphic traps responsible for oil and gas accumulation in conventional reservoirs. Continuous accumulation of unconventional gas in a shale bed is also shown.

originate from facies change or geologic unconformity that provides a barrier to flow and leads to the entrapment of petroleum.

Traps are usually overlain by an impervious caprock or seal rock that deters further migration or seepage of oil and gas from the reservoir. In some cases, the seal is provided by the change in rock facies.

Important differences also exist as to how the conventional and unconventional sources of petroleum are stored in the reservoir.

The petroleum system

Petroleum industry professionals view the entire process of hydrocarbon generation, migration, and accumulation, including the geologic elements that play a part in the above, as an integrated petroleum system. The system refers to the various elements and processes dating back from the origin of petroleum basins in ancient times to the accumulation of oil and gas in reservoirs that are explored and produced today. Note that a large petroleum basin usually has multiple petroleum systems. In various parts of a basin, formation of source rock, migration, and entrapment of petroleum into different reservoirs may occur millions of years apart as the earth processes are continuous.

The elements of a petroleum system for conventional reservoirs are summarized as follows [5]:

- Source rock – where oil and gas originates from the organic matters contained in the rock under elevated temperature and pressure.
- Migration pathway – includes pore channels, microfractures, faults, and joints through which oil escapes from the source rock to the reservoir rock; the principal driving forces are pressure and buoyancy.
- Reservoir rock – where oil and gas are stored, and subsequently produced after discovery.
- Seal rock – an impervious geologic formation that deters the flow of oil and gas from the reservoir.
- Trap formation – a geologic feature, either stratigraphic or structural or a combination, which provides a trapping mechanism to store oil and gas. However, unconventional reservoirs do not have a trap definition in the conventional sense.
- Overburden rock – imparts necessary pressure for compaction of organic-rich sediments and geologic formations.

Petroleum system processes are as follows:

- Generation – petroleum is generated in source rocks under appropriate conditions, including elevated pressure and temperature, over a long period in the geologic time scale.
- Migration – oil is eventually expulsed from source rock under pressure and migrates to the reservoir by hydrodynamic and other forces. Oil and gas migration, however, are chiefly associated with conventional reservoirs.
- Accumulation – petroleum fluids accumulate in present day reservoirs under a suitable trapping mechanism where seal rock plays a critical role.

A typical geologic time scale for the events and processes associated with petroleum system is presented in Table 2.6.

Table 2.6 **The petroleum system [6]**

Sequence of events in geologic time scale	Millions of years ago						
	50	100	150	250	300	350	400
Source rock deposition							▓
Reservoir rock						▓	
Seal rock					▓		
Overburden rock				▓	▓		
Formation of geologic trap				▓			
Generation, migration, and accumulation				▓			
Preservation of petroleum	▓	▓	▓	▓			

Notes: Source rock and reservoir rock are the same in certain unconventional reservoirs. Migration does not occur in unconventional reservoirs.

Comparison between conventional and unconventional reservoirs: source, migration, and accumulation

While the petroleum system addresses all the elements and processes that are in play for conventional reservoirs, certain key aspects of the petroleum system are not found in unconventional reservoirs. In unconventional reservoirs such as shale gas reservoirs, accumulation of petroleum is continuous over a large area with no trap definition. Source rock and reservoir rock are in the same geological formation. In unconventional petroleum reservoirs, migration of petroleum takes place over a short distance, as in shale oil reservoirs, or does not occur at all, as in shale gas reservoirs. Furthermore, migration is controlled by diffusion. In contrast to conventional reservoirs, the ability of unconventional reservoirs to transport fluids is significantly lower due to the ultralow permeability of the rock matrix. Permeability, a critical property of rock, which indicates the ability of rock to transmit fluid, is treated in Chapter 3.

The contrasting features between conventional and unconventional reservoirs are highlighted in Table 2.7 [5]. A shale gas reservoir is used as an example of an unconventional reservoir.

Reservoir heterogeneities

Reservoir rocks are heterogeneous in composition and properties that are of interest to reservoir engineers. There are many types of heterogeneities encountered in geologic formations that affect the performance of the reservoir. In petroleum basins, alternating sequences of shale, sandstone, and carbonate layers are usually encountered due to the repeated encroachment or transgression of the ancient sea into the land, followed by the retreat or regression of water. The cycle may continue over a long period of time leading to the formation of many distinct geologic strata. The grading of rock

Table 2.7 Contrasting features between conventional and unconventional reservoirs

Element/process	Shale gas reservoir (unconventional)	Conventional oil and gas reservoir
Proximity to mature source rocks	Close	Close or distant
Migration of petroleum	Gas trapped in place	Migration may occur over a long distance
Reservoir trap	No evidence of trap in conventional sense	Existence of a structural, stratigraphic, or combination trap
Hydrocarbon charge area	Pervasive through a large area	Relatively limited charge area
Resource in place	Large	Relatively small
Ability of rock to transport fluid	Ultralow	Usually higher by orders of magnitude
Recovery potential	Relatively low	Relatively moderate
Gas−water contact	Not well defined	Well defined
Usual occurrences of water	Updip from hydrocarbon	Downdip from hydrocarbon
Reservoir pressure anomaly	Overpressured reservoirs are commonplace	Pressure anomalies are relatively few

Source: Adapted from Ref. [7].

grains in the vertical direction indicates the transgression/regression cycle. Transgression of the sea is associated with the finer grains deposited upward in a geologic bed or layer. Conversely, coarser grains deposited in the upward direction indicate an environment where the sea has regressed.

A geologic formation may exhibit facies change where the composition of rock may change from one rock type to another. For example, formations in many petroleum reservoirs transition from sand to predominantly shale in the lateral direction. Facies change is an indicator for a change in the depositional environment. Typically, sand particles are deposited in shallow water or coastal environments while silt and clay are deposited in lakes and relatively deep waters. Again, marine organisms can be deposited in deep seas. The occurrence of facies change in a geologic formation may define a boundary for fluid flow and affect the performance of reservoirs.

During transportation, the smaller sized sediments travel longer, and are deposited only in a very low energy environment such as deep sea. Again, well-sorted grains in rock, where most grains are of similar size, indicate long transport of sediments by water or other agents. Size and sorting of grains in reservoir rocks significantly influence the characteristics of reservoirs, including porosity and permeability of rock, which in turn affect the storage and flow capacity of petroleum. The properties of rock, including porosity and permeability, are discussed in Chapter 3.

Case study: Basin and Petroleum Systems Modeling in Alaska [6]

The petroleum industry has developed robust computer models to explore the new frontiers of oil and gas by simulating the formation of sedimentary basins in prehistoric times, maturation of source rocks, and the migration of petroleum. The models are also capable of predicting the location of hydrocarbon accumulation and the estimation of oil and gas volumes. In essence, the models simulate all the aspects of the petroleum system over a geologic time scale, including the deposition of sediments, burial, effects of pressure and temperature, formation of kerogen, generation of oil and gas, migration, and accumulation. The goal is to reduce the significant risk involved in oil and gas exploration in many parts of the world where exploration is cost intensive. In the early 1980s, an exploratory well, known to be the most expensive well in the industry at the time, was drilled in Mukluk prospect in Alaska where drill cuttings showed extensive stains of oil but no commercial quantities of petroleum could be found. It was concluded that oil was present in the structure but escaped due to an ineffective seal or the entire structure was tilted due to certain geologic events [6]. In essence, a critical component of the petroleum system was missing in the decision-making process.

In the following years, a study based on a computer model was conducted for the vast oil region of North Slope in Alaska [6]. Geologic, geophysical, and log information obtained from 400 wells located over an area of 106,000 square miles was utilized in the model. The geological setting is complex, having five source rocks and multiple petroleum systems. Computer aided analysis of overburden rocks facilitated the visualization of burial history and maturation of source rocks. The percentage of kerogen transformed into petroleum over the geologic time scale was estimated by collecting various data related to the source rock, including bed thickness, total organic carbon and hydrogen index. Based on available data related to burial pressure, thermal maturation, and fluid flow, the model simulated the expulsion of fluids from source rock, and subsequent migration and entrapment. The migration pathways predicted the locations where petroleum is likely to be discovered; further exploration was targeted in the areas predicted by the study.

Summing up

It is important to have a clear understanding of the depositional environment and natural events that shape petroleum reservoirs through geologic times. Sedimentary rock types, structural and stratigraphic characteristics, and reservoir heterogeneities including the presence of faults and fractures are directly influenced by various processes and events that occur in nature. Reservoir and source rocks for the storage of petroleum are:

- Sandstone
- Limestone and dolomite, referred to as carbonate rocks
- Shale

An overview of petroleum reservoir rock types indicates that conventional reservoirs are mostly composed of sandstones and carbonate rocks, which are limestones and dolomites. The formations are often interbedded with shale, the latter being the most abundant type of rock in sedimentary basins. About 60% of the world's production of oil and gas is based on carbonate rocks, while sandstone reservoirs account for about 30% of production. In recent times, unconventional reservoirs produce from ultratight shale formations as new technologies in horizontal drilling and fracturing are introduced to unlock the potential. There are about 600 sedimentary basins in the world, of which 26 are major producers of oil and gas.

According to the organic origin theory of petroleum, the origin of oil and gas can be traced to marine organisms and woody plant materials that were deposited along with sediments in marine, shallow marine, deltaic, lagoon, swamps, desert, and various other environments. The deposition process continued through prehistoric times measured in tens of millions of years. The type of rock was determined by depositional environment and the nature of organic material. In a high energy environment such as deserts and deltas, larger particles were deposited, which led to the formation of sandstone. The smaller particles were carried further to low energy locales before deposition, such as lakes and deep marine environments, to form shale beds. In warm shallow seas, shells and algae were precipitated leading to the formation of limestone. In swampy land, woody plant materials were deposited that resulted in the formation of coal.

The precipitated sediments were buried and compacted over a long time, and were subjected to intense temperature and pressure that exist at depths. As a result, rocks were formed out of the sediments due to compressional, thermal, and other effects. The pores of rock contained organic matter that was deposited along with the sediments and were "cooked" to transform into kerogen, and subsequently to oil and gas as found in present day reservoirs. Kerogen is a dark, waxy matter and a precursor to oil and gas. There are three types of kerogen found in the "source rock" of petroleum depending on the hydrogen to carbon (H/C) ratio. Type I kerogen is based on algal material reworked by bacteria and microorganisms. It has an H/C ratio of 1.25 or more and produces oil. The depositional locales for Type I kerogen include lakes and marine environments. Type II kerogen, having an H/C ratio of less than 1.25, produces both oil and gas. It is rooted in plankton remains reworked by bacteria. Type III kerogen originates from woody plant matter in nonmarine and marine environments, and produces gas and coal. The H/C ratio is the least of 3, 1.0, or less.

Hydrocarbons are generated by either of the three processes depending on the intensity of thermal energy:

- Diagenesis, producing kerogen and bitumen at 125°F or below. Biogenic gas is also produced at low temperature coupled with bacterial action.
- Catagenesis, producing oil and gas between 125°F and 275°F.
- Metagenesis, producing dry gas between 225°F and 400°F.

The temperature ranges cited above are provided as a guide only. Worldwide occurrences of petroleum, combined with the temperature range of catagenesis as well as geothermal gradient of sedimentary basins, suggest that there is an "oil window," i.e., the depth range where the petroleum reservoirs are most likely to exist. World-wide statistics indicate that oil reservoirs are discovered at depths between 5,000 ft. and 10,000 ft. in large numbers; however, some heavy oil reservoirs produce from much shallower depths. Although oil can be generated from kerogen at significant depths and migrate upward, oil reservoirs below 12,000 ft. are not common. Dry gas reservoirs may be discovered at further depths than oil as suggested by the temperature range for metagenesis. Below 15,000 ft., the temperature is so intense that the environment is apparently unfavorable to the production of petroleum in significant quantities.

Earth scientists and other professionals view the entire process of generation, migration, accumulation, and entrapment of petroleum as part of the "petroleum system." Petroleum reservoirs are only found where all the essential elements of the petroleum system are at work as follows:

- Source rock for petroleum − oil and gas originate under elevated temperature and pressure due to burial and compaction.
- Migration pathway − petroleum fluids, driven by pressure and buoyancy, migrate through pores, channels, fractures, nonsealing faults and others to present day reservoirs.
- Reservoir rock − oil and gas finally accumulate in reservoir rock.
- Seal − an impervious geologic formation that deters further migration of oil and gas.
- Trap − structural, stratigraphic, and certain other geologic features act as traps to store petroleum.

The processes that are part of the petroleum system include the generation, migration, and accumulation of petroleum fluids. However, the petroleum system of unconventional reservoirs may differ from that of conventional reservoirs just summarized. For shale reservoirs, there is no or little migration of petroleum. Oil and gas are produced from source rock rather than reservoir rock. Furthermore, there is no obvious trapping mechanism evident in unconventional reservoirs. Conventional reservoirs are limited in extent and bounded by features like a geologic structure or hydrodynamic boundary. Shale gas is pervasive over a large extent of ultratight shale formation. There are other distinctions. While conventional reservoirs are found with oil−water and gas−water contacts, no such contact is found in shale gas reservoirs. Exploration of conventional reservoirs requires intensive efforts. Once discovered, the reservoir is produced by traditional methods. On the other hand, unconventional reservoir locations are already known in many regions of the world. However, producing them economically is the challenging part.

This chapter also presents a modeling effort of the petroleum system in Alaska. Geologic, geophysical, and log information obtained from 400 wells located over an area of 106,000 square miles was utilized in the model. The model simulated various aspects including generation, migration, and accumulation of oil in probable locations. Modeling studies are conducted with an objective of aiding oil and gas exploration. In certain petroleum regions of the world, exploration is difficult and cost intensive due to accessibility, climate, and other issues. The study was conducted after drilling one of the most expensive exploratory wells in the industry. The drill cuttings showed extensive oil stains but no oil was found, suggesting that oil had migrated elsewhere due to certain regional forces. Hence, a firm grasp of the processes involved in the petroleum system is of paramount importance in the exploration and production of petroleum.

Questions and assignments

1. How did oil and gas originate? Discuss the natural processes leading to the formation of oil reservoirs.
2. Why is the study of reservoir rocks, structures, and stratigraphy important?
3. Did various rocks deposit in the same environment? How are the rocks distinguished from each other?
4. Why are unconventional shale reservoirs produced economically only in recent times?
5. How old are the petroleum reservoirs? Did the reservoirs potentially undergo any subsequent changes in structure and characteristics?
6. Describe the significance of the oil window in the exploration of petroleum reservoirs.
7. What are the main differences between oil and gas with respect to their origin?
8. What is a petroleum system? What happens when any of the elements of a petroleum system are not present in nature?
9. Is there any distinction between the origin of conventional and unconventional reservoirs?
10. Based on the literature, describe in detail the origin and formation of an offshore oil field. Was the development and production of the reservoir influenced by its origin?

References

[1] Reservoir rocks and source rock types, classification, properties and symbols. Available from: http://infohost.nmt.edu/~petro/faculty/Adam%20H.%20571/PETR%20571-Week-3notes.pdf [accessed 14.06.13].
[2] Petroleum systems, source, generation and migration; 2008. Available from: http://www.ogs.ou.edu/pdf/PetSystemsA.pdf.
[3] GSA geologic time scale. Geologic Society of America. Available from: http://www.geosociety.org/science/timescale/timescl.pdf [accessed 10.12.13].
[4] Encyclopedia Britannica. Available from: http://www.britannica.com [accessed 20.01.14].
[5] The Bakken – an unconventional petroleum and reservoir system. Final Scientific/Technical Report. Colorado School of Mines; 2012.
[6] Schlumberger. Basin and petroleum system modeling. Available from: https://www.slb.com/~/media/Files/resources/oilfield_review/ors09/sum09/basin_petroleum.ashx [accessed 05.01.14].
[7] Zou C. Unconventional petroleum geology. Elsevier; 2012.

Reservoir rock properties

3

Introduction

Rock and fluid properties, along with reservoir characteristics, provide the basis for analysis, development, and production of petroleum reservoirs throughout their life cycle. In order to accomplish the above, reservoir engineers collect and analyze relevant rock properties data from wide-ranging sources on a regular basis, including geological, petrophysical, geophysical, geochemical, well logging, drilling, well testing, and production data. Reservoir simulation studies may also aid in estimating various rock characteristics based on observed reservoir performance. The overall objective is to gather a detailed description and build a conceptual and comprehensive model of the reservoir, which leads to the formulation of a strategy to produce the reservoir optimally and maximize recovery.

Information related to the rock and reservoir is quite limited in the early stages of development of a petroleum reservoir as only a few wells are drilled. At this point, reservoir engineers may rely heavily on regional trends of geology and their experience in developing similar reservoirs. Once more wells are drilled, cored, and logged; a detailed picture begins to emerge about rock properties and reservoir heterogeneities affecting reservoir performance.

This chapter describes the characteristics of reservoir rocks that play a vital role in reservoir engineering and provides answers to the following:

- What are the key rock characteristics that reservoir engineers must be familiar with?
- How do rock properties influence reservoir performance?
- How are rock properties shaped in the first place? Can the properties change once the rock is formed?
- Do depositional environment and rock type determine rock characteristics?
- Do rock properties vary significantly from one location to another in the same geologic formation? Are reservoirs inherently heterogeneous?
- On what scale do the rock heterogeneities occur?
- Are rock properties influenced by the fluids contained in the rock?
- What are the common methods for measuring rock properties?
- How are data related to the rock properties used in reservoir analysis? Does it require the involvement of cross-functional teams?

The chapter concludes with a case study related to the analysis of rock and reservoir characteristics obtained from a number of basins in North America in relation to well production trends.

Reservoir Engineering. http://dx.doi.org/10.1016/B978-0-12-800219-3.00003-6

Properties of conventional and unconventional reservoir rocks

Rock properties of petroleum reservoirs are broadly classified as static and dynamic [1]. Static properties are shaped by the depositional environment in ancient times and various geologic events that occurred during the postdepositional period. The depositional environment influencing the type and characteristics of rock was treated in the previous chapter. The static rock properties include:

- Porosity
- Pore size and distribution
- Pore throat diameter
- Permeability
- Rock compressibility

Porosity and permeability, described in detail later in the chapter, are the two fundamental properties of rock responsible for storing and producing petroleum, respectively. Porosity relates to the microscopic void spaces in rock where oil and gas are accumulated, and permeability indicates the ability of the reservoir fluids to flow through continuous pathways or conduits that exist in rock.

Dynamic properties of rock are influenced by the interaction between rock and fluid properties in the reservoir. Of interest to the reservoir engineers are:

- Relative permeability
- Fluid saturation
- Capillary pressure
- Wettability

Dynamic rock properties such as relative permeability to oil may change significantly at a reservoir location with time as fluid saturation changes during production. Detailed information of fluid saturation, among others, is required to estimate oil and gas volumes present in the reservoir.

For certain unconventional reservoirs that produce from ultratight source rock, the geochemical and geomechanical properties of rock are also of prime interest. The geochemical properties indicate how much hydrocarbon is stored in source rock and its level of maturity. The most commonly sought after properties include the following:

- Total organic carbon (TOC)
- Vitrinite reflectance (VR)

The geomechanical properties indicate how effective hydraulic fracturing will be to produce the tight formation through a network of natural and artificial fractures.

- Young's modulus
- Poisson's ratio
- Fracture stress

Porosity of rock

Typical reservoir rocks have a microscopic network of pores where reservoir fluids are stored. The porosity of a rock is defined as the volume of pore or void spaces present in the rock divided by the bulk volume of the porous rock. The bulk volume of rock is comprised of both pore volume and the volume of solid rock matrix, which is referred to as grain volume.

$$\text{Porosity of rock}, \% = \frac{\text{Volume of porous spaces in rock}}{\text{Bulk volume of rock}} \times 100 \qquad (3.1)$$

For a rock having a bulk volume of 1.0 ft.^{3} and a pore volume of 0.12 ft.^{3}, the porosity of the rock would be 12%. The grain volume would be $1.0 - 0.12 = 0.88$ ft.^{3}. Knowledge of porosity is required to estimate the total volume of oil and gas in place in a reservoir.

Absolute and effective porosity

Not all the pores are interconnected with each other in reservoir rocks, which lead to the concept of effective porosity. It is distinguished from absolute porosity as follows:

$$\text{Absolute porosity}, \% = \frac{\text{Volume of all pores and voids in rock}}{\text{Bulk volume of rock}} \times 100 \qquad (3.2)$$

$$\text{Effective porosity}, \% = \frac{\text{Volume of interconnected pores in rock}}{\text{Bulk volume of rock}} \times 100 \qquad (3.3)$$

Effective porosity is less than the absolute porosity as the effective porosity accounts only for the interconnected pores. Knowledge of effective porosity is important as oil and gas flow can occur only through an interconnected porous network. Any fluid contained in dead-end pores does not contribute to production.

Primary and secondary porosity

The porosity of rock can further be classified as primary and secondary. Primary porosity relates to the pores that initially developed in rock. Secondary porosity may develop due to the various geological and geochemical processes that may occur following deposition. Postdepositional events responsible for the development of secondary porosity include:

- Leaching and dolomitization of carbonate rocks by certain chemical solutions
- Formation of vugs or cavities
- Development of microscopic fissures and fractures

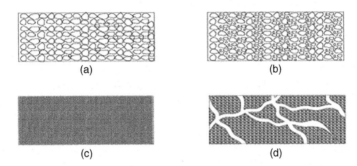

Figure 3.1 Microscopic view of reservoir rock types: (a) well-sorted grains, (b) poorly sorted grains, (c) fine grains as in shale, and (d) solution channels in carbonates.

Secondary porosity is commonly observed in carbonate rocks, although certain sandstones may be found to have porosity that developed in later stages. Many of the petroleum reservoirs are found in dolomite formations. Secondary porosity in rocks often contributes to the reservoir heterogeneities and may introduce additional uncertainties in reservoir analysis. Factors like grain size and sorting, which affect rock porosity as well as the development of secondary porosity, are presented in Figure 3.1.

Range of porosity in petroleum reservoirs

Porosity values in conventional oil and gas reservoirs typically range between 5% and 25%. Larger grains in rock lead to higher values of porosity. Primary porosity is generally higher in sandstones than carbonates. However, porosity values around 30% or more are observed in certain carbonate reservoirs where the development of secondary porosity is quite significant. Certain gas reservoirs with somewhat lower porosity, however, are produced on a commercial scale. In general, higher porosity rocks are associated with good flow characteristics, in addition to the obvious fact that these rocks can store relatively large volumes of oil and gas.

Unconventional reservoir rocks such as shale have porosity values at the lower end of the spectrum. In most shale gas reservoirs, porosity is relatively low, often ranging between 2% and 6%. However, porosity of shale up to 10% has been reported in certain cases. Pore throat diameters in shale are usually much smaller than what is found in conventional reservoir rocks. The size of pore throats can be in nanometers in shale.

Very "tight" sandstone and carbonate reservoirs with limited porosity (usually in single digits) may be regarded as unconventional reservoirs requiring nontraditional approaches to produce on a commercial scale.

Cutoff porosity and net formation thickness

In producing petroleum reservoirs, a lower limit of porosity (and permeability) exists below which oil production is not economically significant. The reasons are that the volume of oil contained in low porosity rock is limited, and the rock is not conducive to flow due to relatively low permeability that is generally associated with low porosity.

The limiting value is known as cutoff porosity. Typical porosity cutoff points are found to be around 5% in conventional oil reservoirs. Hence, only the portion of the geologic formation showing greater porosity is considered in reservoir performance predictions. An implicit fact is that the geologic intervals with higher porosity have better permeability. However, it must be mentioned that many tight and unconventional reservoirs have lower porosity and are often produced through a network of natural and induced fractures.

The concept of cutoff porosity leads to the introduction of net thickness as opposed to gross thickness of a geologic formation in estimating oil and gas reserves. Net thickness represents the portion of the hydrocarbon-bearing formation that can be produced by conventional means where porosity is relatively high. Typical values of the net to gross thickness ratio are in the range of 0.65–0.85.

Fracture porosity

Certain geologic formations, including petroleum reservoirs, are found to have fractures, fissures, and joints, which form due to the various stresses that act on the rock throughout geologic times. The porosity of fractures in oil and gas-bearing rock is generally very small. Literature review suggests that fracture porosity in petroleum reservoir rocks ranges between 1% and 3% or even less. Fractured reservoirs, both conventional and unconventional, are significant producers of oil and gas due to the very high conductivity of the fractures.

Measurement of porosity

Properties of rock, including porosity, change from one reservoir location to another. Again, porosity of same rock type in various geologic layers or facies within a formation may be quite dissimilar due to changes in depositional environment over geologic time. Porosity at various well locations is obtained from petrophysical studies based on log and core analyses. Suitably averaged values of porosity may be used to analyze the reservoir where the rock is assumed to be relatively homogeneous. However, in detailed reservoir studies, mathematical algorithms are utilized to compute porosity values at reservoir locations where data are unavailable.

The oil and gas industry employs a wide range of methods and tools for measuring the porosity and other important properties of reservoir rock. The traditional methods for measuring rock porosity include coring and logging. With the advent of technology, other methods were introduced, including the measurement while drilling (MWD) and nuclear magnetic resonance (NMR).

Porosity based on core samples

In the laboratory, the absolute or total porosity of a core sample can be obtained by first noting the original bulk volume of the core sample; the core is then crushed to obtain the matrix or grain volume. The difference between the two yields pore volume and porosity of the sample. In equation form:

$$\text{Absolute porosity}, \% = \frac{\text{Bulk volume} - \text{grain (crushed) volume of core sample}}{\text{Bulk volume of core sample}} \times 100 \quad (3.4)$$

The effective porosity is determined by filling a dry core sample with a fluid of known density. The increase in weight of the fluid-filled core and the density of the fluid are used to compute the volume of fluid that entered the core. Since the fluid can enter only the interconnected pores of the rock, effective porosity of the core can be calculated as the fraction of fluid volume over the volume of the core. Hence, the corresponding equation to determine the effective porosity of the sample is expressed as:

$$\text{Effective porosity}, \% = \frac{\text{Volume of interconnected pores in core}}{\text{Bulk volume of core sample}} \times 100 \quad (3.5)$$

Bulk volume and grain volume of a core sample can also be measured by displacement method where a core sample is immersed in water and the volume of water displaced is noted (Figure 3.2).

The effective porosity of a core can also be measured by an instrument called a porosimeter as shown in Figure 3.3. A dry core sample is placed in a vacuum chamber. An inert gas, such as helium, is then allowed to flow into the chamber. The resulting increase in the volume of gas, which represents the connected pore volume of the sample, is calculated by noting the increase in pressure in the chamber and then applying Boyle's law.

A helium pycnometer can be used to measure porosity of shale where pore throats are extremely small and measured in nanometers (10^{-9} m).

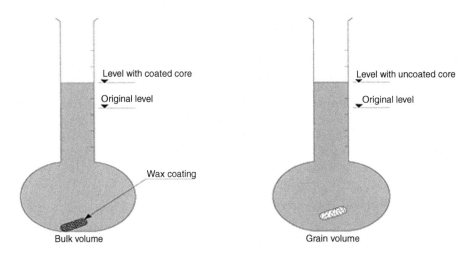

Figure 3.2 Measurement of bulk volume and grain volume or core by displacement method.

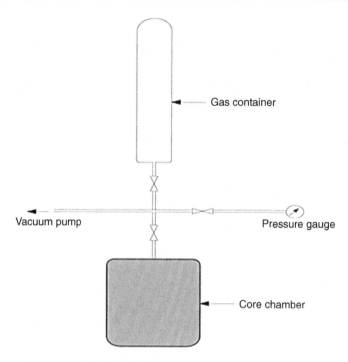

Figure 3.3 Measurement of effective porosity.

Logging

The logging tools that are widely used to determine porosity are acoustic, density porosity, and neutron porosity. The subsurface tools that are introduced in boreholes to measure various properties of rock typically consist of a source and one or more receptors. The source is used to emit some type of energy into the formation, for example, acoustic, electric, or nuclear, which bounces back to the receptors with a signature based on rock characteristics and fluid saturations in the reservoir.

Acoustic or sonic logs operate on the fact that sound waves travel relatively slowly through fluid saturated porous rocks as compared to the speed at which they can travel through the solid rock having the same lithology.

Density tools emit gamma rays that collide with the electrons of rock and the fluids present in rock pores, leading to the determination of the electron density of the subsurface formation. The electron density is proportional to the bulk density of fluid-filled rock. Since the densities of sandstone, limestone, dolomite, and water are known, the measured bulk density of the formation is used to estimate the porosity values across the formation thickness.

Similarly, neutron porosity tools emit high energy neutrons in subsurface formation that are slowed down by colliding with the nuclei of the formation materials. The loss of energy of the emitted neutrons is most pronounced during the collision with hydrogen atoms contained in the pore fluids, namely, oil and water. Porosity interpretations are adjusted for natural gas-filled pores as gas has significantly less hydrogen density. Furthermore, porosity corrections are also needed for any shale volume present in the formation, since petroleum fluids trapped in shale are not easily producible.

Nuclear magnetic resonance

The NMR technique is based on the fact that the NMR signal is proportional to the quantity of hydrogen nuclei present in the fluid in rock pores. The response from the NMR tool indicates how frequently the hydrogen nuclei of the formation fluid collide with the grain surface. In larger pores, the frequency of collision is relatively less.

Logging while drilling

The logging while drilling (LWD) operation was introduced in the late 1980s. Various tools and sensors are incorporated in the borehole assembly during drilling of a well. Tools that measure porosity, such as sonic and neutron porosity, are included. LWD has the advantage in that real-time data are obtained while the formation is relatively undamaged due to the invasion by drilling fluids, among other effects.

Permeability

Permeability of rock indicates how easily the reservoir fluids can move within the porous network. Relatively high values of rock permeability lead to good well productivity and better recovery efficiency from the reservoir as a general rule. In most unconventional reservoirs, however, rock matrix permeability is found to be extremely low rendering traditional methods of production ineffective. Technological innovations enable oil and gas recovery from unconventional resources on a commercial scale where natural or artificial conduits for flow of fluids play a crucial role.

Darcy's law

The definition of permeability is based on an empirical correlation developed by French engineer Henry Darcy [2]. In 1856, Darcy conducted an experiment where water was allowed to flow through a porous sand bed under a known hydraulic head (Figure 3.4). The flow rate of water was found to be proportional to the hydraulic head of water. The constant of proportionality is known as the hydraulic conductivity of the porous medium.

Mathematically, the hydraulic conductivity of a porous medium between points 1 and 2 can be expressed as follows:

$$q = \frac{KA\Delta h}{L} \tag{3.6}$$

where q = volumetric flow rate, m^3/s; K = hydraulic conductivity of the porous medium (sand bed), m/s; A = cross-sectional area of flow, m^2; Δh = hydraulic heads between points 1 and 2, m; L = length of the porous medium between points 1 and 2, m.

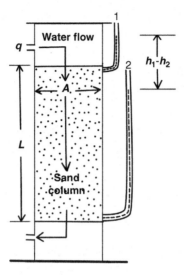

Figure 3.4 Schematic of hydraulic conductivity measurement apparatus. A similar setup was used by Darcy.

The Equation (3.6) is valid for steady-state laminar flow of fluid in a homogeneous medium, meaning that the properties of porous medium, including hydraulic conductivity, are uniform. It must be borne in mind that the geologic formations are not homogeneous. Moreover, unsteady state fluid flow is encountered due to shutting and reopening of wells in a reservoir. Steady-state and unsteady-state fluid flow are described in Chapter 9.

Darcy's law is found to be valid for other fluids, such as reservoir oil, gas, and formation water when Equation (3.6) is extended to include the viscosity of fluid. Hence, Equation (3.6) can be modified to account for the effect of fluid viscosity on flow rate as follows:

$$v = \frac{q}{A} = -\left(\frac{k}{\mu}\right)\left(\frac{\delta p}{\delta L}\right) \tag{3.7}$$

where v = fluid velocity, cm/s; k = average rock permeability, Darcy (D); μ = fluid viscosity, cp; $\delta p/\delta L$ = pressure gradient that drives the fluid, atm/cm.

Equation (3.7) states that the permeability of a porous medium is a function of volumetric flow rate of fluid through the porous medium, length of the porous medium, cross-sectional flow area, and pressure gradient under which the flow of fluid occurs. Fluid flows in the opposite direction of increasing pressure, hence a negative equation appears in the expression. In an inclined plane, Equation (3.7) is modified as follows:

$$v = \frac{q}{A} = -\left(\frac{k}{\mu}\right)\left(\frac{\delta p}{\delta L} - 0.433 y \cos\alpha\right) \tag{3.8}$$

where y = specific gravity of flowing fluid (water = 1.0); α = inclination of dipping bed measured from vertical direction.

There are important points to infer from Equation (3.8):

- All other factors remaining the same, a high value of rock permeability would generally lead to better flow rate (and production) from the reservoir.
- A higher hydraulic head or pressure differential is needed to increase fluid flow rate in the porous media.
- Water is less viscous than oil and will tend to flow with more ease in the reservoir.
- Gas will flow at a significantly higher velocity than oil as it has much lower viscosity.
- Note that the porosity term does not enter Darcy's law explicitly.

Unit of permeability

A porous medium would have a permeability of 1 D when a fluid having viscosity of 1 cp flows through a cross-sectional flow area of 1 cm^2 at a rate of 1 cm^3/s under a pressure gradient of 1 atm/cm. Since Darcy is a rather large unit of permeability, smaller units of permeability are expressed in millidarcies, microdarcies, and nanodarcies. Table 3.1 presents the typical permeability range in conventional and unconventional reservoirs.

Table 3.1 **Range of permeability in conventional and unconventional reservoirs**

Permeability scale	Symbol	Conversion	Notes
Darcy	Darcy or D		Conventional reservoirs. Certain carbonate reservoirs with significant secondary porosity, fractures, vugs, and solution cavities may have a permeability value over a Darcy. Besides, some sandstone formations are also highly permeable.
Millidarcy	mD	10^{-3} D	Conventional reservoirs. Permeability typically ranges from a few millidarcies to a few hundred millidarcies. Permeability of "tight" reservoirs is only a small fraction of mD. Unconventional reservoirs. Typical coalbed methane reservoirs may have permeability between 1 mD and 100 mD.
Microdarcy	μD	10^{-6} D	Unconventional reservoirs. Tight gas sands have permeability in microdarcies. Shale reservoirs have matrix permeability in hundreds of nanodarcies.
Nanodarcy	nD	10^{-9} D	
Picodarcy	pD	10^{-12} D	Virtually impermeable

In oilfield units, Darcy's law can be expressed as follows:

$$q = 1.127 \times 10^{-3} \frac{kA}{\mu L} \Delta P \tag{3.9}$$

where, q is bbl/d, k is in mD, ΔP is in psi, A is in ft.^{2}, L is in ft., and μ is cp.

Radial permeability

Fluid flow pattern around a vertical well is predominantly radial (Figure 3.5). Hence, a value of radial permeability can be calculated based on the knowledge of well dimension, reservoir characteristics, and pressure. The above is accomplished by conducting a transient well test that relates to a fairly large area of the reservoir. Radial permeability obtained by well test, as opposed to core permeability, is affected by the large-scale reservoir heterogeneities and the presence of multiple fluid phases. The equation for radial permeability is in the following form:

$$q = 7.08 \times 10^{-3} \left[\frac{kh(p_e - p_w)}{\mu B_o \ln(r_e / r_w)} \right] \tag{3.10}$$

where q = well rate, STB/day; k = average permeability, mD; p_e = reservoir pressure at outer radius of drainage area, psi; p_w = pressure at wellbore, psi; r_e = outer radius of drainage area, ft.; r_w = wellbore radius, ft.; B_o = oil formation volume factor, rb/STB.

The oil formation factor is a measure of change in volume of oil as it is produced. It is discussed in Chapter 4.

Equation (3.9) can be derived by writing Equation (3.9) in radial coordinates as follows:

$$q = \frac{kA}{\mu(\delta p / \delta r)} \tag{3.11}$$

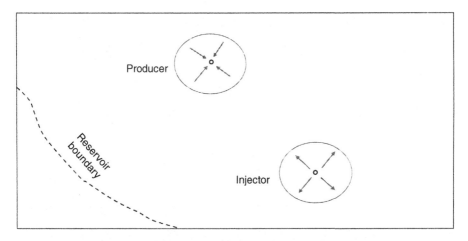

Figure 3.5 Radial flow of reservoir fluid around producer and injector.

Equation (3.11) is then integrated between p_e, p_w, and r_e, r_w to calculate the radial flow rate. As the pressure of fluid decreases with decreasing value of r, the negative sign is eliminated from Equation (3.11).

Measurement of permeability

The permeability of porous medium as calculated by using Equations (3.9) and (3.10) is referred to as absolute permeability as only one fluid is flowing. The most common method of measuring absolute permeability is flooding a core sample in the laboratory with a single-phase fluid (either brine or oil or gas) until a steady-state flow condition is attained. The fluid saturates the core completely. The core is first cleaned, dried, and placed in vacuum chamber to expel all the air in pores and be free of any contaminants. The attainment of the steady-state fluid flow condition in the core sample is indicated by the same fluid flow rate at the inlet and outlet of the core. A constant pressure drop of the flowing fluid across the core is also observed when the steady-state condition is attained. The experiment is repeated at various flow rates and inlet pressures, and a straight line is drawn through the experimental points. The slope of the line is a function of core permeability. Known parameters such as core dimensions and the viscosity of the fluid are used in the analysis to determine the permeability of the core (Figure 3.6).

The air permeability of a core is determined as follows:

$$k_{air} = \frac{q_a P_a \mu L}{p_m \Delta p A} \tag{3.12}$$

where q_a = flow rate of air through the core, cc/s; p_a = atmospheric pressure, atm; $p_m = (p_1 + p_2)/2$, atm; p_1, p_2 = pressure at the two ends of the core, atm.

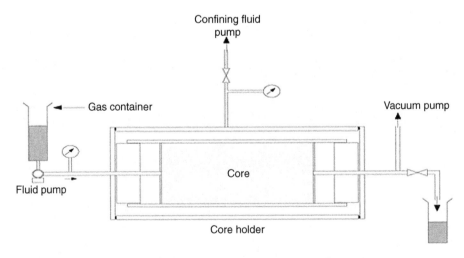

Figure 3.6 Measurement of absolute permeability of core sample.

It must be stressed that the value of permeability obtained in the laboratory is likely to be affected by various factors, including the contrast between reservoir and laboratory environment, namely, pressure and temperature, integrity of rock during coring, and core handling procedure from field to the laboratory. For example, core permeability is affected significantly due to the microfractures that may develop during coring or the core may be exposed to contaminants during handling.

Various methods are employed to determine rock permeability. It is important to recognize that each method may represent rock permeability in a different scale, ranging from about an inch, as in core analysis, to thousands of feet, as in well testing.

Measurement of ultralow permeability by pressure decay method

Ultralow permeability, as encountered in shale reservoirs, is measured in the laboratory by the pressure decay method [3]. The sample having permeability in nanodarcies is crushed to a specific range in size so that the representation of the rock is retained. The crushed particles of spherical shape have large surface area allowing diffusion of gas. The particles are then placed in a sealed chamber and a pressure pulse is transmitted through the particles. The resulting pressure decay of the pulse with time is caused by the diffusion process. The results are then analyzed to measure porosity and permeability of the rock sample. The procedure is advantageous in the sense that any microcracks that are usually present in core samples do not affect the measurements as crushed particles are used. However, the procedure requires a good understanding of the processes involved during pressure decay, and the results need to be calibrated with permeability of similar samples obtained by other methods.

Klinkenberg effect

During the measurement of core permeability based on the flow of gases through the core, it is observed that the apparent permeability values obtained by using various gases are not the same. Furthermore, the values are higher than that obtained by using liquid. The phenomenon is attributed to slippage effect of gas and known as the Klinkenberg effect [4]. The liquid permeability is correlated to gas permeability by the following equation:

$$K_{liquid} = \frac{k_{gas}}{1 + b / P_m} \tag{3.13}$$

where b = Klinkenberg factor; p_m = mean flowing pressure, atm.

The Klinkenberg effect is apparent during the production of CBM, an unconventional resource of gas. As gas is produced from the coalbed, matrix permeability decreases initially due to the increase in *in situ* stress and closure of coal seams. However, at a further decline in reservoir pressure, the Klinkenberg effect combined with the shrinkage of coal matrix results in permeability enhancement. The topic is described in Chapter 22.

Non-Darcy flow

Fluid flow in porous media cannot always be represented by Darcy's law. A common example is the flow of fluid in the immediate vicinity of gas wells where the velocity of gas is quite high and creates turbulence. Consequently, the associated pressure drop is greater than what can be predicted by applying Darcy's law. Hence, it is referred to as non-Darcy flow. The effect of non-Darcy flow is dependent on the rate of flow of gas. Hence, Darcy's law, presented in Equation (3.7), is modified by introducing a nonlinear term that accounts for the additional pressure drop. Forchheimer [5] proposed the following equation that predicts flow rate as a function of pressure gradient:

$$-\left(\frac{dp}{dL}\right) = \left(\frac{\mu_g}{k}\right)\left(\frac{q_g}{A}\right) + \beta\rho_g\left(\frac{q_g}{A}\right)^2 \qquad (3.14)$$

where dp/dL = pressure gradient, atm/cm; μ_g = viscosity of gas, cp; q_g = flow rate, cc/s; ρ_g = density of gas, gm/cc; k = permeability of porous medium, D; β = non-Darcy flow coefficient, atm-sec^2/gm.

The non-Darcy flow coefficient varies from reservoir to reservoir. It is either determined by laboratory core studies or estimated from published correlations. Certain studies indicate that the non-Darcy flow coefficient is inversely related to porosity and permeability, and the relationship is nonlinear.

Fracture permeability

Many conventional and unconventional reservoirs produce predominantly through a network of fractures as the rock matrix permeability is low to ultralow. Permeability of the fractures can be several orders of magnitude higher than matrix permeability, ranging from hundreds of millidarcies to a Darcy or more. An equation that correlates fracture permeability with fracture width is the following:

$$k_{fracture} = \frac{h^2}{12} \qquad (3.15)$$

The equation is valid for any compatible units, such as the fracture thickness in m and fracture permeability in m^2. Fracture permeability is usually determined by well testing.

Dual porosity reservoir

Petroleum reservoirs having fractures and highly conductive channels are referred to as a dual porosity or dual porosity–permeability system. This is due to the fact that the values of fracture porosity and permeability are quite different in scale as compared to matrix porosity and permeability. In addition to conventional fractured reservoirs, dual porosity is common in many unconventional reservoirs, including shale gas and

CBM reservoirs. The net result is that a significantly different reservoir performance, including production rates, is observed than what is expected from a single porosity reservoir system.

Correlation between porosity and permeability

In many sandstone reservoirs where grains have fewer impurities and are well sorted, a good correlation can be obtained between porosity and permeability. It has been observed that a straight line can be drawn through the points when the log of permeability is plotted against porosity as follows:

$$\text{Log}_{10} k = m(\phi) + c \tag{3.16}$$

where m = slope of the straight line, c = intercept on the y-axis.

Note that the values of m and c would vary from reservoir to reservoir and in many cases, from one geologic layer to another within the same reservoir depending on depositional environment.

Referring to grain size, rocks made of coarser grain exhibit relatively high permeability for the same range of porosity, as depicted in Figure 3.7a. However, in carbonate formations with significant secondary porosity and fractures, permeability often varies widely for the same porosity without exhibiting a definitive trend. In unconventional shale reservoirs, matrix permeability is ultralow, and accurate measurement of permeability and identification of a definitive trend between porosity and permeability can be challenging.

Figure 3.7b highlights the trend between porosity and permeability in various formations from different geologic ages in the Cortes Bank area in offshore Southern California.

Permeability anisotropy

Permeability in vertical direction is generally found to be lower than that of horizontal permeability, sometimes by an order of magnitude or more. This occurs as a result of the alignment of the grains of rock during deposition influenced by flow of water, wind, etc. Grains are observed to be flaky and sediments are laminated. Multiple depositional sequences that take place through geologic times also play a part that results in contrasting horizontal and vertical permeability (Figure 3.8). The ratio of horizontal to vertical permeability is an important factor to consider in reservoirs experiencing water coning in oil wells. High vertical permeability may lead to severe coning and premature water breakthrough. Water slumping during waterflood can also be encountered due to high vertical permeability resulting in poor oil recovery.

It is further observed that rock permeability can be oriented in a certain direction in a reservoir. Depending on the direction of waves or wind that prevailed during deposition of sediments, the grains may be oriented in a specific direction. For example, permeability measured in a southeast direction may not be the same as in a northwest

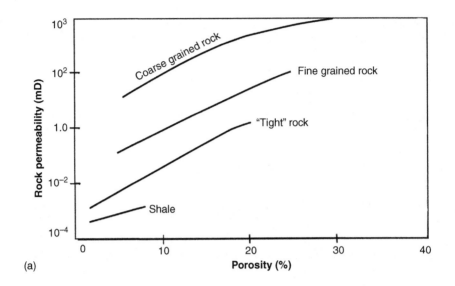

(a)

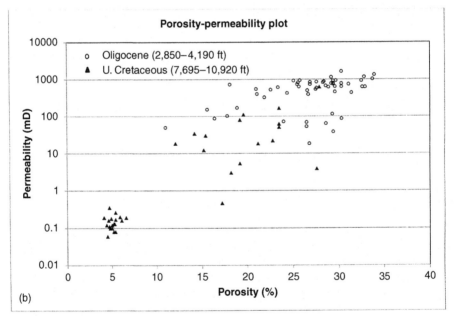

(b)

Figure 3.7 (a) Porosity versus permeability trend in various geologic formations. For a given porosity, rock permeability can vary widely depending on a number of factors, including grain size (plot is not to scale). (b) Porosity–permeability relationship in mostly sandstone formations located at various depths in offshore Southern California. The geologic ages of the formations vary significantly. Data are obtained from a single well. Note the variation in permeability by several orders of magnitude.
Source: Ref. [13].

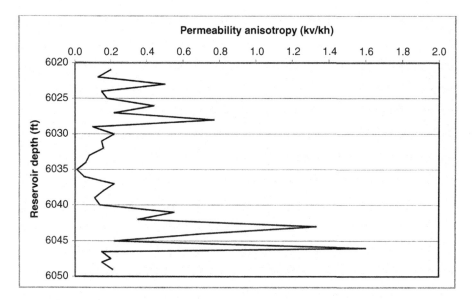

**Figure 3.8 Plot of vertical permeability over horizontal permeability with depth in a
heterogeneous formation.** k_v/k_h varies between 0 and 1.6 within a 30 ft. interval.

direction in a reservoir. These reservoirs are said to have directional permeability.
Again, the rock characteristics result from the conditions that prevailed during deposition of sediments causing the grains to align in a specific direction. Consequently, the
maximum permeability is also observed in that direction. The net result is the flow
of reservoir fluids in a certain preferential direction, potentially leaving significant
portions of oil underground and causing early breakthrough of water in wells that are
located in the path of flow.

The above phenomenon is referred to as permeability anisotropy and may play a
crucial role when an external fluid is injected to enhance oil recovery.

High permeability streaks

Another common occurrence of reservoir heterogeneity is the presence of high
permeability streaks in the reservoir as depicted in Figure 3.9. These are thin intervals of geologic formation that are highly conductive and extend laterally from
well to well. High permeability streaks may pose significant challenges during
water or gas injection into a reservoir resulting in early breakthrough of injected
fluid through the producers and undermining the enhanced oil recovery efforts
(Figure 3.10).

Any prediction of reservoir performance based on limited information on reservoir
heterogeneities may not be adequate, and should only serve as a starting point for the
reservoir engineers in analyzing the reservoir.

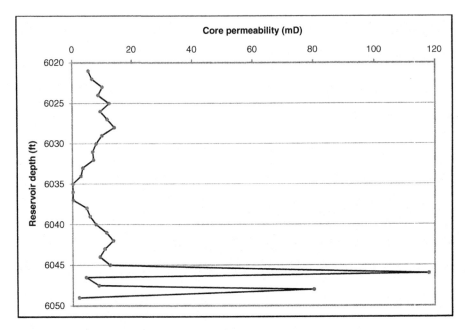

Figure 3.9 Depiction of permeability profile of a stratified reservoir. Note that the two layers are separated by thin impervious shale at 6036 ft. Furthermore, high permeability streaks are suspected at the bottom.

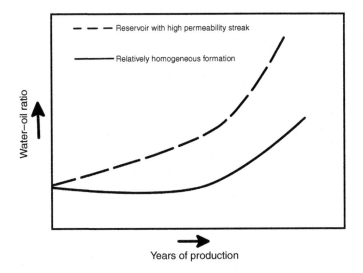

Figure 3.10 Comparison of performance between homogeneous and heterogeneous reservoirs. A high water–oil ratio is encountered early in the life of the reservoir when high permeability streaks are present.

Factors affecting rock permeability and porosity

Sorting and orientation of grains, size of pore throats, tortuosity of pore channels, degree of cementation between grains, and presence of impurities among other factors play critical roles in shaping the permeability of rock. Many of the above factors also determine the porosity of rock. Various effects on rock porosity and permeability are summarized in Table 3.2.

The variation of rock permeability in a reservoir is more pronounced than that of rock porosity. Again, the presence of microfractures, cavities and other heterogeneities contribute to the variations in permeability. In heterogeneous formations, rock permeability can vary by one order of magnitude or more, which can significantly affect reservoir performance.

Effect of reservoir depth

In a study of CBM reservoirs in three basins where depth varied from less than 100 ft. to 10,000 ft., a clear trend in permeability reduction was established with increasing reservoir depth. Formation permeability was quite high at shallow depths of 500 ft. or less, ranging between 100 mD and 1 D. However, permeability was reduced rapidly below 4000 ft. to less than 1 mD. At greater depths, the effects of *in situ* stresses are significant, resulting in the reduction of permeability.

However, studies of the Permian Basin and the Gulf of Mexico have shown that reservoir permeability generally increases with depth. Interestingly, porosity of rock was observed to decrease with depth. Various rock properties, including porosity and permeability, are dependent on the depositional environment prevalent at the time of the formation of rock. Hence, a generalized correlation of rock properties with depth cannot be made.

Formation compressibility

The compressibility of formation is a measure of the rate of change of pore volume with change of reservoir pressure. As oil and gas are produced, reservoir pressure declines resulting in the reduction of bulk as well as pore volume of rock due to the enormous pressure exerted by overlying geologic strata. Rocks are usually slightly compressible; hence, the decrease in pore volume is quite small. Unconsolidated sandstone and other rocks in certain reservoirs are known to be highly compressible in comparison to others.

Formation compressibility is an important parameter in reservoir analysis due to a number of reasons, which include:

- During the primary production of a reservoir, the phenomena of reduction in pore space along with the expansion of reservoir fluids contribute to the driving energy.
- Changes in pore volume occur as reservoir pressure declines due to formation compressibility. Hence, estimation of oil and gas in place is affected.
- The flow of fluids through the compressible rock may decrease noticeably once the reservoir pressure decreases following production of oil and gas. This occurs as rock porosity and permeability are reduced.

Table 3.2 Permeability and porosity affected by rock characteristics

Characteristic	Porosity	Permeability	Notes
Grain size	Ideally, grain size does not have an impact on porosity as long as all grains in rock are uniform.	Smaller grain size leads to narrower pore channels in rock leading to low permeability.	Conglomerates are coarse grained, followed by sand, silt and shale in order of decreasing grain size. Sandstone grains range from very coarse (1–2 mm) to fine (1/8–1/16 mm). Shale is very fine grained, the size of grain being smaller than 1/256 mm. As grains get smaller, pore space and pore throat openings diminish in size.
Sorting of grains	In rocks with poorly sorted grains, larger pores are occupied by smaller grains leading to lower porosity.	Pore channels are smaller and tortuous, leading to lesser permeability.	Clean and well-sorted sandstones have relatively good porosity and permeability.
Cementation	Relatively high degree of cementation results in smaller porosity.	Similarly, highly cemented rock leads to lesser permeability.	
Pore throat diameter	Rocks having smaller pore throats have relatively low porosity.	While the porosity can be similar, presence of smaller pore throats may result in significantly lower permeability in rock by several orders of magnitude.	Ranges from about 2 mm in sandstones in conventional reservoirs to a fraction of mm in tight formations. It is extremely small in ultra-tight tight shale, ranging from micrometers down to nanometers.
Solution and leaching	The postdepositional process of leaching of carbonates leads to secondary porosity in rock.	Rock permeability is enhanced.	Solution effects are mostly found in carbonates. However, sandstones may exhibit such effects.
Presence of fractures	Fracture porosity is significantly lesser than matrix porosity of reservoir rock.	Fracture permeability is larger than matrix permeability by one or more orders of magnitude.	Reservoir acts like dual porosity and dual permeability system.

Formation compressibility is expressed mathematically as follows:

$$c_f = -\frac{1}{V_\varphi}\left(\frac{\delta V_\varphi}{\delta P}\right)T \tag{3.17}$$

where c_f = formation compressibility, 1/psi or psi^{-1}; V_φ = pore volume of rock, ft.^3; P = pressure exerted on formation, psi.

Subscript T denotes that the pore volume change with pressure takes place under isothermal conditions.

Values of formation compressibility are given in units per pounds per square inch, and typically reported in the range of 3–$12 \times 10^{-6} \text{ psia}^{-1}$ for sandstones and carbonates.

Based on laboratory studies, correlations are developed and available to estimate sandstone and limestone compressibility [6]. It is found that formation compressibility is inversely proportional to porosity; however, the relationship is nonlinear.

Sandstone:

$$c_f = \frac{97.32 \times 10^{-6}}{(1 + 55.8721\varphi)^{1.42859}} \tag{3.18}$$

Limestone:

$$c_f = \frac{0.853531}{(1 + 2.47664 \times 10^6\,\varphi)^{0.9299}} \tag{3.19}$$

The above correlations are valid in 2–33% porosity range, with maximum error of 2.6% for sandstone and 11.8% for limestone.

Rock compressibility and bulk compressibility

Note that both the rock matrix and pore volumes are affected by change in pressure. Rock matrix compressibility, c_r, and the bulk compressibility, c_b, of rock are defined in a similar manner. Rock matrix compressibility, simply referred to as rock compressibility, is not the same as pore compressibility or formation compressibility.

Changes in rock porosity due to compressibility

As reservoir rocks are compressible, a reduction in porosity occurs due to reduction in pore pressure. The change in porosity can be estimated as:

$$\phi = \phi_0 \exp[c_f(p - p_0)] \tag{3.20}$$

where p_0 = original pressure, psi; ϕ_0 = original porosity, fraction.

Surface and interfacial tension

In porous media, oil, gas, and water coexist as distinct fluids due to the immiscibility of one fluid with another. The attractive force between the molecules of oil is different than that of water or gas. At the interface or boundary of two immiscible fluids, a very thin film develops due to interfacial tension as the pressure exerted by each fluid is not the same. Studies have reported the thickness of the film in the order of 10^{-7} mm. The physical force that develops at the surface of liquid in contact with an immiscible gas or air is referred to as surface tension in the literature. The net result of interfacial tension or surface tension in rock pores is to affect the fluid flow characteristics in terms of the following:

- Fluid flow rate in porous medium
- Pressure of individual fluid phases
- Preference of one fluid flowing over others in the reservoir

The effects of surface and interfacial tension ultimately reflect on the reservoir performance. Interfacial tension also affects other dynamic properties of rock, including wettability, capillary pressure, and relative permeability of rock to oil, gas, and water as described in the following sections.

Fluid saturation

Most reservoir engineering analysis requires knowledge of oil, gas, and water saturation in geologic formation. Of interest are the following:

- The variation of fluid saturation in the reservoir from one location to another
- Changes in saturation with time as well production continues
- The effects of external fluid injection on the saturation of reservoir fluids

Oil saturation is the ratio of pore space occupied by oil over the total pore space; the rest of the pore space is occupied by either gas or water or both. Similarly, gas saturation is the fraction of pore space occupied by the gas phase. Fluid saturation is also reported in percent in the literature. It is obvious that the saturation of all fluid phases present in fluid-filled pores would add up to 100%.

In conventional oil reservoirs, typical values of oil saturation range between 65% and 85%. Certain oil reservoirs have a gas cap with measurable gas saturation. Dry gas reservoirs generally have high saturation of natural gases, namely, methane, ethane, and heavier components. Examples of oil, gas, and water saturations are noted in Table 3.3.

Fluid saturation in a core sample can be measured directly by heating and evaporating the fluids from the core and then condensing the vapors in a graduated tube. The procedure is referred to as the Dean–Stark method. Heat is imparted by vaporizing toluene (Figure 3.11). Oil and water, being immiscible, are separated in the tube and the volume of water is measured to determine water saturation in the core. Once the water volume and change in the mass of the core before and after the procedure are known, the oil volume that was originally present in the core can be determined.

Fluid saturation in a geologic formation is routinely determined by resistivity log as a well is drilled. Resistivity log is described later in the chapter.

Table 3.3 Examples of oil, gas, and water saturations in petroleum reservoirs

Reservoir type	Oil saturation $(S_o, \%)$	Gas saturation $(S_g, \%)$	Gas condensate saturation $(S_{gc}, \%)$	Connate water saturation $(S_w, \%)$
Oil without a gas cap	65–85	0	0	15–35
Oil with a gas cap	60–70	5–15	0	20–30
Dry gas	0	70–85	0	15–30
Gas condensate	0	40–60	20–40	20

Note: The numbers provided in the table are approximate and should be viewed as a general guide only.

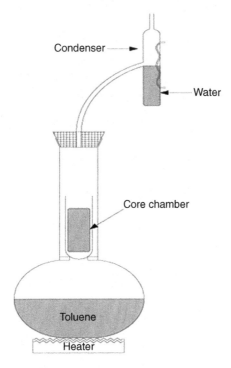

Figure 3.11 Measurement of fluid saturation by the Dean–Stark method.

Irreducible water saturation and movable oil saturation

During the life cycle of a reservoir, not all the oil is producible. Hence, engineers are interested in estimating the movable oil saturation under primary and various enhanced production mechanisms described later in the book. At the abandonment of the reservoir, residual oil saturation indicates the oil volume left behind. Hence, the movable oil saturation can be expressed as:

$$S_{om} = S_{oi} - S_{or} \qquad\qquad\qquad\qquad (3.21)$$

In highly heterogeneous reservoirs, the amount of oil left behind can be significant following primary or secondary recovery. With the advent of new technologies, attempts are made to identify the areas of high residual oil saturation in the formation and produce further by drilling new wells or recompleting existing wells. The topic is presented in Chapter 20. However, due to the effects of wettability and interfacial tension, certain volumes of oil cannot be produced from the reservoir. The minimum saturation of oil that will remain in the porous rock is estimated by conducting laboratory experiments. Wettability and interfacial tension are treated later in the chapter.

Saturation of oil, gas, and water in rocks strongly influences various other properties, including the flow characteristic of the individual fluids described later.

Sorption

In conventional reservoirs, oil and gas remain in the free state in rock pores. However, in certain unconventional reservoirs such as coalbed and shale, gas may remain in the absorbed state in the miniscule pores. Hence, sorption characteristic of unconventional rock is a critical parameter in evaluating an unconventional reservoir. The amount of adsorbed gas is dependent on pore size, type of organic material, mineral composition, and thermal maturity of the rock. Studies indicate that about 15–80% of total gas in shale may remain in the adsorbed state. However, in coalbed formations, virtually all of methane accumulation is found in the adsorbed state.

As the reservoir pressure declines, gas is desorbed from rock and often flows to the wellbore through a network of fractures. The amount of gas desorbed can be estimated by placing the core in a canister and noting the changes in saturation with decreasing pressure. The gas desorption process is modeled by Langmuir isotherm described in Chapters 12 and 22.

Wettability of reservoir rock

The wettability of rock indicates the tendency of one immiscible fluid to spread in the presence of another fluid on the surface of rock. The property of rock is demonstrated by the fact that oil and water tend to spread and adhere to rock surfaces differently (Figure 3.12). The wettability of a fluid is identified by the contact angle of the fluid droplet with the solid surface. When the rock is water-wet, droplets of water spread out to a greater surface area leading to a contact angle with the surface of less than 90°. However, the contact angle would be greater than 90° on oil-wet rocks. There are certain reservoirs that exhibit intermediate or mixed wettability, where the contact angle is about 90°.

The following points are noted with regards to wettability characteristics of rock:

- Wettability is a function of interfacial tension between dissimilar fluids in pores; it is also a function of interfacial tension between fluid and pore surface.

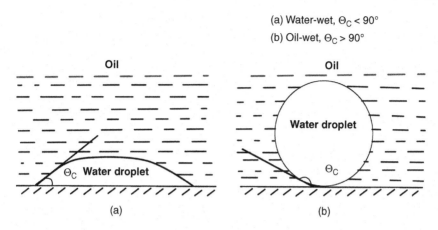

Figure 3.12 Depiction of (a) water-wet and (b) oil-wet surface. Depending on wettability characteristics, a droplet of water contacts the surface differently in the presence of oil. The contact angle determines the wetting preference of the droplet.

- Wettability is influenced by the type of minerals in the rock matrix and by the composition of the fluids, oil, and water in the pores of rock.
- Wettability of reservoir rock may be altered once it comes in contact with injected water. Certain chemical compounds are mixed with injected water to alter wettability in a favorable direction and facilitate recovery.

The majority of oil reservoir rocks are known to be water-wet, meaning that water will spread dominantly over oil in contact with the pore surface. Before the migration of oil into the reservoir, formation water filled the pores, and migrating oil could not expel the water completely as the latter adhered to the pore surface due to water-wet characteristics of the rock. However, oil-wet reservoirs as well as reservoirs exhibiting intermediate wettability are not uncommon.

Rock wettability and waterflood performance

Oil recovery by waterflooding a reservoir is dependent on the wettability of rock. In water-wet reservoirs, injected water displaces oil efficiently as oil has little tendency to adhere to the pore surface. However, in oil-wet reservoirs, performance of water-flood may be less than satisfactory. A relatively large volume of oil may remain in the reservoir due to the oil-wetting characteristics of the rock. Various laboratory investigations and field experience have confirmed the phenomenon.

Methods of measurement of wettability

In the simplest of the methods, wettability of rock is determined by placing a drop of water on the rock surface in the presence of oil and measuring the contact angle between water and rock. The above is shown in Figure 3.12. In a modified method, two parallel plates are used. Water is allowed to advance through a drop of oil placed in between the plates and the contact angle is measured. The process is thought to be similar to displacement of oil by injected water in the reservoir. Wettability is also determined by the Amott test and US Bureau of Mines method. The Amott test is based on the displacement of oil by water in a core and vice versa. The US Bureau of Mines method utilizes a centrifuge for the displacement of one fluid by the other in a core. Values of pressure and saturation obtained from the study are used to determine the wettability of core sample.

Capillary pressure

When two immiscible fluid phases, such as oil and water, are present in a porous medium a pressure differential is observed between the two fluid phases that can be expressed as capillary pressure.

A generalized expression for capillary pressure, as it relates to fluid phases in porous media, is the difference between the pressure exerted by the nonwetting phase and the pressure exerted by the wetting phase pressure.

The magnitude of the capillary pressure in a porous medium is influenced by fluid saturations, interfacial tension between the two fluid phases, and the radius of the pore and pore throat, among other factors.

$$p_c = p_{nw} - p_w \qquad (3.22)$$

where p_{nw} = nonwetting phase pressure, psi; p_w = wetting phase pressure, psi.

In an oil–water system, water is generally the wetting phase, and oil is the nonwetting phase. Hence, the oil–water capillary pressure can be expressed as follows:

$$P_{c,wo} = P_o - P_w \qquad (3.23)$$

where $P_{c,wo}$ = capillary pressure at the water–oil interface, psi; p_o = pressure exerted by the oil phase, psi; p_w = pressure exerted by the water phase, psi.

Furthermore, in a gas–water system, water is the wetting phase. Hence, the capillary pressure equation can be written as:

$$P_{c,gw} = p_g - p_w \qquad (3.24)$$

where $P_{c,gw}$ = capillary pressure at the gas–water interface, psi; p_g = gas phase pressure, psi.

Drainage and imbibition

During drainage in a core, the wetting phase fluid is replaced by a flowing nonwetting phase. In water-wet rock (as shown in Figure 3.13), water saturation is reduced as a consequence of the drainage process, while the saturation of the nonwetting phase,

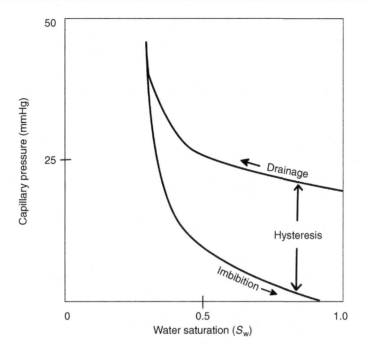

Figure 3.13 Capillary pressure as function of water saturation. During drainage and imbibition, hysteresis takes place as shown. The difference in capillary pressure between drainage and imbibition is exaggerated for illustration only.

oil, is increased. This process can be viewed as the desaturation of the wetting phase in porous media.

In contrast, the imbibition process involves an increase in the saturation of the wetting phase. During imbibition, the wetting fluid phase is allowed to imbibe into the core, thereby increasing its saturation.

Hysteresis effect

This phenomenon, referred to as the hysteresis effect, points to the fact that the capillary pressure in a porous medium would be influenced by the history of saturation changes. While studying a petroleum reservoir, it is important to know whether fluid saturation is either decreasing or increasing to determine the capillary pressure, in addition to the value of fluid saturation.

Methods of measurement

There are a number of methods available to measure capillary pressure. These are: centrifuge method, porous diaphragm method, mercury injection method, and Leverett method. Some of the methods are outlined in the following.

In the centrifuge method, a core saturated with oil is placed in a centrifugal chamber that is rotated at a specific speed until the oil is expelled from the core. The space in the core, filled with oil phase previously, is now filled with gas present outside the core. The centrifugal action at a specific speed is continued until all the oil is expelled. The procedure is repeated several times and the oil saturation is noted based on the pore volume and the oil volume that is expelled and collected. Capillary pressure is then calculated based on the density of oil and gas, rate of centrifugal rotation, axis of rotation, and the dimensions of the sample.

The porous diaphragm method is based on the principle of selective transport of wetting phase fluid through a semipermeable disk. A core sample saturated with brine is placed on a porous disk having much lower permeability than that of the sample. The fluid contained in the core is then displaced by a fluid under increasing pressure. Following each increment in pressure, the amount of displaced fluid is measured until equilibrium is attained.

Mercury injection is a rapid method of determining capillary pressure of a core sample. It is suited to measure the capillary pressure of an irregularly shaped sample in particular. The pore size distribution of rock can also be analyzed. However, the sample cannot be reused for further analysis.

Capillary number

The capillary number indicates the magnitude of viscous forces, represented by fluid viscosity multiplied by its velocity, in relation to forces due to surface or interfacial tension. It is a dimensionless group. In terms of effective permeability to water, rock porosity, and interfacial tension, capillary number can be expressed as follows:

$$N_{ca} = \left(\frac{C k_w \Delta p}{\phi \sigma_{ow} L} \right) \tag{3.25}$$

where N_{ca} = capillary number, dimensionless; C = a constant; k_w = effective permeability to water; ϕ = rock porosity, fraction; σ_{ow} = interfacial tension between oil and water.

In enhanced oil recovery operations, capillary number is an important number to consider. Where the value of capillary number is higher, viscous forces dominate and the effect of interfacial tension between fluids in the rock pores is reduced, thereby augmenting recovery. In typical reservoir conditions, capillary number varies from 10^{-8} to 10^{-2}. Effective permeability to water is described in the following section.

Effective permeability

Effective permeability of rock to a fluid phase (oil, gas, or water) in porous medium is a measure of the ability of that phase to flow in the presence of other fluid phases. For example, effective permeability to oil is a measure of its flow capability in the presence of water, and in some cases, in the presence of both water and gas phases. The same definition of effective permeability applies for gas, indicating its ability to flow in the presence of oil or water or both. Effective permeability to fluids is not the same as the absolute permeability of rock. The absolute permeability reflects 100% saturation of rock by a single fluid whereas the effective permeability to a fluid phase is based upon the presence of two or three fluid phases in porous medium. In laboratory studies, both oil and water phases are allowed to flow through the core to determine effective permeability to individual phases.

Effective permeability is dependent on fluid saturation. Consider an oil reservoir where water injection has just commenced to produce an additional volume of oil. The following observations can be made regarding the dynamic behavior of fluid flow:

1. The effective permeability to oil phase is at maximum initially. Connate water is immobile; hence, the effective permeability to water phase is zero.
2. However, as water is injected and oil is produced, the saturation of oil gradually decreases to residual oil saturation and the effective permeability to oil eventually reduces to zero.
3. During the final stage, effective permeability to water is at maximum.

Furthermore, the following should be noted:

• Not all the oil stored in rock pores will produce due to the forces arising out of surface tension, interfacial tension, wettability characteristics, and other factors.
• The sum of the effective permeability to oil and water is less than the absolute permeability of rock due to the effects of interfacial tension between the fluid phases.
• Since the values of oil and water saturation change with time at a given location of reservoir during production, the values of effective permeability to oil and water also change. A similar observation can be made about gas reservoirs.

Relative permeability

Relative permeability of rock to a reservoir fluid (oil, gas, or water) is defined as the ratio of the effective permeability of the respective fluid phase to the absolute permeability of the rock.

$$\text{Relative permeability to oil} = \frac{\text{Effective permeability to oil phase}}{\text{Absolute permeability of rock}} \tag{3.26}$$

$$\text{Relative permeability to gas} = \frac{\text{Effective permeability to gas phase}}{\text{Absolute permeability of rock}} \tag{3.27}$$

$$\text{Relative permeability to water} = \frac{\text{Effective permeability to water phase}}{\text{Absolute permeability of rock}} \tag{3.28}$$

Using symbols, we can write:

$$k_{ro} = \frac{k_o}{k} \tag{3.29}$$

$$k_{rg} = \frac{k_g}{k} \tag{3.30}$$

$$k_{rw} = \frac{k_w}{k} \tag{3.31}$$

where k_{ro} = relative permeability to oil, ratio; k_o = effectively permeability of rock to oil, mD; k = absolute permeability of rock, mD.

In Equations (3.29)–(3.31) the subscripts o, g, and w are used to indicate oil, gas, and water permeability. Note that the relative permeability is a ratio of two permeability values and does not have any units.

When two or more fluid phases are mobile in a reservoir, relative permeability is the single most important characteristic that controls the production of the flowing phases. Typical scenarios where relative permeability plays a crucial role include the following:

- Primary recovery of oil and gas from a conventional oil reservoir
- Secondary recovery of oil and injected water during waterflooding
- Tertiary recovery of oil and injected chemicals during alkaline flood
- Primary production of CBM along with water stored in seams

Primary, secondary, and tertiary production from a petroleum reservoir is described in Chapters 16 and 17. Production of CBM, an unconventional resource, is discussed in Chapter 22.

In order to determine relative permeability, the values of effective permeability and absolute permeability are determined in the laboratory. Correlations are also available in literature to estimate the relative permeability of oil, gas, and water. The values of relative permeability range between 0 and 1. Hence, it serves as a common standard in reservoir studies regardless of the magnitude of permeability of a specific reservoir. Reservoir engineers seek detailed knowledge of relative permeability to fluids in predicting reservoir performance when designing secondary and enhanced petroleum recovery operations. Such operations involve the displacement of reservoir fluid (oil) by injected fluid (water or gas). In order to build realistic reservoir models, relative permeability data are required from cores obtained at various well locations and geologic layers. In the relative permeability figure, each curve represents an individual fluid phase, namely, oil, gas, or water (Figure 3.14).

Examination of typical relative permeability curves, an example of which is presented in Figure 3.14, reveals the following:

- The relationship between the relative permeability and phase saturation is nonlinear.
- For oil phase, a relative permeability value of 0 is encountered at limiting (end point) saturation where oil ceases to flow in porous medium. The limiting saturation for the oil phase is known as the residual oil saturation.
- In the case of the water phase, the end point saturation where water is immobile is referred to as the irreducible water saturation.

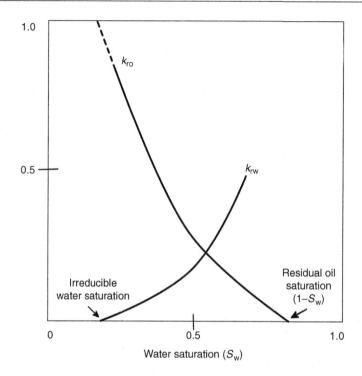

Figure 3.14 Relative permeability of rock to water and oil phases in porous medium.
The plot represents two-phase relative permeability, as both oil and water are present. Note
that, at $k_{ro} = 0$, $S_{or} = 1 - S_w$.

- Values of k_{ro} and k_{rw} between 0 and 1 indicate simultaneous flow of oil and water.
- When $k_{ro} = 1$, only oil phase is flowing; when $k_{rw} = 1$, only water phase is flowing.

Table 3.4 summarizes the relationship between relative permeability and saturation of fluid phase for an oil reservoir under water injection without any free gas present.

In Table 3.4, the following notations are used: S_{oi} = initial oil saturation at the start of waterflood; S_{or} = residual oil saturation at the end of waterflood; $S_{w,irr}$ = irreducible water saturation at the start of waterflood.

Note that the determination of end point saturations, $S_{w,irr}$ and S_{or}, are of interest to reservoir engineers to determine the ultimate recovery potential of the reservoir. The above is required in future planning and economic analysis. End point saturation can be determined in a relatively straightforward manner without conducting the measurements of relative permeability or capillary pressure.

Oil-wet, heterogeneous, and unconventional reservoirs

The shape of the relative permeability curve depends on wettability, heterogeneity, and other rock properties. The relative permeability characteristics of oil and water are

Table 3.4 **Relationship between relative permeability and saturation**

Stage of waterflood	Oil phase saturation (S_o)	Oil phase relative permeability (k_{ro})	Water phase saturation (S_w)	Water phase relative permeability (k_{rw})
At the start of waterflood	S_{oi}	1	$S_{w,irr}$*	0
During waterflood	$S_{oi} < S_o < S_{or}$	$0 < k_{ro} < 1$	$S_{w,irr} < S_w < 1 - S_{or}$	$0 < k_{rw} < 1$
At the end of waterflood	S_{or}*	0	$1 - S_{or}$	1

* End point saturation.

influenced by the wettability of rock. In water-wet rock, there is preferential tendency of the water phase to adhere to the pore wall; hence, the irreducible water saturation is higher. In oil-wet rock, however, wetting tendency of the oil phase is greater in comparison to the water phase. Hence, the irreducible water saturation is relatively low. The net effect is that the relative permeability curve of water in the oil-wet system is that it shifts to the right of what is observed for the water-wet system. Similarly, the residual oil saturation tends to be higher in oil-wet rocks, shifting the oil relative permeability curve to the left. In heterogeneous formations having high permeability streaks as well as in fractured formation, the water relative permeability curve may be steep, indicating that water breakthrough at the producers may occur rather prematurely.

Relative permeability characteristics of rock are not only important in conventional reservoirs during secondary recovery as displacement of oil occurs by water or gas injection. In unconventional CBM reservoirs, two-phase flow of water and methane occur during primary production [7]. Figure 3.15 depicts the typical relative permeability curves of water and gas during the production of CBM.

Apart from direct measurements in the laboratory, various correlations exist in the literature to estimate relative permeability values based on fluid saturation in conventional reservoirs. These correlations are frequently used in reservoir models in the absence of field data. Some widely known correlations are presented in the following [8].

Well-sorted sand (unconsolidated)

1. Oil–water relative permeabilities:

$$k_{ro} = (1 - S^*) \tag{3.32}$$

$$k_{rw} = (S^*)^3 \tag{3.33}$$

2. Gas–oil relative permeabilities:

$$k_{ro} = (S^*)^3 \tag{3.34}$$

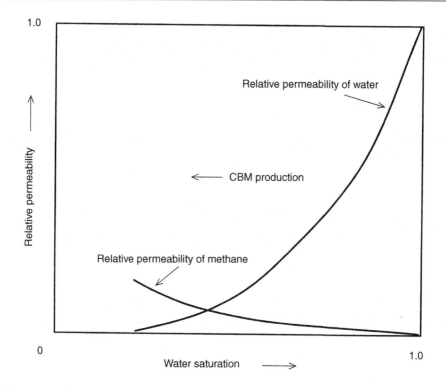

Figure 3.15 Relative permeability curves of gas and water depict the primary production of CBM.

$$k_{rg} = (1 - S^*)^3 \tag{3.35}$$

Poorly sorted sand (unconsolidated)

1. Oil–water relative permeabilities:

$$k_{ro} = (1 - S^*)^2 (1 - S^{*1.5}) \tag{3.36}$$

$$k_{rw} = (S^*)^{3.5} \tag{3.37}$$

2. Gas–oil relative permeabilities:

$$k_{ro} = (S^*)^{3.5} \tag{3.38}$$

$$k_{rg} = (1 - S^*)^2 (1 - S^{*1.5}) \tag{3.39}$$

Cemented sandstone, oolitic limestone, and vugular rocks

1. Oil–water relative permeabilities:

$$k_{ro} = (1 - S^*)^2 (1 - S^{*2})$$

(3.40)

$$k_{rw} = (S^*)^4$$

(3.41)

2. Gas–oil relative permeabilities:

$$k_{ro} = (S^*)^4$$

(3.42)

$$k_{rg} = (1 - S^*)^2 (1 - S^{*2})$$

(3.43)

where

$$S^* = \frac{S_o}{1 - S_{wc}}, \text{gas–oil system,}$$

(3.44)

$$S^* = \frac{S_w - S_{wc}}{1 - S_{wc}}, \text{oil–water system, and}$$

(3.45)

S_{wc} = connate water or irreducible water saturation, fraction.

Laboratory measurements of relative permeability

Two-phase relative permeability is obtained by either the unsteady-state or steady-state method. In the unsteady-state method, the core is initially saturated with brine completely and the absolute permeability of the core sample is determined. The core is then flooded with oil. As a result, brine is displaced from the core by oil except what is retained due to the irreducible water saturation. The process is somewhat similar to oil migration into the reservoir. At this point, the fluid saturations in the core reflect the initial values of oil and gas saturation encountered in the reservoir. Next, water is injected into the core at a predetermined rate to displace the oil; the resulting changes in pressure and the volume of oil produced with time are noted. The displacement process is very similar to water injection into oil reservoirs where injected water displaces oil. The relative permeability values of the fluid phases at various points in the flow period, i.e., at various fluid saturations, are determined by using the Buckley–Leverett equation. The equation is described in Chapter 9.

The steady-state method involves injection of both the fluid phases simultaneously into the core at predetermined and constant rates and achieving the steady-state flow condition. The pressure gradient and production rate of each fluid phase will remain constant over time as the steady-state condition is achieved. Fluid saturation of the

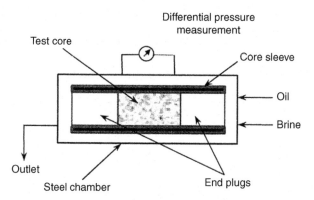

Figure 3.16 Determination of relative permeability of a core sample by the Penn State method. The test core is put in a rubber sleeve and placed in a steel chamber. Two other cores are placed at the ends of the core in order to minimize the end effects.

individual phases is then measured by weighing or by measuring the resistivity of the core sample. Modern techniques of determining fluid saturation in cores involve computer aided tomography scanning, NMR scanning and X-rays. The procedure is repeated at several times at various injection rates of the two fluids. It is important to note that the individual fluid saturations obtained are not the same as the injection ratio of the fluids. There are several methods to measure the relative permeability based on steady-state flow, including the Penn State method, the Hassler method, the Hafford method, and the dispersed feed method (Figure 3.16). Certain permeability measurement devices are capable of simulating reservoir pressure and temperature.

Geochemical properties of rock

Various characteristics of source rocks are of interest to assess petroleum generation, including TOC and vitrinite reflectance (R_o). Rocks considered to be viable sources of petroleum usually have TOC ranging between 2% and 10%. TOC is usually higher in shale than in carbonates. Values of TOC greater than 5.0 for shale and 2.0 for carbonates are considered to be good to excellent sources for the generation of hydrocarbons. VR is an indicator of the thermal maturity of source rocks, i.e., the extent of heat energy or temperature the rock is subjected to to generate petroleum. Vitrinites, abundant in source rocks, originate from plant cell walls and woody matter. Values of VR between 0.6 and 1.1 indicate generation of oil in source rocks; however, values greater than 1.1 suggest intense heating of source rock leading to the generation of natural gas rather than oil. R_o values greater than 3.0 indicate the formation of graphite only. The above properties of rock are routinely determined for various unconventional reservoirs (Table 3.5).

Geomechanical properties of rock

In certain unconventional reservoirs, including shale gas reservoirs, good fracturing characteristics are of critical importance to produce economically. Apart from

Table 3.5 **Total organic carbon (TOC) and kerogen quality [9]**

TOC (wt, %)	Kerogen quality
<0.5	Very poor
0.5–1.0	Poor
1.0–2.0	Fair
2.0–4.0	Good
4.0–12.0	Very good
>12.0	Excellent

the presence of natural fractures, ultratight rock must have sufficient brittleness so that the artificial fractures having the desirable characteristics can be created. Furthermore, the fractures must remain open for long periods of time so that the well remains productive. Hence, geomechanical properties of rock that are most significant in ultratight shale reservoir stimulation and production are discussed briefly in the following.

Young's modulus

Young's modulus is defined as the stress over strain for a material, including rock. Stress represents the pressure applied to the material while strain is a measure of deformation due to the applied pressure. In equation form, Young's modulus can be expressed as follows:

$$E = \frac{\sigma}{\varepsilon} \tag{3.46}$$

$$E = \left(\frac{F/A}{\Delta V / V_o} \right) \tag{3.47}$$

where E = Young's modulus, psi; σ = stress, psi; ε = strain, ratio; F = force applied on material, lb_f; A = area on which pressure is applied, in^2; V_o = original volume of material, in^3; ΔV = change in volume, in^3.

Hooke's law states that stress is proportional to strain. When values of stress are plotted against strain for a material, a straight line can be drawn through the points. The slope of the straight line is the Young's modulus.

A relatively high value of Young's modulus indicates that the rock resists deformation, which leads to brittleness and hence good fracturing characteristics. On the other hand, a low value implies that the material is ductile and would deform with relative ease. Typically, shales that are good candidates for fracturing have Young's modulus in the order of 10^6 psi. Methods of measurement of Young's modulus include dipole sonic log and microseismic studies.

Poisson's ratio

Poisson's ratio is defined as the ratio of lateral strain over longitudinal strain, which is the result of the application of stress on a material. Consider a solid cylinder subjected to stress in a longitudinal or vertical direction. The effects are:

- Increase in cylinder diameter in lateral direction
- Decrease in cylinder length in longitudinal direction

Mathematically, Poisson's ratio can be expressed as:

$$v = \left(\frac{\Delta d / d_o}{\Delta L / L_o} \right) \tag{3.48}$$

where v = Poisson's ratio, ratio; Δd = change in cylinder diameter, in; d_o = original diameter, in; ΔL = change in cylinder length, in; L_o = original length, in.

Typical values of Poisson's ratio in shale are in the order of 10^{-1}. A relatively low Poisson's ratio is viewed as favorable for fracturing.

In situ *stress*

Knowledge of *in situ* stresses is also essential to understand fracture propagation and characteristics, including the orientation of naturally occurring fractures. Horizontal wells are drilled transverse to the principal direction of stress in rock in order to maximize fluid flow through the wellbore.

Storativity and transmissibility

Reservoir rock characteristics are evaluated in terms of storativity and transmissibility to indicate the storage and flow potential of petroleum fluids, respectively. The two parameters combine various rock and fluid properties as in the following:

$$\text{Storativity} = \text{Porosity} \times \text{total compressibility} \times \text{thickness}$$

$$\text{Transmissibility} = \frac{\text{Permeability} \times \text{thickness}}{\text{Fluid viscosity}}$$

Storativity indicates the amount of fluid that will be released from the porous medium when there is a unit drop in reservoir pressure. The unit of storativity is pounds per square inch (psi^{-1}). Storativity of rock is directly proportional to effective porosity, net thickness, and total compressibility. Larger values of the three properties lead to greater storage of petroleum. Transmissibility is directly proportional to reservoir permeability and net thickness, and inversely proportional to fluid viscosity. The unit of transmissibility is mD-ft./cp. High values of rock permeability along with greater formation thickness and relatively low fluid viscosity lead to large volumetric flow in the porous medium that is ultimately produced through the wells. On the contrary, low reservoir permeability and viscous oil reduce transmissibility of the rock and are hence detrimental to production.

Reservoir quality index

Reservoir quality index, based upon the porosity and permeability of rock, is indicative of how much hydrocarbon is stored in the geologic formation and how well it will produce. The reservoir quality index is defined as:

$$RQI = 0.0314 \times \left(\frac{k}{\phi} \right)^{1/2} \tag{3.49}$$

In a broader perspective, reservoir quality is evaluated in terms of rock properties, along with pertinent geologic features such as lateral continuity and number of fluid flow units present. Reservoir drive mechanisms, described in Chapter 11, also contribute to reservoir quality. Poor reservoir quality, often associated with unconventional reservoirs, leads to technical and management challenges, implementation of new tools and innovative technologies, and higher investments. That said, a large number of conventional reservoirs producing currently in various parts of the world are of poor reservoir quality. Reservoir quality is again discussed in Chapter 6.

Well logging: a brief introduction

Well logging, as it applies to the petroleum industry, is used to identify oil and gas intervals, and quantify properties of reservoir rock by placing various types of sensors in the borehole. The rock characteristics include, but are not limited to, lithology, geologic structure, porosity, fluid saturation, and degree of drilling fluid invasion. The sensors are electric, electromagnetic, acoustic, neutron, gamma ray, and other tools that can send and receive signals into the geologic formation. Depending on various properties of rock and conditions surrounding the borehole, the emitted signals are transformed in character and attenuated in strength, which are captured by the sensors. The signatures are then analyzed to evaluate the formation properties that are of interest for producing oil and gas. In addition, caliper logs are run to determine the size of bores.

Certain logging and imaging tools aid in reservoir characterization by providing information on fractures and faults. Downhole imaging tools are used for various purposes, including the detection of rock fractures and high permeability "thief" zones. Well logging is an invaluable tool in determining how a well will be completed to produce effectively, and how the future wells will be drilled to develop and manage the reservoir.

Logging is broadly classified as either open hole or cased hole. Open hole logs, as the name implies, are run in the open borehole of a newly drilled well to measure various rock properties, including the identification and saturation of hydrocarbon intervals. Once the casing is set in the borehole, cased hole logs are used. One of main uses of cased hole logs is to determine the integrity of the casing and identify any damage. Once a well is drilled, a suite of logging devices on a string is lowered into the open hole to collect downhole data. The information received by the sensors can be recorded downhole in memory mode or at the surface in real time. During the 1970s, MWD technology was introduced, where the logging tools are attached to the drill string and data collected by the tools are sent to the surface continuously by using mud pulse technology.

The credit for introducing well logging techniques goes to the Schlumberger brothers who founded the company in the earlier part of the twentieth century. The logging tool was initially developed for detecting metal ore deposits, and was adopted for the oil and gas industry later. In 1927, a resistivity logging tool was first used downhole in Alsace, France. A few years later, the spontaneous potential (SP) log was introduced to identify the permeable zones having hydrocarbons. The gamma ray log, which measures natural radioactivity of a formation, was introduced by Well Surveys, Inc. in 1939. The log is particularly useful in identifying shale beds and works in cased holes. In the late 1940s, induction log was developed, which works in nonconductive oil-based mud environments.

Common logging tools and techniques used in the oil and gas industry are as follows:

- Resistivity logs: Water is a better conductor of electricity than oil and gas. Petroleum fluids are much more resistive to electricity than formation water having measurable salinity. Resistivity logs operate on this principle. The resistivity tool basically consists of two electrodes. The first electrode sends electric current into the fluid-filled formation, and the current flows back to the second electrode located at the other end of the tool forming an electric circuit. Depending on the conductivity of the formation fluid, the intensity of current varies as the tool is slowly pulled toward the surface. Oil zones are indicated by relatively high resistivity. Common resistivity tools include, but are not limited to, dual laterolog and microspherically focused log. Additionally, dual induction logs are also employed to determine fluid type and saturation, and are equipped with induction coils to measure the conductivity.
- SP log: One of the earliest logging tools in the industry, the SP log is used to measure the potential difference between the borehole and the surface by lowering an electrode into the borehole and measuring the difference in potential with a reference electrode at the surface. As a permeable formation is encountered by the tool, noticeable deflection in electrochemical potential occurs, which is dependent upon the clay content of the formation and water salinity.
- Density logs: Density logging tools are used to measure the bulk density of the formation. The tool is based on a radioactive source. As the tool is run against the formation, resulting gamma ray count due to Compton scattering and photoelectric adsorption is analyzed to determine the bulk density of rock. The latter can be used to evaluate formation porosity.

- Acoustic log: A sonic log is based on the travel time of sound in rock, which is influenced by rock porosity, lithology, and texture. A transmitter emits sound waves, and the time taken by the sound waves to travel from the transmitter to the receiver, which is also located in the tool, is recorded to evaluate the rock characteristics.
- Gamma ray log: This is based on a tool that measures the natural radioactivity of the formation. Shale beds contain radioactive potassium in clay, and are distinguished from sandstone layers, which are chiefly composed of nonradioactive quartz particles. Shale also contains uranium and thorium in an adsorbed state.
- Caliper log: Caliper logs are used to obtain a borehole profile, including the diameter and shape of the borehole. The tool consists of two arms that are pressed against the bore wall as the tool travels upward. The arms are connected to a potentiometer at the surface. Any change in borehole geometry is recorded.
- Spectral noise logging: This acoustic tool is used to determine well integrity and identify production or injection intervals, among others. The tool operates by recording noise generated by fluid flow in the subsurface system, including any leaks.
- Dipmeter logs: The logging tool is used to aid in characterizing the reservoir. Dipmeters determine the orientation of geologic beds as well as the orientation of faults and fractures by using imaging techniques.
- MWD: With the advent of new technology, the MWD tool was introduced in the drilling industry in the 1970s and is capable of sending real-time data from the subsurface formation, including rock porosity, density, fluid pressure, borehole trajectory, etc. A suite of logging tools is involved in MWD, including electric and acoustic logs. Radioactive sources are also included. The information is sent to the surface by mud pulse telemetry, which is based on transmitting pressure pulses through a mud column. MWD, also referred to as logging while drilling, provides real-time information about the location and direction of a lateral section during horizontal drilling.

Reservoir heterogeneity

Reservoir heterogeneities can be microscopic, macroscopic, or megascopic, meaning that rock properties vary from micro- to field scale. Common examples of microscopic heterogeneity include the variations in pore size and pore throat diameter, sorting of grains, presence of impurities, and the tortuous nature of miniscule channels in rocks, resulting in a wide range of permeability from one core sample to another. Field-scale heterogeneities include stratification within a formation, variations in formation thickness, pinchouts, and facies change. In certain cases, the entire reservoir can be represented by an equivalent homogeneous model when heterogeneities are present only on a microscopic level. Many reservoir models and analyses are rooted in the assumption that the geologic formation is homogeneous, having uniform rock properties in all directions. Some of the common heterogeneities present in oil and gas-bearing formations are listed in Table 3.6.

However, when large-scale heterogeneities are present, including multiple layers, fractures, and compartments, detailed treatment of the heterogeneities is warranted in analyzing the reservoir and predicting its performance. In between the two extremes in scale, rock heterogeneity occurs on an interwell scale, where the reservoir quality between two adjacent wells cannot be assumed for other wells. An example is the

Table 3.6 **Common heterogeneities in reservoir rocks**

Rock type	Heterogeneities in rock	Possible effects
Sandstone	Poorly sorted grains, presence of impurities including siltstone and mudstone, existence of fractures, etc.	Degradation in reservoir quality. However, fractures usually enhance productivity of tight sandstones.
Carbonate	Presence of vugs, channels, solution cavities, and fractures	Fractures and channels may enhance primary production. However, unexpected water breakthrough is common during water injection.
Shale	Presence of laminations and fractures. Significant variations in geochemical, geomechanical, and petrophysical properties from region to region	Unpredictable well performance, although the wells may be located in the same general area

early water breakthrough at a producer during water injection while the other producers located in the same area continue to have dry oil production.

The effects of field-scale heterogeneities on a petroleum reservoir are as follows:

- Uncertainty in the estimates of oil and gas in place
- Uncertainty in estimating petroleum reserves
- Uncertainty in identifying reservoir boundaries
- Unknown variables in field development
- Unexpected decline in well productivity
- Complexity in understanding reservoir dynamics
- Premature breakthrough of oil or gas during waterflooding and enhanced oil recovery
- Challenges in optimizing recovery of oil and gas
- Variations in fluid properties in the reservoir
- Unexpected influence of adjacent aquifers
- Uncertainty in locating future wells
- Requirement of relatively large number of wells
- Difficulties in planning a water injection project
- Difficulties in selecting an enhanced recovery process
- Difficulties in building and validating meaningful reservoir models
- Need for well recompletion and workover
- Need for intensive data collection and rigorous analysis

Some reservoirs are more heterogeneous than others, depending on rock type, depositional environment that existed in prehistoric times, and postdepositional geologic events. In most cases, rock properties vary both in the horizontal as well as in the vertical direction. Reservoir heterogeneities are commonly analyzed in terms of porosity and permeability variations in rock, and by the overall structural makeup of the reservoir. Petroleum reservoirs are considered to be heterogeneous where there is a wide variation in permeability and a distinct correlation between porosity and permeability cannot be established, among other factors. In reality, virtually all petroleum reservoirs exhibit heterogeneous rock characteristics, meaning that rock properties vary from one well to another, and from microscopic to megascopic scale.

Some of the common tools and techniques to identify and characterize reservoir heterogeneities are presented in the following.

A case study has been presented in the following section, which correlates the formation heterogeneities encountered in various petroleum basins of the United States with well production pattern. Reservoir characterization is discussed in Chapter 6.

Case Study: Reservoir Heterogeneity and Well Performance

The United States Geological Survey (USGS) conducted a detailed study of formation characteristics and rock heterogeneities of a large number of reservoirs that affect the addition of petroleum reserves [10]. The study correlated the past production trends of various fields in the light of known heterogeneities. The conclusions of the study are important as the geologic characteristics of a reservoir determine the following, among others:

- Petroleum reserves and updates as reservoirs are produced
- Number and location of stepout wells
- Delineation of reservoir boundaries
- Infill drilling potential
- Selection of enhanced recovery methods once natural reservoir energy is insufficient for production
- Variations in well production trends within the same reservoir
- Fluid dynamics that affect well rates and type of fluids produced
- Continuity of geologic formation
- Effects of facies change on reservoir production

The petroleum basins studied include: (i) Gulf of Mexico, (ii) Powder River Basin, (iii) Denver Basin, (iv) Fort Worth Basin, (v) Anadarko Basin, (vi) Permian Basin, (vii) Midland Basin, and (viii) Piceance-Unita Basin. The formations included both conventional and unconventional sources of petroleum. Reservoir characteristics are presented in Table 3.7. It gives a fairly good idea of the typical range of porosity and permeability in a large number of petroleum basins across the United States. The reservoirs were divided into subcategories depending on contrasting rock characteristics that control production behavior.

Of the formations studied, the Ellenburger Karst formation of the Permian Basin, having a porosity of 2–7% and permeability ranging between 2 mD and 750 mD, showed the largest variations in production trends of wells. A large variation in productivity of wells is tied to a high degree of reservoir heterogeneity and less predictability in the growth of future reserves. The study also highlights the need for appropriate technology and further investigation in producing efficiently from certain formations. It is quite interesting to note that, of all the basins studied, the highest number of wells as well as the largest production of oil is obtained from the heterogeneous Ellenburger Karst formation (Table 3.8).

Table 3.7 **Tools and techniques to identify and characterize rock and reservoir heterogeneities**

Scale of heterogeneity	Tools and techniques	Notes
Microscopic	Core and thin section studies	Rock properties on microscopic scale require upscaling in various reservoir studies
Interwell/ macroscopic	Well logging; pressure transient testing; tracer studies; microseismic studies	
Field	Geological and geophysical studies; pressure transient testing	Production data analysis, reservoir surveillance, reservoir simulation, and others are used extensively to understand a heterogeneous reservoir

Case Study: Role of Rock Fracture Properties in Unconventional Shale Gas Development [12]

This topic showcases several important aspects of modern-day reservoir engineering. Many of the concepts introduced in the following are discussed in detail in subsequent chapters of the book. As noted earlier, shale has ultralow permeability in nanodarcies, which is impossible to produce on a commercial scale without inducing a large conductive fracture network. Well productivity gets better when the network of fractures covers large reservoir volume, fracture density is high, induced fractures connect to the naturally occurring fractures in shale, and fractures remain open for long periods of time to facilitate the flow of gas.

The technology of shale gas reservoir development involves horizontal drilling combined with multistage fracturing. Horizontal wells are drilled 10,000 ft. or more in the lateral direction in ultratight formations to contact the reservoir volume as much as possible. The wells are then hydraulically fractured in multistages to facilitate the flow of natural gas through highly conductive pathways. Drilling, fracturing, and completion of the wells are capital intensive, costing several millions of dollars per well. The management must ascertain whether such ventures are economic. Hence, the analysis begins with the modeling of shale gas production from horizontal wells with multistage fracturing. Reservoir models, based on reservoir description, integrated data analysis, and flow equations, attempt to replicate the processes and events associated with fluid flow and predict reservoir performance. Reservoir model simulation, horizontal wells, multistage fracturing of shale formations, and economic analysis are described in Chapters 15,18,22,and 24, respectively.

Table 3.8 Reservoir characteristics of various petroleum basins in the united states [11]

Basin location	Geologic formation	Depositional environment	Geologic age	Lithology	Maximum gross (ft.)	Porosity (%)	Permeability	Cum. oil production (MMbbl)	Cum. gas production	Notes
Gulf of Mexico Basin	Frio	Fluvial or deltaic – shallow marine	Oligocene	Sandstone	15,000	10–35	8–3500 mD	534.4		Least heterogeneous among the formations studied
Gulf of Mexico Basin	Smackover	Marine carbonates	Upper Jurassic		1,000	2–35	<1 mD to several darcies			
Gulf of Mexico Basin	Norphlet	Eolian sands	Upper Jurassic	Sandstone	100	As much as 20% in onshore reservoirs and 12% in deeper offshore reservoirs	Generally high; as much as 500 mD			
Anadarko and Denver Basins	Morrow	Fluvial or deltaic (shallow marine)	Pennsylvanian	Sandstone	1,500	4–22	<1 mD to several darcies	74.7		Low to moderate heterogeneities
Powder River Basin	Minnelusa	Eolian sands	Pennsylvanian–Permian	Sandstone with minor shale and carbonate	1,200	12–24%, as high as 47% in certain cases	10–830 mD, as high as 3,200 mD	586		Relatively high heterogeneity
Fort Worth Basin	Barnett shale	Marine shale	Mississippian	Shale with some carbonates	650	<6	Ultralow, in nanondarcies			Unconventional reservoir

Williston Basin	Bakken	Marine shale	Devonian–Mississippian	Marine shale, siltstone–sandstone	140	Typically 3–10%	<.01–109 mD		
Permian Basin	Ellenburger	Marine carbonates	Ordovician	Dolomitized mudstone	1,500	1–14	<1–750 mD	1155.8 (karst), 65 (ramp)	The Ellenburger group was divided into ramp and karst. The latter was found to be the most heterogeneous formation studied
Midland Basin	Spraberry	Submarine sands	Permian	Sandstone (Turbitide), with minor black shales, silty dolostones, and argillaceous siltstones	1,000	Usually 5–15%, as high as 18%	Generally low; <1–10 mD		
Uinta-Piceance Basin (Utah, Colorado)	Wasatch	Nonmarine fluvial–deltaic	Paleocene–Eocene	Sandstone	5,000	Up to 15% in shallower depths (<4,000 ft.); 10% or less at greater depths	Very low; <0.1 mD. As high as 40 mD in some cases	89.9	Relatively high heterogeneity

The objectives of the study include, but are not limited to:

- Detailed understanding of reservoir geology and rock quality that directly influences production potential
- Geological screening that points to the placement of wells with highest potential
- Identity of bypassed reserves
- Optimization of horizontal well length
- Optimization of spacing between the wells
- Design of multistage hydraulic fracturing
- Well completion
- Increased drainage area and better gas recovery

Productivity of the reservoir is dependent upon rock geological, geochemical, and geomechanical properties described earlier in the chapter. In the context of induced and natural fractures, the models simulate the effects of the lateral extent of fractures, fracture spacing, density, connectivity, and the orientation of the fractures on production of gas. As a well produces, the effects of drawdown of gas leading to depletion in reservoir pressure and closure stress on the fractures are also evaluated. In order to validate the model, the result of simulation is matched with the historical production data before production forecasts are made. Fracture data as obtained by conducting microseismic studies are incorporated in the model. The data include fracture density, intensity, and complex configuration of fractures.

The integrated model is based upon the following:

- 3D seismic study
- Fracture mapping and modeling
- Well pressure testing
- Production history matching
- Drilling and formation evaluation

The goal is the optimization in well design for increased ultimate recovery and net present value.

Summing up

Rock properties in conventional and unconventional reservoirs are as shown in Table 3.9.

Questions and assignments

1. What are two most important properties of reservoir rock and how do they characterize a petroleum reservoir?
2. How do the dynamic properties of rock differ from the static properties? Explain with several examples.

Table 3.9 **Important rock properties of petroleum reservoir rocks**

Rock properties	Oilfield units	Definition	Typical range in petroleum reservoirs	Notes
Pore volume	ft.^{3}	Volume of void space in rock, including microscopic pores and channels in rock	About 5–30% of total volume of rock in oil and gas reservoirs	
Grain volume	ft.^{3}	Volume of rock matrix or solid portion of rock	n/a	Pore volume and grain volume of a core add up to its bulk volume
Porosity	Fraction, %	Ratio of pore volume over bulk volume, i.e., pore volume plus grain volume	5–35%	Rock porosity can be less than 5%, but these formations are not economically producible
Absolute porosity	Fraction, %	Ratio of all voids in rock over bulk volume	5–35%	
Effective porosity	Fraction, %	Ratio of interconnected pore volume over bulk volume	5–35%	
Secondary porosity	Fraction, %	Porosity that develops due to geochemical and other processes following the initial development of rock pores		Secondary porosity and the presence of cavities or vugs can be significant in carbonate rocks
Dual porosity		Refers to two different porosities in matrix and fractures in fractured formation		
Permeability	mD	Measure of ability of fluid through porous rock	Nanodarcies to several darcies	Reservoirs having ultralow permeability are produced by horizontal drilling and multistage fracturing
Effective permeability	mD	Measure of ability of one fluid phase to flow through porous rock over other fluid phase(s)		

(Continued)

Table 3.9 **Important rock properties of petroleum reservoir rocks** *(cont.)*

Rock properties	Oilfield units	Definition	Typical range in petroleum reservoirs	Notes
Relative permeability	Fraction	Ratio of effective permeability over relative permeability	0–1.0	
Fracture permeability	mD	Refers to permeability in the fractures of rock	Darcies	
Compressibility	psi^{-1}	The compressibility of formation is a measure of the rate of change of pore volume with change of reservoir pressure.	In the order of 10^{-5}–10^{-6} psi^{-1} for most formations	
Saturation of oil, gas and water in rock	Fraction, %	Measure indicates the ratio of volume of a fluid present in pore volume of rock.	0–1.0	
Interfacial tension		Forces arising at the interface of two immiscible fluids		
Surface tension		Forces arising between the surface of rock pores and the fluid contained in the pores		
Sorption	scf/ton	The process of trapping of gas in near liquid state in micropores of coal or shale. Sorption can be either physical, due to weak molecular attractions, or chemical, due to strong chemical bonding.	From tens to hundreds of $ft.^3$ per ton of rock	Significant amount of natural gas is trapped in micropores of coal and shale due to sorption.
Wettability	Degrees	Indicates tendency of one fluid to wet rock surface in the presence of another	0–180°	
Capillary pressure	psi	Measure of the difference between pressure exerted by wetting and nonwetting phases in contact with rock		

(Continued)

Table 3.9 **Important rock properties of petroleum reservoir rocks** *(cont.)*

Rock properties	Oilfield units	Definition	Typical range in petroleum reservoirs	Notes
Total organic carbon	%	Measure of organic carbon content in source rock of petroleum	1–12% or more	
Vitrinite reflectance		Indicator of the thermal maturity of source rocks, i.e., the extent of heat energy or temperature the rock is subjected to to generate petroleum	0.6–2.5	
Young's modulus		Ratio of stress over strain for a material, including rock. Stress represents the pressure applied to the material while strain is a measure of deformation due to the applied pressure.		
Poisson's ratio		Ratio of lateral strain over longitudinal strain, which is the result of application of stress on a material		

3. How do reservoir engineers obtain information regarding the rock properties? List the sources of various types of data.
4. Distinguish between absolute porosity and effective porosity. Why do the two porosities differ? What porosity value is needed to estimate recoverable reserves?
5. What is Darcy's law? Describe the assumptions and limitations of Darcy's law as applied to petroleum reservoirs. What is its significance in characterizing fluid flow in porous medium?
6. What is the unit of permeability? How it is defined? Describe the range of permeability encountered in petroleum reservoirs.
7. Describe the effects of various reservoir parameters on fluid flow characteristics, including rock permeability, fluid viscosity, and pressure gradient.
8. Besides Darcy flow, what other type of fluid flow is observed in porous medium? What are the effects of non-Darcy flow?

9. Distinguish among absolute permeability, relative permeability, and effective permeability. What are their units?
10. What is the significance of relative permeability during oil and gas production? Explain with examples.
11. Draw oil–water relative permeability diagrams for well-sorted and poorly sorted sandstones based on the correlations presented in the chapter. Why are the relative permeability curves different in the two cases?
12. Define connate water saturation, irreducible water saturation, movable oil saturation, and residual oil saturation. How are the saturations determined?
13. What are end point saturations and why are these important? Describe the effects of various fluid saturations in reservoir performance.
14. What is wettability? How does the wettability of rock affect reservoir performance?
15. What is capillary pressure? Explain drainage and imbibition processes. Is the capillary pressure observed during the two processes the same?
16. Does capillary pressure affect the flow of oil and water? Explain.
17. What is capillary number? Why is it significant in analyzing reservoir performance?
18. Why is interfacial tension important in reservoir studies? What other rock properties are affected by surface and interfacial tension?
19. Distinguish between pore compressibility and matrix compressibility. How does rock compressibility affect a reservoir? What is the effect on production when reservoir rock is highly compressible?
20. Based on a literature review, describe the three-phase relative permeability correlations and prepare a diagram showing the relative permeabilities of individual fluid phases.
21. What types of well logging tool are commonly used in the petroleum industry? Conduct a literature review and prepare a table describing the role of various logging techniques in describing and developing a reservoir.
22. Define storativity, transmissibility, and reservoir quality. How can these properties be obtained from a reservoir?
23. What rock properties are important in unconventional reservoir development?
24. How does the storage of unconventional gas differ from that of conventional gas?
25. What is reservoir heterogeneity? How does reservoir heterogeneity affect reservoir performance and assets? Based on a literature review, describe a heterogeneous reservoir and the considerations made to develop the reservoir.

References

[1] Satter AS, Iqbal GM, Buchwalter JL. Practical enhanced reservoir engineering assisted with simulation software. Tulsa, OK: Pennwell; 2008.
[2] Darcy H. Les fontaines publiques de la ville de Dijon. Paris: Victor Dalmont; 1856.
[3] Chertov MA, Suarez-Rivera R. Modeling pressure decay permeability for tight shale characterization, AGU Fall Meeting, 2011.
[4] Klinkenberg LJ. The permeability of porous media to liquids and gases. API Drilling and Production Practice; 1941.
[5] Forchheimer P. Wasserbewegung durch boden. Zeit Ver Deutsch Ing 1901;45:1781–8.
[6] Newman GH. Pore-volume compressibility. J Petrol Technol 1973;25(2):129–34.
[7] McKee CR, Bumb AC, Bell GJ. Effects of stress-dependent permeability on methane production from deep coalseams. Paper SPE 12858, presented at the Unconventional Gas Recovery Symposium, Pittsburgh, Pennsylvania, May 1984.

[8]   Coalbed methane: principles and practices, Halliburton, 2008. http://www.halliburton. com/public/pe/contents/Books_and_Catalogs/web/CBM/CBM_Book_Intro.pdf.

[9]   Wyllie MRJ, Gardner GHF. The generalized Kozney–Carmen equation – its applications to problems of multi-phase flow in porous media. World Oil; 1958.

[10] Boyer C, Kieschnick J, Suarez-Rivera R, Lewis RE, Waters G. Producing gas from its source, Oilfield Review; Autumn 2006, pp. 36.

[11] Fishman NS, Turner CE, Peterson F, Dyman TS, Cook T. Geologic controls on the growth of petroleum reserves, USGS Bulletin 2172-1; 2008.

[12] Shale Field Development workflow, http://www.halliburton.com/public/solutions/contents/Shale/Presentations/FINAL_Ron%20Dusterhoft.pdf [accessed 08.12.14].

[13] Paul RG, Arnal RE, Baysinger JP, Claypool GE, Holte JL, Lubeck CM, Patterson JM, Poore RZ, Slettene RL, Sliter WV, Taylor JC, Tudor RB, Webster FL. Geological and operational summary, southern California deep stratigraphic test OCS-CAL No. 1, Cortes Bank area offshore southern California, Open-File Report 76-232, https://pubs.er.usgs. gov/publication/ofr76232 [accessed 15.06.14].

Reservoir fluid properties

4

Introduction

Reservoir fluid properties, along with rock properties described in the previous chapter, determine how a petroleum reservoir would be developed, engineered, and managed. Petroleum fluids range, as encountered in reservoirs throughout the world, from dry natural gas to ultraheavy oil. Naturally occurring petroleum varies widely in viscosity, gravity, composition, and phase behavior, which leads to the formulation of unique strategies to develop and produce the reservoirs effectively. The most common classification of petroleum reservoirs is based on the type of hydrocarbons it stores and produces. Petroleum reservoirs are classified as follows:

- Dry gas
- Wet gas
- Gas condensate
- Light oil
- Black oil of intermediate composition
- Heavy oil
- Extra heavy oil, bitumen

Dry gas has the lightest hydrocarbons and is least viscous. Naturally, it has maximum mobility in the porous medium. Some fraction of wet gas condenses at the surface under stock tank conditions. Gas condensate reservoirs are distinguished by the fact that certain hydrocarbon components remain in gas phase at high pressure, but condense out as droplets within the reservoir when reservoir pressure is reduced.

In light crude, hydrocarbons having reduced molecular weight are found in large proportions. Oil gravity and viscosity are comparatively reduced. Oil having low viscosity flows with relative ease in porous media. One important characteristic of light oil is that the volatile components are liberated from the liquid into vapor phase as the reservoir pressure is reduced. In the domain of heavy oil reservoirs, heavier hydrocarbon components are relatively abundant in crude oil. Both oil gravity and viscosity increase as heavier hydrocarbons of higher molecular weight are in large proportions. Oil becomes less mobile at higher viscosity. Ultraheavy oil and bitumen hardly flow in porous media unless oil viscosity is reduced by thermal recovery or other methods.

Oil, gas, and formation water are referred to as the three fluid phases in porous media. This chapter describes the important properties of reservoir fluids and aims to answer the following:

- What are the important fluid properties in reservoir engineering?
- What factors affect the reservoir fluid properties?
- How are fluid properties used in reservoir analyses?
- How are reservoirs classified according to fluid properties and what strategies may be adopted to produce the reservoirs?

Reservoir Engineering. http://dx.doi.org/10.1016/B978-0-12-800219-3.00004-8

Utilization of petroleum fluid properties data

Most fluid properties can be correlated to each other and are dependent on the prevailing pressure and temperature to a varying degree. Hence, the properties are also referred to as pressure–volume–temperature (PVT) properties; the acronym stands for pressure, volume, and temperature. Knowledge of reservoir fluid properties like viscosity, gravity, composition, and phase behavior aid the reservoir engineers to understand the following:

- How easily the reservoir fluids would flow toward the wellbore under operating pressure
- How fluid properties affect well rates
- How wells will be designed and operated to achieve maximum productivity
- To what extent oil or gas would change in volume once brought to the surface
- How equilibrium among various fluids occurs in the geologic formation
- If there would be a change in fluid phase (form liquid to vapor or vice versa) as the reservoir pressure declines
- How the fluid properties and any change in fluid phase would affect ultimate recovery

Various fluid properties are required in virtually every aspect of reservoir engineering studies. Some of the important roles that fluid properties play in reservoir engineering are as follows:

- Volumetric estimates of oil and gas in place
- Classical material balance analysis
- Insight into reservoir drive mechanisms
- Estimation of well rates
- Reservoir simulation and prediction of reservoir performance
- Determination of applicable enhanced oil recovery processes

Properties of reservoir oil

The oil properties that are of primary interest to reservoir engineers include the following:

- Specific gravity
- Viscosity
- Compressibility
- Bubble point pressure
- Solution gas–oil ratio
- Producing and cumulative gas–oil ratio
- Oil formation volume factor
- Two-phase formation volume factor

The phase behavior of petroleum is discussed in the next chapter. Fluid phase behavior, including any changes from liquid to vapor or vice versa, is dependent upon the composition of the fluids, as well as the reservoir pressure and temperature.

It is observed that reservoir fluid properties may vary from one geologic layer to another where the layers are not in communication. In certain reservoirs, highly viscous tar mat is encountered at the periphery of the reservoir.

Specific gravity of oil and API gravity

The specific gravity of crude oil is defined as the ratio of the oil density over the density of the water, both measured at the same reference temperature and pressure. Specific gravity measurements are usually based on 60°F temperature. Specific gravity is a ratio of two densities, hence it has no units. The specific gravity of oil is commonly expressed as API gravity. As defined by the American Petroleum Institute, the API gravity is computed as follows:

$$API = \left(\frac{141.5}{\gamma_o} \right) - 131.5 \tag{4.1}$$

where

γ_o = specific gravity of oil, a ratio.

Note that the API gravity is inversely proportional to the specific gravity of fluid. Heavier crude having higher specific gravity would lead to a lower API gravity. API gravity can range approximately from 40° for light crude oil to 10° for heavy crude oil. Equation (4.1) suggests that oil would sink in water when API gravity is less than 10°, which is a characteristic of extra heavy oil and bitumen.

Typical values of API gravity of crude oil as sold in the world market are given in order of increasing specific gravity in Table 4.1 (decreasing API gravity).

As oil, gas, and formation water have different specific gravities, three distinct zones can be observed in a typical petroleum reservoir where all three fluid phases are present. Gas having the least specific gravity is encountered at the top part of the geologic formation. Gas is underlain by oil due to gravity. Finally, formation water remains at the bottom as it is heavier than oil and gas. The relative location of oil, gas, and water zones determine where a well will be drilled, how it will be completed, and whether a reservoir pressure maintenance operation would be necessary to maximize recovery.

Depending on rock and fluid properties, gas–oil and oil–water interfaces may not be sharp and a transition zone may exist between two fluid phases (Figure 4.1). Knowledge of transition zone is quite important in effectively producing certain reservoirs having long transition zones.

Table 4.1 API gravity of commercially available crude oil

Crude oil	°API
West Texas Intermediate	39.6
Brent (North Sea)	38
Nigerian Bonny Light	35–37
Saudi Light	33.3
Russian Export Blend	32
Dubai	31

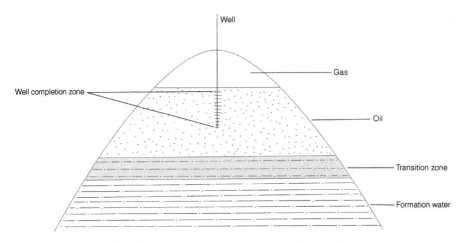

Figure 4.1 Vertical equilibrium of gas, oil, water, and oil–water transition zone in a typical petroleum reservoir. The producing well, completed in oil zone, is designed to avoid water and gas production.

Oil viscosity

Viscosity of oil indicates how easily it will flow in the reservoir. It is a measure of the internal resistance to flow. The unit of viscosity is the centipoise as commonly used in reservoir calculations. Viscosity data are required to calculate fluid flow rate in the reservoir and is considered to be one of the most important properties of reservoir fluids. Oil is more viscous than water; hence, water is more mobile in porous media in comparison to oil. Consequently, one of the most prevalent issues in reservoir management is to combat unwanted water production, which flows better than oil. It is further noted that gas may be produced in excessive quantities due to very low viscosity in certain oil wells, and engineering solutions are needed to reduce the gas–oil ratio as well.

Light oil having low viscosity flows quite easily through porous media in comparison to heavy viscous oil. An examination of Darcy's law, presented in Chapter 3, indicates that a more volumetric flow rate would be achieved when the fluid viscosity is less, given all other conditions are the same. Viscous crude would require more energy to flow towards the wellbore than low viscosity oil. The heaviest and most viscous of hydrocarbon deposits usually require unconventional methods of recovery. The range of viscosity and API gravity for various types of oil are shown in Table 4.2.

As noted earlier, viscosity, specific gravity, and other PVT properties of petroleum fluids depend on the relative abundance of light or heavy hydrocarbon components. Viscosity is also a function of reservoir pressure and temperature. As oil is produced and reservoir pressure declines, oil viscosity reduces somewhat as long as no gas evolves from the liquid phase. However, once the light components are liberated with further decline in pressure below the bubble point, oil becomes more viscous. Bubble point of petroleum fluid is described later in the chapter. Figure 4.2 presents the change in oil viscosity with reservoir pressure.

Table 4.2 Viscosity, API gravity, and method of recovery of oil

Type of oil	Viscosity (cp)	Gravity (°API)	Recovery method
Light	0.7–5.0	38–42	Conventional
Intermediate	6–12	22–38	Conventional
Heavy	12–100	18–22	Conventional
Extra heavy	100–10,000	<20	Unconventional
Oil sands/bitumen	>10,000	7–9	Unconventional

Note: Ranges of oil viscosity and gravity cited in the table are approximate and provided as a guide only.

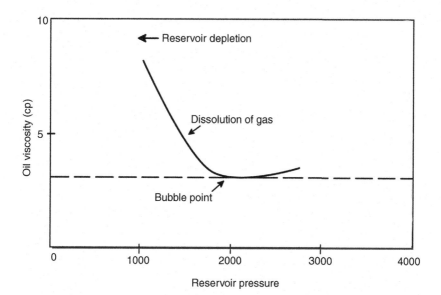

Figure 4.2 Change in oil viscosity before and after the evolution of volatile components.

Oil viscosity is reduced with the increase in temperature. Hence, thermal methods are employed in order to enhance recovery from heavy oil and oil sands. Thermal enhanced oil recovery methods are discussed in Chapter 17.

Various laboratory methods are available to measure the viscosity of a crude sample obtained from the field. A literature review suggests that a number of correlations exist to estimate the viscosity of oil.

Viscosity of oil having no volatile components remaining in the liquid phase is referred to as dead oil viscosity. The dead oil viscosity can be estimated by the following correlation when the API gravity is known [1]:

$$\mu_{od} = C(T - 460)^{-3.444}[\log(y_o)^a] \tag{4.2}$$

where $C = 3.141 \times 10^{10}$; $a = 10.313[\log(T - 460)] - 36.447$; y_o = oil gravity, °API.

The viscosity of "live" oil at the bubble point, where the fluid contains volatile hydrocarbons in a dissolved state, can be estimated when the solution gas–oil ratio is known. The solution gas–oil ratio, explained later in the chapter, is a measure of the volume gas dissolved per unit volume of oil. Chew and Connally [2] proposed the following correlation:

$$\mu_{ob} = (10)^a (\mu_{od})^b \tag{4.3}$$

where μ_{ob} = oil viscosity at the bubble point where volatile components are in solution, cp; $a = R_s[2.2(10^{-7})R_s - 7.4(10^{-4})]$; $b = 0.68/10^c + 0.25/10^d + 0.062/10^e$; $c = 8.62(10^{-5})R_s$; $d = 1.1(10^{-3})R_s$; $e = 3.74(10^{-3})R_s$.

The changes in oil viscosity as a function of API gravity and solution gas–oil ratio are presented in Figure 4.3.

Above the bubble point pressure of oil, the following correlation can be used to estimate the viscosity [3]:

$$\mu_o = \mu_{ob} \left(\frac{p}{p_b}\right)^m + 1 \tag{4.4}$$

where p_b = pressure at bubble point, psia; $m = 2.6p^{1.187} \exp(a)$; $a = -11.513 - 8.98 \times 10^{-5}p$.

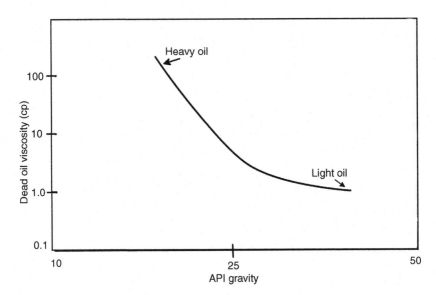

Figure 4.3 **Viscosity of oil as a function of API gravity and gas–oil ratio under typical conditions.**

Isothermal compressibility

Oil compressibility is a measure of change in volume as a result of change in prevailing pressure. It is defined as the rate of change in the volume of crude oil per unit change in pressure divided by the volume of oil. Mathematically, compressibility at a given pressure and temperature can be expressed by:

$$C = \left[-\frac{1}{V} \left(\frac{\partial V}{\partial p} \right) \right] \tag{4.5}$$

where c = fluid compressibility, psi^{-1}; V = oil volume, bbls; p = fluid pressure, psi.

Note that the compressibility oil in Equation (4.5) is measured at a reference temperature, T.

Oil is slightly compressible. As reservoir pressure declines, oil undergoes slight expansion in volume as long as volatile hydrocarbons are not liberated from the liquid phase. The unit of oil compressibility is the inverse of pounds per square inch (psi^{-1}). Values of oil compressibility can typically range from 5×10^{-6} psi^{-1} to 12×10^{-6} psi^{-1} or more. If one million barrels of oil in a reservoir are found to have a compressibility of 12×10^{-6} psi^{-1} and a drop of 100 psi in reservoir pressure occurs, the volume of oil would expand by 1200 barrels. The result would be obtained by noting that:

$$\begin{aligned}\text{Change in oil volume} = \text{Original oil volume} \times \text{oil compressibility} \\ \times \text{change in reservoir pressure}\end{aligned} \tag{4.6}$$

Total and effective compressibility

The total compressibility of the system accounts for the compressibility of the fluid phases present in the system as well as the formation compressibility. Hence, total compressibility, c_t, can be expressed as:

$$c_t = c_f + c_o S_o + c_g S_g + c_w S_w \tag{4.7}$$

where c_t = total compressibility, psi^{-1}; c_f = formation compressibility, psi^{-1}; c_o = compressibility of oil, psi^{-1}; S_o = water saturation, fraction; c_g = compressibility of gas, psi^{-1}; S_g = gas saturation, fraction; c_w = compressibility of formation water, psi^{-1}; S_w = water saturation, fraction.

The gas compressibility term drops out of the equation in the case of an oil reservoir where no free gas is present.

The effective compressibility of a fluid phase is obtained by dividing the total compressibility by the saturation of that phase in porous media. Hence, in an undersaturated oil reservoir where a free gas phase is not present, the effective compressibility of the oil phase can be expressed as follows:

$$c_e = \frac{c_f + c_o S_o + c_w S_w}{1 - S_w} \qquad (4.8)$$

Compressibility of oil and gas influences the flow characteristics in the porous medium. Moreover, certain unconsolidated formations have significant compressibility affecting the recovery of petroleum.

Bubble point pressure

The bubble point of reservoir oil is an important fluid property the reservoir engineers seek. Simply stated, it is the pressure where the volatile components present in oil begin to "bubble up." The pressure at which these bubbles of light hydrocarbons first appear is referred to as the bubble point for the fluid system.

Reservoir performance changes significantly when the reservoir produces below the bubble point, including oil and gas rates at the wells. Consider the production from an oil reservoir that does not have any gas cap at discovery. Above the bubble point, only the oil phase is present in the reservoir along with formation water. However, as the reservoir is produced and pressure declines, phase change takes place and light hydrocarbons are liberated from the oil. From this point onward, both oil and gas are produced at the wellbore, and flow of gas may eventually dominate over the production of oil. Many reservoir management strategies involve the maintenance of reservoir pressure above the bubble point pressure by water injection and avoid the production of gas.

Volatile oil with abundance of light hydrocarbon components has relatively high bubble point pressure, and the gas phase begins to evolve relatively early during depletion. The bubble point is relatively low for heavy oil. The bubble point pressure in conventional oil reservoirs ranges between 1800 psi and 2600 psi in typical cases.

Standing [4] proposed the following correlation to estimate the bubble point of oil based on the properties of oil and gas dissolved in oil:

$$p_b = 18.2 \left[\left(\frac{R_s}{\gamma_g} \right)^{0.83} (10)^a - 1.4 \right] \qquad (4.9)$$

where R_s = solubility of gas at the bubble point, scf/STB; γ_g = specific gravity of gas under surface conditions; $a = 0.00091(T - 460) - 0.0125(^\circ API)$; T = temperature, $^\circ R$; API = API oil gravity.

Note that the above correlation is subject to certain limitations in the presence of impurities.

The bubble point pressure of a liquid phase is also referred to as the saturation pressure because the liquid is completely saturated with dissolved gas above this pressure.

Solution gas–oil ratio

In a reservoir, oil is in liquid phase with a certain quantity of gas dissolved in it. As we have seen earlier, when the reservoir pressure declines due to production of oil, lighter hydrocarbons begin to evolve out of solution and form a gas phase.

Solution gas–oil ratio is indicative of the amount of gas dissolved in reservoir oil. It represents the volume of gas that would dissolve per unit volume of oil under reservoir pressure and temperature; however, the volumes of oil and gas are expressed in standard pressure and temperature, scf and STB, respectively. Hence, the solution gas–oil ratio can be expressed as follows:

$$R_s = \frac{\text{Volume of dissolved gas in oil in reservoir, scf}}{\text{Reduced volume of oil following liberation of gas, STB}} \tag{4.10}$$

where R_s = solution gas–oil ratio, scf/STB.

Oil having high solution gas–oil ratio is rich in volatile components and exhibits relatively high bubble point pressure.

Marhoun [5] proposed the following correlation to estimate the solution gas ratio of reservoir oil based on reservoir pressure, temperature, and specific gravity of oil and gas:

$$R_s = [a\gamma_g^b \gamma_o^c T^d p]^e \tag{4.11}$$

where γ_g^b = specific gravity of gas, dimensionless; γ_o = specific gravity of stock-tank oil, dimensionless; T = temperature, °R; $a = 185.843208$; $b = 1.877840$; $c = -3.1437$; $d = -1.32657$; $e = 1.398441$.

Producing and cumulative gas–oil ratio

The gas–oil ratio at the well is defined as:

$$\text{Gas – oil ratio (GOR)} = \frac{\text{Volume of gas produced in scf}}{\text{Volume of oil produced in STB}}$$

When a gas cap exists on the top of the oil zone, the producing gas volume is based on the free gas flow from the gas cap as well as the solution gas that evolves from the crude oil under declining reservoir pressure. Therefore, the producing gas–oil ratio is greater than the solution gas–oil ratio.

The cumulative gas–oil ratio is the cumulative volume of gas produced over the cumulative volume of oil produced from a reservoir. As the reservoir is produced, cumulative gas–oil ratio increases with time.

Oil formation volume factor

The oil formation volume factor is a measure of the degree of change in oil volume as it is produced from the reservoir and brought to surface conditions. In the subsurface formation, pressure as well as temperature is significantly higher than stock-tank conditions. As oil is produced, it undergoes shrinkage or reduction in volume due to the liberation of dissolved gas. The effect is greater in the case of highly volatile oil due to the abundance of light hydrocarbons.

The oil formation volume factor is defined as follows:

$$B_o = \frac{\text{Volume of oil plus dissolved gas under reservoir pressure and temperature, rb}}{\text{Reduced volume of oil under stock tank pressure and temperature following liberation of gas, stb}} \qquad (4.12)$$

Depending on the relative abundance of volatile components, the formation volume factor may typically range from 5.0 for highly volatile oil to a value close to 1.0 for heavy oil. Oil having a formation volume factor of 2.0 indicates that the volume of oil will be reduced to half when produced. Heavy oil, on the contrary, has a relatively low formation volume factor meaning that oil volume is not significantly reduced under surface conditions, as shown in Figure 4.4. Formation volume factor of highly volatile liquid condensate is presented in Figure 4.5, where significant change in formation volume factor occurs as the reservoir pressure declines.

Oil is slightly compressible. Hence, the oil formation volume factor increases slightly with decline in reservoir pressure due to expansion of the liquid phase as long as the reservoir produces above the bubble point. However, as the bubble point is reached and the reservoir begins producing below the bubble point, reduction in oil volume is observed due to the evolution of the gas phase. Consequently, the oil formation volume factor increases with the decrease of reservoir pressure.

It is further noted that the formation volume factor can be calculated above the bubble point if the oil compressibility and certain other fluid properties are known. The equation is as follows:

$$B_o = B_{ob} \exp[-c_o(p - p_b)] \qquad (4.13)$$

where B_{ob} = formation volume factor at bubble point, rb/STB; p_b = bubblepoint pressure, psia.

Below the bubble point pressure, however, the effect of liquid expansion becomes relatively small compared to shrinkage of oil as the lighter hydrocarbons are liberated and form a vapor phase.

The formation volume factor can be estimated based on the following Petrosky–Farshad correlation when certain fluid properties are known:

$$B_o = 1.0113 + 7.2046(10^{-5}) \left[R_s^{0.3738} \left(\frac{\gamma_g^{0.2914}}{\gamma_o^{0.6265}} \right) + 0.24626(T - 460)^{0.5371} \right]^{3.0936} \qquad (4.14)$$

where R_s = solution GOR, scf/STB; γ_g = specific gravity of gas, ratio; γ_o = specific gravity of oil, ratio; T = reservoir temperature, °R.

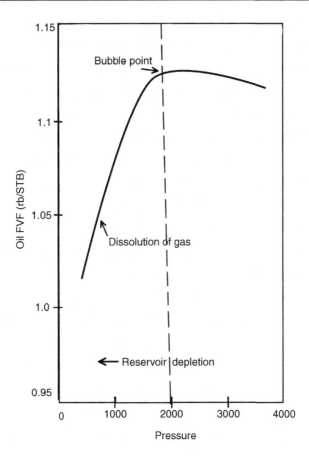

Figure 4.4 Formation volume factor of oil as a function of pressure. The value is unity when no more volatiles are present in the liquid phase.

Two-phase formation volume factor

The two-phase formation volume factor takes into account the oil formation volume factor as well as the formation volume factor of dissolved gas expressed in rb/stb. It can be expressed as follows:

$$B_t = B_o + (R_{si} - R_s)B_g \qquad (4.15)$$

where B_t = two-phase formation volume factor, rb/STB; B_o = oil formation volume factor, rb/STB; R_{si} = initial solution gas–oil ratio, scf/STB; R_s = solution gas–oil ratio, scf/STB; B_g = gas formation volume factor, rb/scf.

Above the bubble point pressure, only the oil phase exists, and the two-phase formation volume factor simplifies to the single-phase formation volume factor for oil.

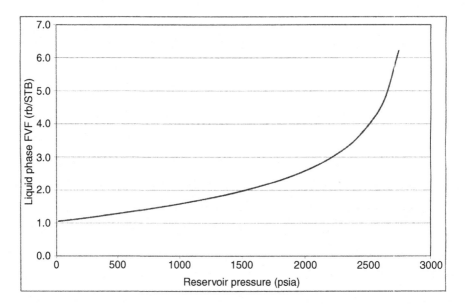

Figure 4.5 Plot of formation volume factor of condensate liquid versus reservoir pressure. Due to the presence of highly volatile hydrocarbons, formation volume factor is quite high at reservoir conditions. Courtesy: Computer Modelling Group.

Below the bubble point, however, the two-phase volume factor may have a significantly high value due to the expansion of gas.

Properties of natural gas

Natural gas is primarily composed of light hydrocarbons compared to oil. Due to the low viscosity of natural gas, it is produced with relative ease from gas and gas condensate reservoirs. Oil reservoirs with a gas cap or the reservoirs that are operating under bubble point pressure also produce natural gas along with oil. Primary interests to reservoir engineers include the compression and expansion characteristics of natural gas under changing reservoir pressure, mobility contrast of gas in relation to oil, and the changes in solubility of gas in oil as reservoir pressure declines, among others. Natural gas properties discussed in this chapter are as follows:

- Ideal gas law
- Real gas law
- Gas compressibility and gas compressibility factor
- Pseudo-reduced pressure and temperature
- Formation volume factor of gas
- Gas viscosity
- Gas density

Ideal gas law

The ideal gas law states that the pressure, temperature, and volume of gas are related to each other. The following equation can be used to express the relationship:

$$pV = nRT \tag{4.16}$$

where p = prevailing pressure, psia; V = volume of gas, ft.^{3}; n = number of pound-moles of gas, lb_m-mol; R = gas law constant, (psia)(ft.^{3})/($°$R)(lb_m-mol); T = prevailing absolute temperature, $°$R.

The value of gas law constant is 10.73 based on the units used in the above equation. It is also noted that T, $°$R = T,$°$F + 460.

Equation (4.16) is based on Boyle's law and Charles's law. The above relates the change in ideal gas volume to the changes in prevailing pressure and temperature, respectively. Furthermore, Equation (4.16) is referred to as the equation of state for an ideal gas.

Real gas law

Under typical reservoir conditions with high pressure and temperature, real gas volume may deviate significantly from that of the ideal gas. Hence, the ideal gas law is modified by introducing gas compressibility factor or gas deviation factor in order to develop an equation of state for real gases, as shown below:

$$pV = z \, n \, RT \tag{4.17}$$

where z = gas compressibility factor, a function of prevailing pressure and temperature.

The gas compressibility factor can be expressed as follows:

$$z = \frac{\text{Actual volume of gas at specific pressure and tempeature}}{\text{Volume predicted by ideal gas law at the same pressure and temperature}} \tag{4.18}$$

Note that the compressibility characteristics are also dependent on its composition. The gas compressibility factor, z, can be determined experimentally by utilizing equation. Standing and Katz [6] published a chart based on experimental results that plots z factor as a function of pseudo-reduced pressure and temperature. The chart is valid for computing z factor for natural gases regardless of their composition as long as the pseudo-reduced pressure and temperature are known. Pseudo-reduced pressure and temperature can be calculated when the pressure and temperature data of a hydrocarbon component at the critical point are available. The vapor and liquid phases of a pure substance at the critical point are indistinguishable. Pseudo-reduced pressure and temperature are defined as follows:

$$P_{pr} = \frac{P}{P_{pc}} \tag{4.19}$$

$$T_{pr} = \frac{T}{T_{pc}} \tag{4.20}$$

where P, T = pressure (psia) and temperature (°R) at which the z factor is calculated; P_{pc}, T_{pc} = critical pressure (psia) and temperature (°R) of the hydrocarbon component.

The values of z factor deviate significantly from ideality ($z = 1.0$) under certain pressure and temperature conditions. The following equation can be used to compute the critical pressure and temperature of a multicomponent mixture such as natural gas:

$$p_{pc} = \sum yi_{pc,i} \tag{4.21}$$

$$T_{pc} = \sum yi_{Tc,i} \tag{4.22}$$

It is noted that the critical pressure and temperature values based on the above equations do not represent the actual critical values for the mixture. Rather, these are used in the estimation of the gas compressibility factor.

Correlations are also available to compute the gas compressibility factor for a specific composition of gas. An equation proposed by Dranchuk and Abou-Kassem [7] can be expressed as follows:

$$z = 1 + A.\rho_r + B.\rho_r^2 - C.\rho_r^5 + D.\exp(-0.721\rho_r^2) \tag{4.23}$$

where $\rho_r = 0.27P_r/(zT_r)$; $A = 0.3265 - 1.07/T_r - 0.5339/T_r^3 + 0.1569/T_r^4 - 0.05165/T_r^5$; $B = 0.5475 - 0.7361/T_r + 0.1844/T_r^2$; $C = 0.1056(-0.7361/T_r + 0.1844/T_r^2)$; $D = 0.6134(1 + 0.721\rho_r^2)(\rho_r^2/T_r^3)$.

Since the compressibility factor z appears in both sides of Equation (4.23), an iterative approach to determine the value of z at a specific P_r and T_r is necessary.

In practice, values of z factor are computed by software applications available in the industry. An example is presented in Figure 4.6 where the compressibility factor of a gas of known composition is plotted over a pressure range. In this specific case, z factor decreases with decreasing pressure between 3500 psia and 2400 psia, reaches a minimum, and then increases as pressure further decreases to atmospheric conditions. The data used in the calculation are as follows.

Viscosity of natural gases

Viscosity is a measure of the internal resistance to flow. The unit of viscosity is the centipoise. Due to considerably less viscosity than oil, gas will flow at a significantly

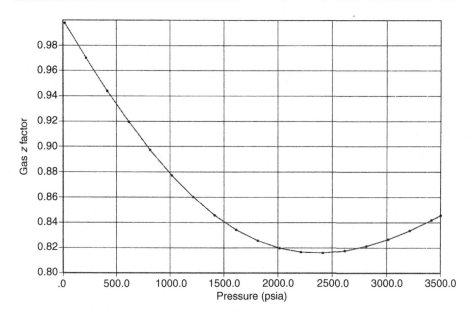

Figure 4.6 Values of gas compressibility factor as function of pressure based on gas composition and reservoir temperature. Courtesy: Computer Modelling Group.

higher rate than oil in porous media. The change in viscosity of the natural gas with pressure is shown in Figure 4.7. The same gas composition and reservoir temperature from the previous example is used.

Gas formation volume factor

Gas formation volume factor is the volume of gas in a reservoir barrel divided by volume of gas under standard conditions, scf.

Standard cubic feet of gas is the volume of gas at 14.7 psi^{-1} pressure and 60°F temperature.

$$B_g = 5.021 \left(\frac{zT}{p} \right)_{res} \tag{4.24}$$

where B_g = gas formation volume factor, rb/Mscf; z = gas compressibility factor; T = reservoir temperature, °R; p = reservoir pressure, psia.

As the above equation suggests, the formation volume is a direct function of gas compressibility factor and varies inversely with reservoir pressure. Since the gas formation volume is a very small number, it is conveniently expressed in rb/Mscf rather than rb/scf.

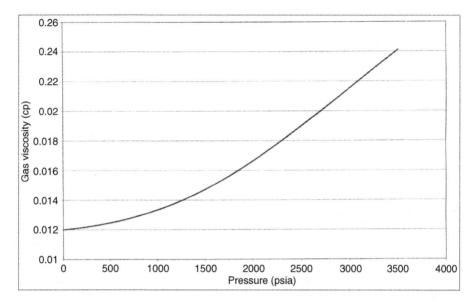

Figure 4.7 Plot of gas viscosity versus pressure. Viscosity of natural gas decreases nonlinearly with decreasing pressure. Courtesy: Computer Modelling Group.

Example 4.1

Calculate formation volume factor at 2600 psig. Use gas composition and reservoir data in Table 4.3.
Reservoir pressure = 2614.7 psia
Temperature = 150°F
Based on Figure 4.6, $z = 0.818$
$B_g = 5.021 \times [0.818 \times (150 + 460)/2614.7] = 0.9582$ rb/Mscf

Isothermal compressibility

Gas compressibility is a function of the rate of change in the volume of the gas per change in gas pressure divided by gas volume at a specified temperature.

Properties of gas condensates

Gas condensates contain an amount of intermediate to heavy hydrocarbon fractions, which condenses out of the vapor phase as droplets in the porous medium when the reservoir pressure declines below its dew point. The formation volume factor of gas condensate is defined as follows:

$$B_{gc} = \frac{\text{Combined volume of gas and condensate in vapor phase measured in rb}}{\text{Volume of condensate produced as liquid measured in stb}} \quad (4.25)$$

Table 4.3 Composition of dry gas used to plot gas compressibility factor

Dry gas component	Percent
Methane	88.1
Ethane	6.0
Propane	2.9
i-Butane	1.9
i-Pentane	1.1

Reservoir temperature: 150°F.

Charts [4] and correlations are available in the literature to estimate the formation volume factor based on the specific gravities of gas and condensate. The heavier components are usually combined and reported as heptanes-plus or C_7+ fractions.

An example of gas compressibility factor and condensate–gas ratio of a gas condensate system is presented in Figure 4.8. The composition of gas condensate is presented in the Table 4.4. For the gas condensate used in the study, the liquid phase essentially drops out of vapor between 2300 psia and 2200 psia, as the plot suggests.

Due to the high volatility of condensate, formation volume factor is quite high at reservoir conditions (Figure 4.5).

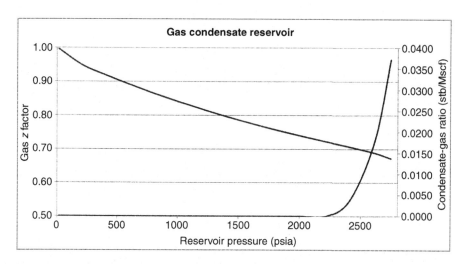

Figure 4.8 Plot of gas compressibility factor and condensate–gas ratio for gas condensate sample. The composition of gas condensate is shown in Table 4.4. Courtesy: Computer Modelling Group.

Table 4.4 **Composition of gas condensate**

Component of hydrocarbon	Percentage
Methane	72.8
Ethane	10.1
Propane	3.9
i-Butane	1.9
n-Butane	1.4
i-Pentane	0.51
n-Pentane	0.5
Hexanes	1.1
Heptanes and higher	7.79

Properties of formation water

Knowledge of formation water properties is needed in various reservoir studies along with oil and gas properties. Properties of formation water are generally dependent on reservoir pressure, temperature, and concentration of salt compounds.

Formation water compressibility

Water compressibility is a function of the rate of change in the volume of the water per change in pressure at a specified temperature divided by water volume.

Formation water viscosity

Viscosity of water depends upon the reservoir temperature, pressure, and the salinity of water. Viscosity of formation water decreases with temperature, and increases the concentration of salt compounds. At a reservoir temperature of 140°F, formation water viscosity increases from 0.46 cp to 0.9 cp as the amount of salt compounds increases from 0% to 26%.

Solution gas–water ratio

Solution gas–water ratio is defined as the volume of dissolved gas in water over water volume. Natural gas has limited solubility in water.

Formation volume factor

Water formation volume factor is the volume water and dissolved gas at elevated pressure and temperature in the reservoir divided by one stock-tank barrel of water under standard conditions. The formation volume factor is quite low, around 1.01 in typical cases.

Laboratory measurement of reservoir fluid properties

Besides using correlations, fluid samples collected from subsurface formation and studies are conducted to evaluate various fluid properties. Laboratory measurements are also made on separated liquid and gas phases obtained from surface facilities, which are recombined in correct amounts to reproduce fluid samples under simulated reservoir conditions.

The usual measurements include the determination of oil specific gravity, viscosity, compressibility, bubble point pressure, solution gas–oil ratio, oil formation volume factor, and gas deviation factor.

Measurement techniques using a PVT cell are:

- Flash vaporization
- Differential vaporization

In flash vaporization (constant composition expansion), the evolved gas is kept in contact with the liquid phase at all times in a closed chamber. The process is similar to what takes place in surface separators.

Objectives of flash vaporization include the determination of bubble point or saturation pressure, specific volume at saturation pressure, coefficient of thermal expansion, and isothermal compressibility of the liquid above the bubble point

In differential vaporization, the gaseous hydrocarbons are removed as soon as they evolve from the solution. This emulates the rapid gas movement towards the wells in porous media immediately following the dissolution of the gas.

Results of the measurements are solution gas–oil ratio, relative oil volume, relative total volume, density of the oil, gas deviation factor, gas formation volume factor, incremental gas gravity, and fluid viscosity as a function of pressure.

Factors affecting reservoir fluid properties

As noted earlier, reservoir pressure, temperature and fluid composition primarily affect the properties of various fluid phases, which are described in the following.

Reservoir pressure

Reservoir pressure is an important factor governing the phase behavior and the properties of the reservoir fluids. As a petroleum reservoir is produced and depleted, one or more of the following are observed, depending on the type of reservoir:

- Dissolution or liberation of gas phase from oil phase in the reservoir
- Changes in oil volume as oil is slightly compressible and more predominant due to the liberation of the gas phase
- Expansion of gas
- Retrograde condensation where the formation of oil droplets occurs from the gas phase

Furthermore, as Darcy's law suggests, the oil and gas production rates depend upon the reservoir pressure and the pressure at the wellbore.

Reservoir pressure is determined at discovery as well as periodically through the production phase of the reservoir; values of reservoir pressure are then used to calculate the fluid properties and conduct the reservoir performance analysis.

Estimation of reservoir pressure

Reservoir pressure is estimated by reservoir depth, and change of pressure with depth.

Reservoir pressure is commonly expressed as gauge pressure and absolute pressure. The unit of gauge pressure is psi^{-1}. Absolute pressure = gauge pressure + atmospheric pressure (usually 14.7 psi). The unit of absolute pressure is psia. It can be shown that the pressure gradient of fresh water is 0.433 psi^{-1}, considering its density (62.4 lb-m/ft.^{3}), units of area (1 ft.^{2} = 144 in.^{2}), acceleration due to gravity (32.2 ft./s^2), and lb-m to lb-f conversion factor.

$$(62.4\, \text{lb-m / ft.}^3) \times (1\, \text{ft.}^2\, / 144\, \text{in.}^2) \times (32.2\, \text{ft. / s}^2) / (32.2\, \text{lb-m ft. / lb-f s}^2) \tag{4.26}$$
$$= 0.433\, \text{psi / ft.}$$

The pressure gradient implies that the fresh water (sp. gr. = 1.0) would exert a pressure of 0.433 ft.$^{-1}$ of depth in the reservoir. Formation water, however, is heavier than fresh water as it contains dissolved solids. The specific gravity of formation is greater than 1.0. Hence, it exerts more pressure per foot of depth in the reservoir. The pressure gradient of formation water is calculated as:

$$\text{Change of pressure of formation water with depth} = 0.433\gamma_w\, \text{psi / ft.} \tag{4.27}$$

where γ_w = specific gravity of formation water, ratio.

When reservoir depth and the specific gravity of formation water or connate water are known, the reservoir pressure can be calculated as follows:

$$p = 0.433\gamma_w D + 14.7\, \text{psia} \tag{4.28}$$

where D = reservoir depth, feet.

Abnormally pressured reservoirs

Due to various geologic events, structural anomalies, and hydrodynamic processes that can occur through ages, certain oil and gas reservoirs exhibit higher or lower pressure than that computed by Equation (4.28). These reservoirs are referred to as abnormally pressured reservoirs. Certain gas reservoirs have been discovered where the gradient is 0.8 psi/ft. Hence, Equation (4.28) is modified to accommodate abnormally pressured reservoir conditions:

$$p = 0.433\gamma_w D + 14.7 + C\, \text{psia} \tag{4.29}$$

where C = correction factor for abnormally pressured reservoir.

Certain unconventional shale gas reservoirs having ultralow permeability are known to be overpressured.

Typical measures of pressure in petroleum reservoirs

Reservoir pressures are measured under various conditions, such as:

- The reservoir fluid pressure in the rock pores is the reservoir pressure or formation pressure.
- The reservoir pressure at discovery without any production is the initial reservoir pressure. It declines continuously with production when there is no support in the form of fluid injection or aquifer influx. Reservoir fluid properties change accordingly affecting recovery.
- The average reservoir pressure is the pressure when all production and injection activities cease and equilibrium is reached throughout the reservoir.
- Abandonment pressure is the pressure when the producing well reaches its economic limit following a decline in rates.
- Flowing bottom hole pressure is the pressure measured at the bottom of a well when oil and gas flow are produced.
- Static bottom hole pressure is the pressure when there is no flow at the well and pressure has reached a stabilized condition. A static condition may be achieved by shutting the well for a considerable period of time.
- Wellhead pressure is the pressure measured at the wellhead. The wellhead pressure in a producing well is less than the bottom hole pressure.
- Fracture pressure is the threshold pressure at which the subsurface formation is fractured by injecting fluid.
- Overburden pressure is the combined pressure exerted by the formation rock and the reservoir fluid.

Note that reservoir pressures recorded at various depths in multiple well locations are corrected to the same datum depth by using the known fluid gradient. The datum is usually taken at the oil–water contact.

Reservoir temperature

Like reservoir pressure, reservoir temperature is also an important factor governing the phase behavior and the properties of the reservoir fluids.

Reservoir temperature, which depends upon the reservoir depth, can be estimated by the following equation:

$$T = T_s + \frac{T_{gradient} \times D}{100} \tag{4.30}$$

where T = reservoir temperature, °F; $T_{gradient}$ = temperature gradient, °F/100 ft.; T_s = temperature at surface, °F.

Change of temperature with depth can vary from 0.8°F to 1.6°F per 100 ft. A value of 1.2–1.4°F per 100 ft. is usually assumed for sedimentary basins. Temperature anomalies may occur due to geothermal processes.

Example 4.2

.Estimate the pressure and temperature of a newly discovered reservoir at a depth of 6200 ft. where very few data are available. Make necessary assumptions.
Based on reservoir data obtained from the region, the following assumptions are made:
  Specific gravity of the formation water = 1.08
  Temperature gradient = 1.3°F/100 ft.
  Mean surface temperature = 62°F
  Further assume that the reservoir is overpressured by 200 psi as the regional data suggest.
  Referring to Equation (4.28), the estimated reservoir pressure is calculated as follows:
  $p = 0.433(1.08)(6200) + 14.7 + 200 = 3114$ psia
  Referring to Equation (4.29), reservoir temperature is estimated as:
  $T = 62 + (1.3/100)(6200) = 142.6$°F

Composition of petroleum fluids

Crude oil and natural gas are composed of many hydrocarbon compounds with a wide range of molecular weights. The lighter and simpler compounds are produced as natural gas after surface separation. Heavier and more complex compounds are produced as crude oil at stock-tank conditions. Example compositions of petroleum fluids of increasing gravity, from dry gas to black oil, are shown in Table 4.5.

The above indicates that the heptanes and heavier fractions are more in proportion with increasing gravity of petroleum fluids.

Summing up

Fluid and rock properties are fundamental to reservoir engineering. Fluid properties are essential in understanding fluid flow characteristics in porous media, designing a well, developing a reservoir, planning waterflood operations, and optimizing ultimate recovery, to name a few.

Some of the important roles that fluid properties play in reservoir engineering include the estimate of hydrocarbon in place by various methods, analysis of fluid flow and well rates, reservoir simulation studies, and determination of enhanced oil recovery methods.

Reservoir classification is usually based on the type of petroleum fluid it mainly produces. The reservoir types are listed in the following in the order of increasing gravity and viscosity:

- Gas reservoirs: Dry gas, wet gas, and gas condensate
- Oil reservoirs: Light oil, black oil, heavy oil, extra heavy oil, and bitumen

Table 4.5 **Compositions and properties of oil and gas**

Components	Dry gas	Wet gas	Gas condensate	Volatile oil	Black oil
Methane	86.6	82.9	75.88	55.22	33.6
Ethane	5.4	6.6	8.3	7.1	4.01
Propane	3.3	3.1	3.5	3.87	1.01
i-Butane	1.8	0.3	0.66	1.12	0.82
n-Butane	0.2	1.5	2.2	1.08	0.33
i-Pentane	0.45	1.35	0.6	0.81	0.43
n-Pentane	0.06	0.71	1.22	1.22	0.22
Hexanes	0.05	2.09	1.5	1.87	1.8
Heptanes-plus			2.5	26.7	57.4
CO_2	0.16	1.2	3.2	0.9	0.07
N_2	1.98	0.25	0.44	0.11	0.31
Total	100	100	100	100	100
API gravity			46	36	25
Color of liquid			Straw	Amber	Green to black
GOR, scf/STB			>5000	1500	350

As petroleum fluids become more viscous, their mobility in porous medium is diminished. Relatively little effort is needed to produce gas that is least viscous. Light and intermediate oil are produced by conventional recovery methods. However, extra heavy oil and bitumen are not mobile at all. These require either thermal or unconventional recovery methods to produce economically.

Oil and gas properties discussed in the chapter are highlighted in Tables 4.6 and 4.7.

Questions and assignments

1. Why are pressure and temperature important for reservoir fluid properties, and how are they estimated?
2. What are the main characteristics of dry gas, wet gas, gas condensate, volatile oil, and black oil?
3. Reservoir engineers are interested in what oil and gas properties?
4. How can solution gas ratio affect oil production?
5. Why are certain extra heavy oil reservoirs viewed as unconventional? How extensive is the occurrence of heavy oil and bitumen in the world compared to all other types?
6. Why does real gas deviate significantly from the ideal gas law?
7. How is the ideal gas law modified to account for the behavior of the real gas?
8. How are pseudo-reduced pressure and temperature used to compute gas deviation factor, z?
9. What are the types of laboratory measurements of reservoir fluid properties?
10. What are the applications of fluid properties in reservoir engineering?

Table 4.6 **Properties of oil**

Property	Description	Typical range	Notes
API gravity	Common unit for oil gravity. Inversely related to specific gravity	See Table 4.2.	Affects vertical equilibrium of oil, gas and water in porous medium
Viscosity	Measure of how easily the fluid will flow	See Table 4.2.	Gas having the least viscosity flows with ease. Water flows preferentially in porous media as it is less viscous than oil. Extra heavy oil and bitumen are so viscous that these require thermal or unconventional methods to recover.
Compressibility	Measure of change in oil volume per unit pressure		Affects fluid flow characteristics in porous media
Bubble point pressure	Pressure at which bubbles of gas begin to come out of liquid phase as a reservoir pressure declines	High for volatile oil and low for black oil; typically ranges between 1800 psi and 2500 psi	Many reservoirs are operated above bubble point pressure to avoid gas production
Solution gas–oil ratio	Measure of the amount of gas dissolved in liquid phase at reservoir condition	High for volatile oil and low for black oil; typically ranges between 250 scf/STB and 1500 scf/STB	Liberation of gas occurs early
Oil formation volume factor	Measure of shrinkage of oil from reservoir condition to stock-tank condition due to dissolution of gas	Low for heavy oil and high for volatile oil; typically ranges between 1.0 and 5.0	
Two-phase formation volume factor	Oil formation volume factor combined with gas formation volume factor	Same as oil formation volume factor above the bubble point; increases significantly below the bubble point due to the liberation of gas.	

Table 4.7 **Properties of natural gas**

Property	Description	Notes
Gas compressibility (z factor)	Ratio of actual volume of real gas over the ideal volume at a specific pressure and temperature	It is a measure of nonideality of real gas and a nonlinear function of pressure, temperature, and composition
Gas formation volume factor	Ratio of volume of gas in reservoir bbls over the volume in scf under standard conditions	Essential in calculating gas in place and reserves
Gas viscosity	Defined in Table 4.6	

References

[1] Glaso O. Generalized pressure-volume-temperature correlations. J Petrol Technol 1980;May:785–95.
[2] Chew J, Connally CA Jr. A viscosity correlation for gas saturated crude oils. Trans AIME 1959;216:270–5.
[3] Vasquez M, Beggs D. Correlations for fluid physical properties prediction. J Petrol Technol 1980. p. 32.
[4] Standing MB. A pressure-volume-temperature correlation for mixtures of California oils and gases. Drilling and production practices. Washington DC: American Petroleum Institute; 1947. p. 285–7.
[5] Marhoun MA. PVT correlation for Middle East crude oils. J Petrol Technol 1988;May: 650–65.
[6] Standing MB, Katz DL. Density of natural gases. Trans AIME 1942. p. 146.
[7] Dranchuk PM, Abou-Kassem JH. Calculation of z factors for natural gases using equations of state. J Cdn Pet Tech July–Sept 1975. p. 34–6.

Phase behavior of hydrocarbon fluids in reservoirs

Introduction

Reservoir pressure changes significantly during the production of oil and gas, which may lead to a change in phase, namely, vaporization of oil or condensation of gas. Phase behavior of the reservoir fluids is studied in detail as it usually has a profound effect on reservoir performance. The phenomenon is dependent on prevailing reservoir pressure and fluid composition. Reservoir oil and gas are typically composed of large numbers of hydrocarbon components with wide-ranging bubble points and dew points. As the pressure declines during production from an oil reservoir and the bubble point is reached, volatile components in oil are liberated and a free gas phase is formed. Since gas is more mobile than oil, the gas–oil ratio at the wells may become quite high, adversely affecting oil production. In a gas condensate reservoir, relatively heavier components may condense out within the reservoir as the dew point is reached following the decline in reservoir pressure. In both cases, certain strategies are undertaken to optimize the recovery of oil and gas.

This chapter describes the phase behavior of various petroleum fluids with the aid of phase diagrams and related reservoir performances, and answers the following questions:

- What is a fluid phase diagram?
- How is fluid behavior explained with a phase diagram?
- Why is a phase diagram needed to study fluid flow in reservoirs?
- What are phase envelope, bubble point curve, dew point curve, and critical point of a reservoir fluid?
- How does the fluid phase behavior affect oil and gas recovery?

Phase diagram

The phase behavior of petroleum fluids is best described by a phase diagram [1,2]. A generalized version of the phase diagram is presented in Figure 5.1. Liquid and vapor phases are represented in a two-dimensional plot where reservoir temperature and pressure are represented in x and y coordinates, respectively. It must be noted that each reservoir fluid, due to varying composition of hydrocarbons and impurities, has unique properties and phase behavior. Hence, a reservoir fluid is represented by its own phase diagram.

Reservoir Engineering. http://dx.doi.org/10.1016/B978-0-12-800219-3.00005-X

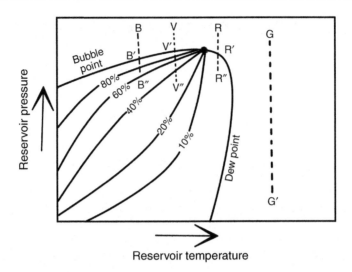

Figure 5.1 A generalized phase diagram showing volatilization of oil and retrograde condensation of gas that affects reservoir performance.

The important aspects of the phase diagram are summarized as follows:

- Single- and two-phase regions: The two regions are distinguished by the phase envelope. Within the phase envelope, petroleum fluid exists in two phases, liquid and vapor. Outside the phase envelope, petroleum fluid exists in single phase, either in liquid or in vapor form.
- Isosaturation lines: The curved lines within the phase envelope, referred to as isosaturation lines, represent the relative percentages of liquid and vapor at the specific pressure and temperature. Along an isosaturation line, liquid and vapor fractions present in the fluid are constant.
- Bubble point curve: The outer periphery of the phase envelope toward the upper left. Fluid at any pressure above the line only exists in liquid form.
- Dew point curve: The outer periphery of the phase envelope toward the lower right. Fluid at any temperature beyond the line only exists in vapor form.
- Critical point: Bubble point and dew point curves meet at critical point C where liquid and vapor phases are in equilibrium and indistinguishable from each other. At the critical point, the properties of liquid and vapor phases are identical. Moreover, liquid and vapor phases are indistinguishable at the critical point. The critical point changes with the composition of hydrocarbons.
- Cricondenbar: The maximum pressure over which fluid can only exist in liquid form.
- Cricondentherm: The maximum temperature over which fluid can only exist as vapor.

It is again noted that the phase diagram is dependent upon the composition of the oil and gas. *In situ* fluid in each reservoir will have its own phase diagram having different bubble point and dew point curves. The shape of the phase diagram for highly volatile oil is quite different to that of heavy oil. In Figure 5.2, phase diagrams for light and heavy oils are shown for comparison.

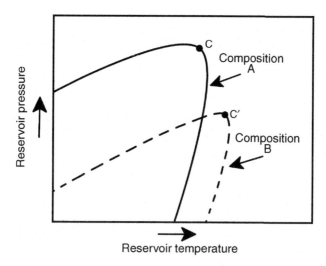

Figure 5.2 Phase diagram of volatile oil (composition A) compared to black oil (composition B). Reservoir fluids have unique compositions; hence, they exhibit unique phase behavior.

Phase diagram based on software application

In pressure–volume–temperature studies and reservoir simulation, phase diagrams are generated by using the equations of state (EOS). The well-known EOS include Peng–Robinson and Soave–Redlich–Kwon. An example of a phase diagram for black oil generated by the Peng–Robinson EOS is presented in Figure 5.3. The composition of black oil used in the study is shown in Table 5.1.

Reservoir types and recovery efficiency

Petroleum reservoirs are commonly classified according to the composition of petroleum fluids. In this section, the phase behavior of each fluid type is summarized with the help of a generalized phase diagram. Performance of a reservoir is dependent on the type of fluid, pressure, and temperature. The recovery efficiencies from each type of reservoir are also discussed.

In the order of increasing presence of heavier hydrocarbon components, petroleum reservoirs are classified as follows:

- Dry gas reservoir
- Wet gas reservoir
- Gas condensate reservoir
- Volatile oil reservoir
- Black oil reservoir
- Heavy oil reservoir

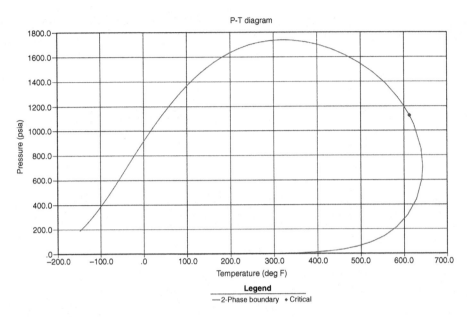

Figure 5.3 Phase diagram of black oil based on the Peng–Robinson equation of state.
Reservoir temperature of 150° is assumed.
Courtesy: Computer Modelling Group.

Table 5.1 Composition of black oil used in phase diagram

Component	Mole (%)
Methane	33.1
Ethane	3.9
Propane	1.2
i-Butane	0.77
n-Butane	0.42
i-Pentane	0.4
n-Pentane	0.18
Hexanes	0.16
Heptanes and higher	59.87
Total	100

- Dry gas reservoir: Dry gas has only the lighter components, and no liquid phase is formed as reservoir pressure declines. The path taken by the dry gas reservoir is shown as G-G' in the phase diagram (Figure 12.1). The line does not enter the two-phase region inside the phase envelope. The reservoir drive mechanism is the expansion of gas. Recovery efficiency from conventional dry gas reservoirs having good porosity and permeability is quite high, in the range of 70–85%, due to the fact that gas is significantly less viscous than oil and water. In unconventional shale gas reservoirs, however, recovery is significantly less due to ultralow rock permeability, typically less than 10%.

- Wet gas reservoir: Wet gas is distinguished by the presence of certain heavier components that are converted to liquid under stock-tank conditions at the surface. However, the gas phase essentially remains as vapor in the reservoir as pressure declines due to production.
- Gas condensate reservoir: In gas condensate reservoirs, gas has relative abundance of heavier components, which condense out in the reservoir as the pressure declines. Gas traces a path where it enters the shaded retrograde condensation region within the two-phase region R–R′. In order to minimize the loss of enriched hydrocarbon components, gas recycling is implemented where the certain amount of produced gas is injected back in the reservoir. The path traced by gas condensate reservoirs is shown by the line RR′. The recovery efficiency of retrograde condensate reservoirs is less than that of dry and wet gas reservoirs.
- Saturated and undersaturated oil reservoirs: An oil reservoir can either be saturated or undersaturated. Initially, the reservoir pressure may be above the bubble point pressure and the petroleum fluid is completely in liquid phase (undersaturated oil reservoir), or at or below the bubble point (saturated oil reservoir). The term undersaturated denotes that the liquid phase is not fully saturated with gas and has the capacity to dissolve more gas. With regards to the phase diagram, the initial point of undersaturated oil reservoir is above the bubble point curve. On the other hand, saturated oil is located on the bubble point curve or within the phase envelope.
- Volatile oil reservoir: Volatile oil is relatively high in lighter hydrocarbon components compared to black oil reservoirs and has higher API gravity (40° or more). Let us first consider a saturated oil reservoir. The path traced by volatile oil is closer to the critical point than heavier oil. In the phase diagram, the path has two distinct portions and characteristics as shown in Table 5.2.
- Black oil reservoir: "Black" oil is less volatile due to the presence heavier hydrocarbons. The path traced by a producing black oil reservoir is further away from the critical point. The path of the black oil is labeled as B–B′–B″. Like in volatile oil reservoirs, recovery above

Table 5.2 **Phase behavior of volatile oil**

Path	Characteristics	Drive mechanism	Typical recovery (%)
V–V′	Steady decline in reservoir pressure unless an external source such as aquifer drive is present	Volumetric expansion of rock and fluid	In single digits, 1–7
	Gas remains in solution		
	Only oil is produced		
V′–V″	Gas phase is liberated from liquid phase	Solution gas or deletion drive	20–35
	Gas becomes mobile beyond critical gas saturation		
	Liquid phase is driven toward the wellbore by gas phase		
	The gas–oil ratio is initially low, increases with time, and finally decreases as most of the liberated gas is produced		

and below the bubble point is based on expansion and solution drive, respectively. As black oil has higher viscosity, recovery can be somewhat less than that of volatile oil reservoirs given all other factors remaining the same.

- Heavy oil reservoir: Heavier and complex hydrocarbons are abundant in heavy oil leading to very high viscosity, in the order of 10,000 cp or more. Heavy oil is much less volatility. The path traced by heavy oil is further to the left of the phase diagram. Recovery is quite low, unless thermal enhanced oil recovery methods are implemented.

Study of gas condensate reservoir performance

The constant volume depletion test is widely used in the industry to evaluate the performance of a gas condensate reservoir. The test replicates the pressure depletion that is encountered in the reservoir during production. A fluid sample obtained from the reservoir is kept in a high-pressure chamber; pressure is gradually lowered by releasing gas and noting the dropout volume of the liquid phase inside the chamber. It is a direct and reliable analysis of fluid phase behavior in a gas condensate reservoir. The dew point of the vapor is determined by observing the first appearance of liquid droplets in the chamber. Subsequently, liquid volumes are noted as a function of depleting pressure (Figure 5.4). Besides dew point, the test delivers a wealth of data, including the compositional changes as a function of pressure, recovery of hydrocarbons, accumulated volumes of liquid condensates, and compressibility factor.

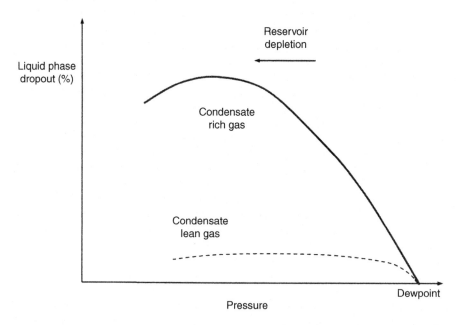

Figure 5.4 Results of constant volume depletion test showing pressure versus retrograde condensation.

Optimization of oil and gas recovery

In volatile and black oil reservoirs, pressure maintenance above the bubble point is widely practiced to optimize recovery. Liberation of gas from oil is avoided so that the mobility of oil in the reservoir is not hampered by the appearance of gas. Recovery from a solution gas drive reservoir is generally less than that of a reservoir where volatile components are kept in a dissolved state by pressure maintenance operations.

For retrograde condensate reservoirs, pressure maintenance is accomplished by recycling the produced gas through injection wells, among other methods, as described below. Maintenance of reservoir pressure above the dew point ensures that the rich hydrocarbon components in gas do not condense out and are not left behind in the reservoir.

Case Study: Review of Gas Condensate Reservoir Performance

As stated earlier, fluid properties and phase behavior play a critical role in the performance of a gas condensate reservoir, more than in any other type of reservoir. In reservoirs where liquid droplets build up in the vicinity of wells due to reservoir pressure below the dew point of reservoir fluid, severe loss in well productivity can be encountered. In certain cases, the loss in productivity can be as high as 80%. Some gas reservoirs are under contract to deliver a fixed quantity of gas per day to the customer, where the issue may become even more serious. A literature review on this topic points to the following [3]:

- Liquid dropout characteristics: Three regions are identified within the drainage area depending on the reservoir pressure. In the outer region located farthest from the well, fluid is initially in single phase as reservoir pressure is above the dew point. Reservoir pressure drops continuously toward the well, and is below the dew point. In the middle region, condensation of heavier hydrocarbon occurs. However, saturation of liquid droplets is below a critical limit and the droplets are not mobile in the pore channels. In the innermost layer, the saturation of liquid droplets is above the critical value and the droplets are mobile. As the reservoir is depleted, the outer layer shrinks and more of the reservoir area experiences condensation of liquid droplets. Liquid droplets are trapped in the formation pores leading to loss of valuable hydrocarbon components.
- Liquid holdup in the wellbore: As gas and condensate are produced, some portion of the liquid may fall back due to gravity leading to liquid holdup in the wellbore. As a high pressure drop is observed around the well, non-Darcy flow may occur, which lowers the apparent permeability of the formation.
- Effect of formation permeability: The adverse effect of gas condensation is dependent on permeability thickness (kh) of the formation. It is usually severe in low permeability reservoirs, which require higher drawdown pressure to produce. High permeability formations, on the contrary, allow the transport of liquid condensate with relative ease. Hence, loss in well productivity may be less severe.
- Remedial measures: As stated earlier, a common approach to produce gas condensate reservoirs is to recycle the produced gas into the reservoir under high pressure in order

to operate above the dew point pressure and avoid any condensation of heavier hydro-carbons in the wellbore. However, when the gas price is high, the approach may turn out to be less attractive. A huff and puff method may also be implemented through a single well, where gas is injected at high pressure to increase the reservoir pressure for a period of time, followed by production through the same well. After a period of pro-duction, the condensate dropout problem reappears as pressure is reduced. Hydraulic fracturing is also used to enhance well productivity as the fracture creates a pathway for enhancing production for a period of time. Eventually, production declines as con-densate builds up around the fractures. Wells drilled in carbonate reservoirs may be acidized to enhance permeability. Additionally, field tests have been conducted involv-ing the injection of solvents into the formation to remove the blockage by condensate and enhance well productivity.

Summary

Reservoir fluids are composed of various hydrocarbon components with wide-ranging bubble points and dew points. As a reservoir is produced, reservoir pressure changes significantly, which may result in the volatilization of lighter components present in oil or the condensation of heavier components in gas. As it has been observed from the early days of oil and gas production, any change of fluid phase affects reservoir performance significantly.

The phase behavior of a reservoir fluid can be best described with the help of a phase diagram, which is unique for each reservoir fluid having different compositions. A phase diagram depicts the state of fluid (single phase or two phase) for a range of pressures and temperatures. Pressure is plotted as ordinate (y-axis) while temperature is plotted as abscissa (x-axis). One of the main characteristics of the phase diagram is the phase envelope. Within this envelope, fluid remains in two phases, namely, oil and gas, while outside the envelope, fluid exists in only one phase, either oil or gas. The phase envelope also marks the bubble point and dew point curves for the system. At a pressure above the bubble point, all volatile components remain in solution. Similarly, heavier hydrocarbons remain in gas phase outside the dew point curve. Bubble point and dew point curves meet at critical points, where the liquid and gas phases are indistinguishable and the properties are identical. Within the two-phase envelope, isosaturation lines can be drawn where the ratio of liquid and gas content is the same.

Oil reservoir performance can be optimized by producing a reservoir above the bubble point, which avoids the liberation of gas phase within the reservoir. This en-sures that oil is produced without any adverse effects of gas on the mobility of oil in the reservoir. Again, gas condensate reservoirs are reinjected with produced gas in order to maintain the reservoir pressure above the dew point, ensuring that all the rich hydrocarbon components do not drop out in the reservoir and are left behind. Other methods are also applied to enhance productivity of gas condensate wells, including the huff and puff method, hydraulic fracturing, acidization, and solvent injection.

Questions and assignments

1. Describe fluid phase behavior as typically observed in various types of petroleum reservoir.
2. How does a phase diagram help in visualizing fluid behavior?
3. What are the main features of a phase diagram? What is critical point?
4. Why is the formation of liquid droplets in a gas condensate reservoir called retrograde condensation?
5. Can the phase diagram change for a reservoir fluid with production? Explain.
6. How would the shape of a phase diagram change between light and intermediate oil.
7. Is a phase diagram critical to develop a heavy oil reservoir? Why or why not?
8. Draw a phase diagram for a typical volatile oil, and show the path traced by the liquid in the reservoir through the two-phase region all the way to the surface. How would you modify the plot if the reservoir is under pressure maintenance and operates above the bubble point?
9. What is the constant volume test? How can it aid in enhancing the recovery of gas and liquid phase from a gas condensate reservoir?
10. Based on a literature review, describe gas recycling operation in a gas condensate reservoir and how it enhances reservoir performance.

References

[1] McCain, William D Jr. The properties of petroleum fluids. Tulsa, OK: Pennwell; 1990.
[2] Satter A, Iqbal GM, Buchwalter JA. Practical enhanced reservoir engineering: assisted with simulation software. Tulsa, OK: Pennwell; 2008.
[3] Fan L, Harris BW, Jamaluddin A, Kamath J, Mott R, Pope GA, Shandrygin A, Whitson CH. Understanding gas condensate reservoirs. World Oil; Winter 2005/2006.

Characterization of conventional and unconventional petroleum reservoirs

Introduction

Reservoir characterization aims at describing the reservoir in sufficient detail in order to optimize well design and placement, completion, fracturing, fluid injection, and oil production. The ultimate goal of reservoir characterization is to add value to the assets, i.e., oil and gas reserves. The goal is achieved by understanding the uniqueness of the reservoir and minimizing potential risks in reservoir development. Knowledge acquired from reservoir characterization studies leads to more reliable reservoir simulation models and prediction of performance. For large and complex reservoirs, reservoir characterization holds the key to successful reservoir management.

This chapter discusses reservoir characterization efforts and answers the following queries:

- What are the objectives of reservoir characterization?
- What type of reservoir studies lead to reservoir characterization?
- What is reservoir quality? What role does it play in reservoir development?
- How does reservoir characterization contribute to reservoir management?
- What information is sought based on reservoir characterization studies?
- What workflow can be implemented to conduct reservoir characterization and add value to reservoir assets?

Objectives

The objectives of reservoir characterization include enhancement of reservoir performance and add to ultimate recovery potential. Based on various reservoir characterization studies, engineers seek the following information, among others:

- Identification of structure, lithology, rock types, facies change, and other factors that contribute to reservoir heterogeneity
- Distribution of porosity, permeability, fluid saturation, hydrocarbon pore volume, and fluid contact throughout the reservoir; the data are used to build realistic reservoir models, quantify reservoir quality, identify pay zones, design and drill wells, and optimize reservoir performance. Upscaling of core data to reservoir scale is necessary
- Reservoir complexities such as the presence of faults, fractures, barriers, channels, and change in rock facies that may affect reservoir performance
- Information leading to the optimization of well design, including the length, trajectory, and number of laterals for horizontal wells

Reservoir Engineering. http://dx.doi.org/10.1016/B978-0-12-800219-3.00006-1

- Mechanical properties of tight reservoir rocks, including Young's modulus, Poisson's ratio, bulk modulus, closure stress, and others that would optimize fracturing. Horizontal wells with multistage fracturing are essential to produce economically from tight and unconventional reservoirs

Reservoir quality

One of the goals of reservoir characterization is the evaluation of reservoir quality. In simple terms, reservoir quality indicates how much hydrocarbon is stored in the formation and how easily it will produce. Porosity, permeability, fluid properties and saturations, geological continuity, formation heterogeneities, number of flow units, reservoir drive mechanisms, and pressure contribute to reservoir quality. Geological aspects that influence reservoir quality are the target for reservoir characterization studies. Poor reservoir quality often leads to engineering challenges, innovative solutions, and higher investments.

Tools, techniques, and measurement scales

Most petroleum reservoirs are inherently heterogeneous and complex. Unconventional reservoirs with ultralow permeability exhibit a high degree of heterogeneity in rock properties. No single tool is adequate in characterizing a reservoir in high resolution. Characterization of heterogeneous reservoirs requires the integration of large amounts of data obtained by various tools and techniques at scales that range from over a kilometer down to a nanometer. Besides scale, the resolution of the tools can be quite different. Some of the tools used in characterizing the reservoir are listed in Table 6.1.

Workflow

All in all, reservoir characterization is an integral part of workflow related to reservoir engineering and management (Figure 6.1). A workflow is outlined in the following:

- Develop earth model based on geology, geophysics, and geochemistry; involves mapping of reservoir quality

Table 6.1 **Reservoir characterization tools [1]**

Reservoir characterization tool	Reservoir scale
Seismic surveys	Several meters to kilometers
Transient well tests	Several meters to a kilometer or more
Studies of outcrops	Less than a meter to hundreds of meters
Well logs	Less than a meter to hundreds of meters
Core analysis	About a millimeter to a meter or more
X-ray–CT scanner	About a millimeter to several centimeters
Micro-CT scanner	Few micrometers to about a centimeter
Scanning electron microscopy	Several nanometers to less than a millimeter
Helium pycnometer	About a nanometer to less than a millimeter

- Develop a dynamic reservoir model based on rock and fluid properties; integrate log and core data
- Review regional trends in characterizing the reservoir
- Design new wells; in case of horizontal wells, design the number of laterals, horizontal length, and the trajectory of horizontal section
- Validate the reservoir models based on past production history
- Simulate the reservoir models to predict performance
- Continue validation of the reservoir models with new production data; update models as necessary

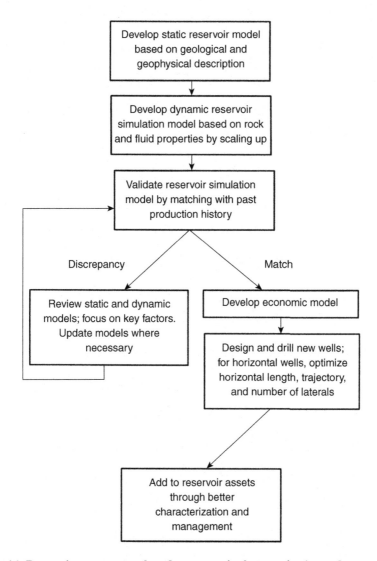

Figure 6.1 Reservoir management based on reservoir characterization and reservoir model validation.

However, reservoir characterization for the entire reservoir is resource intensive in terms of time and effort. When the resources are limited, reservoir characterization studies may focus on specific wells and localized areas.

Unconventional reservoirs

The workflow for developing unconventional reservoirs such as a shale gas reservoir may require a focus on mechanical and geochemical properties of rock, identification of sweet spots, and optimization of multistage fracturing of horizontal wells. Core Lab [2] proposes a series of steps in characterizing and developing ultralow permeability shale reservoirs, some of which are as follows (Figure 6.2):

- Geology: Study of depositional environment, facies, lithology, clay content, clay types, pore structures, and presence of natural fractures on macro- and microscale, among others.
- Geochemistry: Total organic carbon (TOC), vitrinite reflectance, kerogen type and rock evaluation pyrolysis.
- Petrophysical properties: Porosity, permeability, fluid saturations (oil, gas, and water), hydrocarbon-filled porosity, and bound water saturation.
- Geomechanical properties: Young's modulus, Poisson's ratio, bulk modulus, and closure stress; embedment characteristics of proppants to keep the fractures conductive are also included.
- Fracture stimulation design: Rock–fluid compatibility and fracture conductivity of proppant.
- Petrophysical model: Core–log calibration of open hole logs leading to identification of target zones for stimulation.
- Integrated studies: Integration of core and log data, fracture stimulation techniques, and production test results.
- Regional trends: Review of available regional data in characterizing and developing the unconventional reservoirs.

Reservoir characterization scenarios

Reservoir characterization is based on wide-ranging tools and techniques, and often requires an interdisciplinary approach to integrate all available data. For example, characterization of the geologic layers in a conventional oil reservoir may be based on log, core, and well testing studies as well as production history of the field. Identifying the characteristics of a thin geologic interval having high permeability streaks and high water saturation may determine whether a well should be completed in that particular interval. In the event that the well is completed to produce oil from the interval, reservoir characterization may address what design considerations regarding well placement and completion would be necessary to implement.

In another scenario, reservoir characterization, may involve identification of "sweet spots" in an unconventional shale gas reservoir where a horizontal well can be drilled followed by multistage fracturing of the tight formation. The sweet spots are certain localized areas in a pervasive shale formation having favorable petrophysical, mechanical, and geochemical properties. These spots are more likely to produce economically.

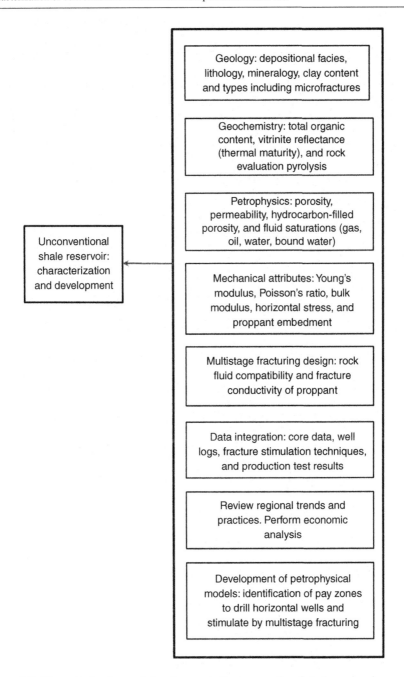

Figure 6.2 Characterization and development of unconventional shale reservoirs.

Case Study: Characterization of a Low Permeability Oil Reservoir in Saskatchewan, Canada [3]

An integrated study involving petrophysical log and core data in a sandstone reservoir was conducted with an objective to enhance the production potential from the low permeability zones of a reservoir. The reservoir is located in southwestern Saskatchewan, Canada. The Upper Shaunavon B reservoir consists of two facies that are characterized by variations in permeability: one with a low permeability in the range of 0.1–10 mD, the other with a high permeability in the range of 10–1000 mD. However, the latter is relatively thin with limited hydrocarbon volume. Recovery from the reservoir was estimated to be quite low (less than 4%). The low permeability facies was considered to be part of the unconventional reservoir for the purpose of economic production. Wells were historically targeted to produce from the high permeability formation with associated production from the low permeability bed. A number of horizontal wells have been drilled in the reservoir and multistage fracturing techniques are utilized to enhance oil recovery.

The reservoir characterization study was based on data obtained from 177 wells, concentrating on the following:

- Identification of flow units
- Reservoir quality
- Connectivity between various layers
- Extent of the oil-saturated reservoir rock
- Pore volumes and production history

The reservoir was characterized by mapping and conducting volumetric analysis. Porosity–thickness (Φh), permeability–thickness (kh), and production bubble maps were prepared (Figure 6.3). Overlaying the bubble maps the Φh and kh maps showed a strong correlation between pore volume and produced volume of oil.

Production data were also analyzed and compared from wells completed in both facies or in one facies. The study identified the presence of six facies in the reservoir of which five are of reservoir quality, implying good porosity and permeability that are producible economically. Large volumes of oil remain untapped due to low permeability of rocks and reservoir heterogeneities, which can be recovered by further implementing horizontal drilling and multistage fracturing techniques.

Case Study: Identification of "sweet spots" in Marcellus Shale [4]

With significant technological advances in the areas of horizontal drilling, completion and fracturing, the petroleum industry is witnessing rapid development of low permeability conventional and unconventional reservoirs in the United States and other countries. Notable among them are shale gas development and production in commercial quantities. Since shale is of ultratight matrix permeability (in microdarcies or less) and limited porosity (usually in single digits),

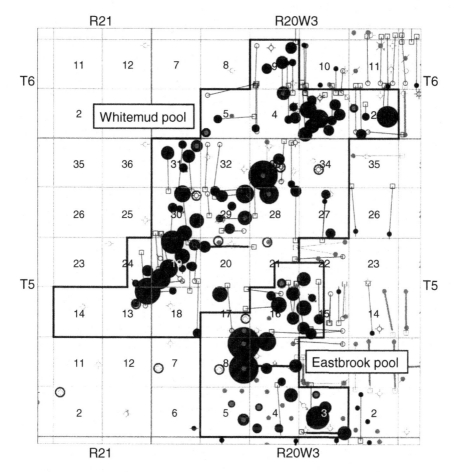

Figure 6.3 Production bubble map; cumulative volumes of oil produced are shown as bubbles. Larger bubbles represent higher production from wells.
Source: Taken from Ref. [3].

the initial production rate may not be satisfactory or may decline rapidly unless the wells are located in "sweet spots." These spots have good reservoir quality and favorable fracturing characteristics that are amenable to economic recovery. Rock properties including porosity and permeability, TOC, thermal maturity, brittleness leading to good fracturing characteristics, along with formation thickness, contribute to the producibility of shale.

Shale gas reservoirs are generally pervasive extending over a very large area. For example, Marcellus shale in the Appalachian Basin extends several hundred miles from New York to Virginia, and is estimated to contain about 500 trillion ft.[3] of natural gas, sufficient to meet the demand of the United States for nearly two decades. However, not all parts of the reservoir support economic production based on the technology currently available. The Marcellus shale is composed of upper and lower shale members with an intervening limestone layer. A study was conducted to identify and map the sweet spots in Marcellus shale.

It is observed that a relatively high porosity of shale combined with high values of TOC may qualify as the sweet spots in the pervasive formation. The reservoir characterization study was based on the following:

- Well logs: Porosity and resistivity log data obtained from thousands of wells that penetrated Marcellus formation
- Geochemical data: TOC data obtained from over 90 wells penetrating the Marcellus formation
- Petrophysical data: Porosity of shale

Reservoir locations with better porosity and relatively high TOC were identified to be likely candidates for sweet spots. Validation of results is obtained by evaluating the available data on gas production trends around the sweet spots.

The contour map of TOC in Marcellus shale is obtained by using the Passey method [5]. The method correlates the TOC with resistivity logs as well as porosity logs that are obtained from sonic, density, and neutron logs. A term $\Delta \log R$ is computed that represents the separation between the deep resistivity curve and the porosity curve. The larger the separation, the higher would be the TOC of shale (Figure 6.4). The above data are used to generate the TOC contours for Marcellus shale.

The calculated TOC contour values were compared against the actual TOC data obtained from core samples as part of a quality assurance procedure. The study indicated that a good match is obtained between the two values.

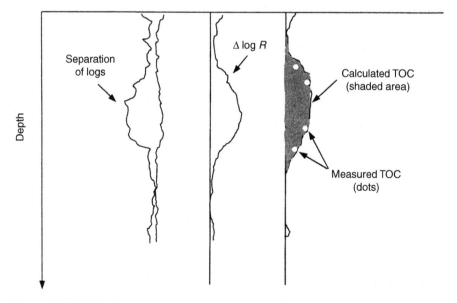

Figure 6.4 Identification of sweet spots in shale having favorable TOC.

Summing up

Reservoir characterization, as the name suggests, aims at obtaining a detailed description or characteristics of a reservoir. The ultimate goal is to add value to reservoir assets by identifying the rock heterogeneities, structural attributes, and flow units of a reservoir that lead to better field development and management. Distribution of porosity, permeability, fluid saturation, hydrocarbon pore volume, fluid contact, structural discontinuities, and facies change are commonly sought parameters in reservoir characterization studies. For unconventional reservoirs such as shale gas, geochemical composition of rock as well as geomechanical characteristics are also important. Reservoir characterization studies require multidisciplinary efforts including, but not limited to, seismic, geological, geochemical, petrophysical, and geomechanical studies. The wide-ranging tools used in reservoir characterization differ in both scale and resolution. Data obtained from field studies may range from over a kilometer down to a nanometer or less.

Reservoir characterization efforts are part of reservoir development and management workflow where static earth and dynamic reservoir simulation models are built and tested against production history of the reservoir. In order to obtain a satisfactory match, iterations are performed to update the models with appropriate values. Once a match is obtained, new wells are designed to optimize production following appropriate economic analysis.

Studies for the characterization of unconventional shale reservoirs include, but are not limited to:

- Depositional environment, facies, lithology, clay content, clay types, and pore structures
- Presence of natural fractures
- TOC, vitrinite reflectance, kerogen type
- Porosity, permeability, fluid saturations (oil, gas, and water), hydrocarbon-filled porosity, and bound water saturation
- Young's modulus, Poisson's ratio, bulk modulus, and closure stress; embedment characteristics of proppants
- Rock–fluid compatibility and fracture conductivity of proppant
- Core–log calibration, integration of core and log data, fracture stimulation techniques, and production test results
- Regional trends in reservoir characteristics and production

Two case studies are presented highlighting the value of reservoir characterization in enhancing reservoir performance as follows:

- Characterization of facies in a low permeability sandstone formation with a goal to enhance productivity by drilling horizontal wells
- Identification of "sweet spots" in Marcellus shale based on integrated log and core studies

Questions and assignments

1. What is reservoir characterization and what are its objectives?
2. What information is usually sought in reservoir characterization studies?
3. Why does reservoir characterization require a multidisciplinary approach? What disciplines are generally involved?

4. List the tools and techniques commonly used in reservoir characterization.
5. How might reservoir characterization aid in the design of a horizontal well in unconventional gas reservoir? Explain.
6. Is there any "best time" in the reservoir life cycle to perform reservoir characterization or is it a continuous process? Include a field example in your answer.
7. How does reservoir characterization lead to better simulation models?
8. How would you characterize source rock? How might it differ from characterizing reservoir rock?
9. Your company is planning to drill five horizontal wells in a newly discovered dolomite reservoir where several layers are interbedded with shale. List the parameters that could of prime importance in locating the future wells.
10. You have been assigned a task to enhance the productivity of a few oil wells that are showing high water cut issues lately. Other wells in the reservoir are producing as expected. What steps would you take to characterize the formation and propose appropriate solutions?

References

[1] Solano NA, Clarkson CR, Krause FF, Aquino SD, Wiseman A. On the characterization of unconventional oil reservoirs. Available from: http://csegrecorder.com/articles/view/on-the-characterization-of-unconventional-oil-reservoirs [accessed 20.02.14].
[2] Tight oil reservoirs of the midland basin: reservoir characterization and production properties; 2014. Available from: http://www.corelab.com/irs/studies/tight-oil-reservoirs-midland-basin.
[3] Fic J, Pedersen K. Reservoir characterization of a "tight" oil reservoir, the middle Jurassic Upper Shaunavon member in the Whitemud and Eastbrook pools, SW Saskatchewan. Marine Petrol Geol 2013;44:41–59.
[4] Logs reveal Marcellus sweet spots, TGS. Available from: http://www.tgs.com/uploaded Files/CorporateWebsite/Modules/Articles_and_Papers/Articles/0311-tgs-marcellus-petrophysical-analysis.pdf [accessed 23.08.14].
[5] Passey QR, Creaney S, Kulla JB, Moretti FJ, Stroud JD. A practical model for organic richness from porosity and resistivity logs. AAPG Bulletin 1990;74/12:1777–94.

Reservoir life cycle and role of industry professionals

7

Introduction

A petroleum reservoir goes through several distinct phases throughout its life. Some reservoirs maintain production on a commercial scale for over 100 years. A reservoir's life cycle consists of exploration, discovery, appraisal and delineation, development, production, and abandonment. Tasks associated with each phase are quite challenging. Professionals from various disciplines, including earth scientists and engineers, contribute to develop and produce the reservoir. Apart from technical and financial considerations, various laws and regulations play important roles in the life cycle of a reservoir. In developing and operating large oil and gas fields offshore, huge investments, in billions of dollars, are often required. Hence, robust reservoir simulation models serve as potent tools to manage the reservoir successfully throughout the life cycle. A case study on the development of an offshore field is presented at the end of the chapter.

This chapter highlights the life cycle of a petroleum reservoir, conventional and unconventional, and answers the following questions.

- What are the phases in the reservoir life cycle?
- How is a petroleum reservoir explored, developed, produced, and abandoned?
- What is the role of industry professionals in the life of a reservoir?
- Is the life cycle for unconventional reservoirs any different?

Life cycle of petroleum reservoirs

Typical life cycle of a reservoir, presented in Figure 7.1, involves exploration, discovery, delineation, development, and production [1]. The various phases in the life of a reservoir are discussed as follows.

Exploration

The petroleum industry conducts oil and gas exploration on a continuous basis to find new horizons for reserves that can be produced economically. The exploration activities, mainly based on geological and geophysical studies, started with relatively shallow inland fields over 100 years ago. With the advancement in technology such as horizontal drilling and offshore platforms, petroleum exploration gradually moved offshore to shallow coastal areas and finally to deep-sea reservoirs. In recent years, unconventional petroleum reserves are explored vigorously as game changing technologies such as multistage fracturing come to the fore. Geologists and geophysicists

Reservoir Engineering. http://dx.doi.org/10.1016/B978-0-12-800219-3.00007-3

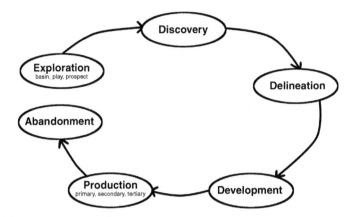

Figure 7.1 Reservoir life cycle.

are involved in exploration and contribute to reservoir description. This includes depth, structure, stratigraphy, fractures, faults, size, aquifer system, and the location of the prospect reservoir. Tools and techniques include geological and geophysical surveys, basin analysis, and others. In certain unconventional reservoirs such as shale gas, geochemical and geomechanical studies are also important in exploration, in addition to traditional exploration for conventional reservoirs.

Exploration of petroleum starts with play. A play is a geologic structure that has recognizable features suggesting possible oil and gas storage and entrapment. Presence of oil and gas accumulation in the same region may provide credence to a play. However, considerable uncertainties and risk are usually associated as a play is identified. A play becomes a prospect when earth scientists gather sufficient evidence to believe that there is a good chance of striking oil or gas in the geologic formation of interest. Prospects are ranked on the basis of risks associated. Risks are based on the quality of source rock including total organic carbon and level of maturity, existence of migration pathways, reservoir quality (porosity and permeability), and presence of cap rock, among others. In the case of conventional accumulations of petroleum, drillable prospects are those where chance of discovery of petroleum appears to be the highest in view of geologic structure (such as anticlines), stratigraphy, reservoir rocks, cap rocks, source rocks, migration pathways, and regional successes.

Some unconventional reservoirs, including shale oil and gas, are exceptions. In these cases, source rock is the reservoir rock. Hence, the occurrence of reservoir rock or migration pathways is not a factor in the exploration of the petroleum deposits.

Discovery

Exploratory drilling may lead to discovery of a new oil or gas field when luck favors. Historical data suggest that the chances of success of finding exploratory wells in the United States are about 30%. Chances improved slightly in the latter part of the twentieth century with the advent of new technologies in exploration. It must be borne in mind that the relatively large geologic plays in various petroleum basins of the world are already explored with diminishing chance of oil and gas finds in very large quantities.

Based on limited geologic information, attempts are made to estimate the initial oil and gas in place and potential reserves. The tools and techniques used include, but are not limited to, petrophysical studies, measurement while drilling, logging while drilling, drill stem test, and reservoir simulation. Uncertainty related to reservoir description is quite high in this phase.

Again, unconventional reservoirs are distinguished from conventional reservoirs in the discovery phase. Extensive deposits of unconventional petroleum are already known in many regions of the world. New wells in unconventional reservoirs are drilled with a high chance of striking oil or gas or both. Shale gas and oil reservoirs are good examples. Shale formations are often found to extend over tens or hundreds of miles and wells are drilled with certainty. However, attaining good recovery from such reservoirs can be technically and economically challenging.

Geologists, drilling engineers, petrophysicists, and reservoir engineers contribute to locating producible formations with pay thickness, porosity, oil saturation, oil–water contact, reservoir pressure, and probable producing rates.

Appraisal

Drilling of additional wells, including appraisal wells, leads to the definition of reservoir size and quality. Any geologic complexities involved may also be brought to light at this stage. A reservoir located in a complex geologic setting, including layering, faults, fractures, barriers, compartmentalization, and facies change, usually requires the drilling of quite a number of wells and multidisciplinary studies for detailed characterization (Figure 7.2). Any uncertainty related to the reservoir geometry and characteristics diminishes to a large extent as more wells are drilled.

Drilling engineers, petrophysicists, and reservoir engineers are again involved. Additional data on reservoir continuity and variations in pay thickness, porosity, oil saturation, and reservoir pressure are collected. Depending on reservoir complexity, one or more wells are cored, which are analyzed in the laboratory for porosity, absolute permeability, relative permeability, and spectrographic characteristics. Oil, gas, and water properties, such as gas solubility, formation volume factor, compressibility, and viscosity, are determined by analyzing the reservoir fluid samples.

Development

Reservoir, drilling, operation, and facilities engineers are mainly involved in developing the field using an economically viable number of wells and spacing between the wells. Development strategy of the reservoir and drilling of the future wells are based on reservoir simulation studies that run large numbers of what-if scenarios in terms of reservoir uncertainties, well locations, and design. The most appropriate strategy is adopted to develop the field. Tight reservoirs having relatively low permeability generally require the drilling of closely spaced wells for economic production. With the

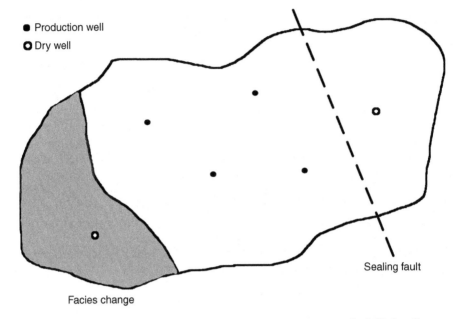

Facies change

Figure 7.2 Reservoir delineation with fault and facies change as newly drilled wells provide detailed information.

advent of horizontal drilling technology and other techniques, reservoirs with complex geology are developed and produced effectively with high recovery that was not attainable before.

Offshore fields require large capital investments in the development of platforms and sub-sea structures that support production, storage, and transport of oil and gas. Often, multiple oil and gas fields are developed at an offshore site within the framework of a large project. Wells are usually horizontal in design to contact large reservoir areas to produce economically. Wells are drilled through slots in a single platform in various directions to reach different parts of the reservoir or more than one reservoir. Practical concerns regarding offshore reservoir development also include the availability of drilling rigs and number of slots in a platform. From the perspective of capital investment, this is the most important phase of a reservoir life cycle.

Development of oil and gas fields in remote locations, deep-sea environments, and highly complex geologic settings may also pose significant technological and economic challenges.

Production

Reservoir production overlaps development as existing wells are produced while new wells are being drilled as per the reservoir development schedule. Production usually occurs in multiple stages. The stages of production are based on primary, secondary, and enhanced oil recovery (EOR) processes. Primary production of oil or gas

Table 7.1 **Production of conventional oil reservoirs**

Reservoir production	Typical recovery (%)	Notes
Primary	20–20	Production by natural drive mechanisms
Secondary	15–25	Mostly waterflood and gas injection
Tertiary	5–15	EOR methods

reservoirs is obtained at the expense of the natural reservoir energy. There are multiple sources of natural energy, including rock and fluid expansion, liberation of dissolved gas, water influx from adjacent aquifers, and gravity. The primary drive mechanisms lead to oil recovery to a varying degree. Reservoir performance under primary production including recoveries is discussed in Chapter 11.

Secondary recovery from oil reservoirs is accomplished by injecting fluids to augment natural energy. Secondary recovery is based on waterflooding, gas injection, or gas–water combination floods. EOR processes include thermal, chemical, and miscible floods. These are employed by using an external source of energy to recover oil that cannot be produced economically by conventional primary and secondary means. Waterflooding of petroleum reservoirs is described in Chapter 16. Major EOR processes are highlighted in Chapter 17. Secondary production and EOR are referred to as part improved oil recovery (IOR) efforts.

Ultimate recovery varies quite significantly from one reservoir to other. However, the recovery estimates shown in Table 7.1 are typical for many reservoirs throughout the world.

Abandonment

Oil or gas fields are abandoned when no more recovery can be obtained economically. Well production rate, reservoir location (onshore vs. offshore), operating costs, market conditions, environmental and other regulations, and other factors may play a critical role in abandoning a reservoir. Common reasons of abandonment include:

- Declining oil and gas production rates, which are not economically sustainable
- Excessive water–oil ratio (WOR) or gas–oil ratio (GOR) at producing wells (see Figure 7.3)
- IOR efforts do not recover the remaining oil economically
- Cost of operation and maintenance is excessive with unfavorable rate of return on investment

Rejuvenation

A discussion of reservoir life cycle would not be complete without the rejuvenation of abandoned or nearly abandoned oil fields that have been witnessed by the petroleum industry. As production from oil and gas wells declines, management of the reservoir becomes more challenging. Revitalization efforts of matured reservoirs as practiced in the industry are discussed in Chapter 20.

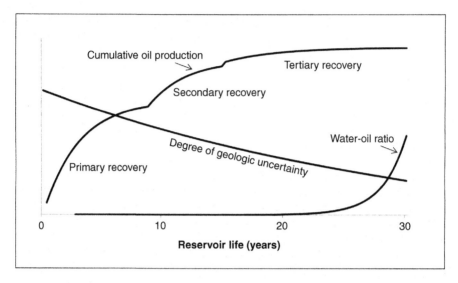

Figure 7.3 Oil and gas production, WOR, and uncertainty throughout the life cycle of a typical oil reservoir.

Unconventional reservoir life cycle

The life cycle of unconventional reservoirs may vary somewhat in certain aspects as compared to the conventional reservoirs. For example, in developing shale gas assets, emphasis is given in locating "sweet spots" for development and production rather than focusing on the vast extent of shale. The unconventional reserves are based on continuous accumulations of gas in shale formation that may spread tens to hundreds of miles, but can be produced economically only from localized sweet spots.

Role of professionals

The role and contribution of petroleum industry professionals are outlined in Tables 7.2 and 7.3.

Table 7.2 Role of multidisciplinary professionals

Life cycle phase	Professionals
Exploration of oil and gas	Geologists, geophysicists, and geochemists
Discovery	Drilling engineers, petrophysicists, and reservoir engineers
Appraisal	Drilling engineers, petrophysicists, geologists, geophysicists, geochemists, and reservoir engineers
Development	Reservoir drilling, operation, and facilities engineers
Production	Production, operation, facilities, and reservoir engineers

Table 7.3 Contributions of professionals: It is teamwork

Professionals	Contributions to reservoir
Geophysicists	Depth to reservoir, structural shape, faulting, boundaries, visualization of reservoirs
Geologists	Origin of hydrocarbon deposits, migration, accumulation, rock types, mineralogy, depositional environments, structures, stratigraphy
Geochemists	Organic content and thermal maturity of source rock
Petrophysicists	Producing zone depths, zone thickness, rock types, rock porosity, reservoir fluid saturations
Engineers	Rock properties, fluid properties, well test (reservoir pressure, temperature, wellbore conditions, faults, effective permeability), injection and production data, material balance calculations to determine original hydrocarbon in place, decline curve analysis, gas cap, aquifer size and strength, primary drive mechanism, production and injection optimization, reservoir simulation, design of secondary and tertiary recovery, and reservoir surveillance, among others.

Case Study: Development of Offshore Oil and Gas Fields Based on Model Studies

Large projects involving multiple offshore fields and reservoirs require significant capital investments. Detailed studies are required to minimize risks and ensure sound economic returns in the face of various uncertainties and financial scenarios. Detailed mathematical models are developed and utilized to build what-if scenarios before the investment is made. Studies related to the development of an offshore asset involving multiple fields attempt to address the following [2]:

- Development sequence of oil and gas fields, which field will be developed first
- How much oil and gas will be produced from each field over a time frame
- Number and drilling sequence of wells in the fields
- Design of offshore platforms and their sizes
- Determination of the optimum connectivity between specific fields and planned facilities

There are multitude of constraints and uncertainties to consider, including the following:

- Reservoir profile needs to be realistic, including oil and gas production rates, water–oil ratio, and gas–oil ratio
- Oil and gas reserves in individual fields
- Availability of drilling rigs for a specific field in view of limited availability
- Availability of platform slots
- Development of a specific field only after the facility is built for the field
- Fiscal considerations and related uncertainties
- Oil pricing and market conditions
- Long planning period and potential for future expansion

The model is simulated for maximizing the net present value of the entire project. The model predicts the number of platforms and other structures to be built in the first year along with the number of wells to be drilled initially. Once several wells are drilled, the degree of uncertainties is reduced regarding initial well production rates and petroleum reserves, thereby enabling the model to predict more accurately in subsequent years (Figure 7.4).

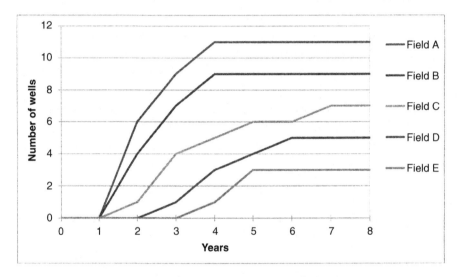

Figure 7.4 Drilling schedule of multiple platforms in the offshore field development phase of the life cycle.

Summing up

Petroleum reservoirs go through a complete life cycle from their inception to their end. Certain reservoirs are known to produce over 100 years. The five distinct phases in reservoir life cycle are exploration, discovery, appraisal and delineation, development, and production and abandonment. Certain phases overlap, for example, the development and production of a large reservoir can proceed concurrently. Depending on the recovery efforts, the production phase may be subdivided into primary, secondary, and tertiary recovery.

• Exploration: Life of a typical oil and gas reservoir begins with exploration of petroleum plays. The plays are the geologic structures that indicate possible oil and gas accumulation, including the existence of source rock, reservoir rock, migration pathway, geologic trap, and cap rock. Plays become prospects when there is sufficient evidence, based on geological and geophysical studies, to believe that oil and gas may exist. Exploratory wells are drilled and petroleum reservoirs are discovered when luck favors. Historical data suggest that approximately 30% of exploration wells strike oil and gas. Certain unconventional reservoirs, such as shale oil and gas, are known to extend over a very large area. However, producing the unconventional reservoirs is a technical challenge. Hence, emphasis is on efficient production rather than exploration in these circumstances.

- Appraisal and delineation: Next, one or more appraisal wells are drilled to evaluate the reservoir size and quality. In this phase, wells are drilled to identify the extent of the reservoir. The appraisal of a petroleum reservoir is commonly accomplished by conducting petrophysical studies and transient well tests. Petrophysical properties are described in Chapter 3. Transient well tests are outlined in Chapter 10.

- Development: Appraisal of reservoirs is followed by development. During this phase, oil and gas wells are drilled to produce the reservoir optimally. The location, placement, and design of wells are decided by conducting reservoir simulation studies and reviewing from what-if scenarios. The experience of reservoir professionals and regional trends are also useful in developing a reservoir. Development of offshore fields with multiple platforms requires huge capital investments and economic optimization. A case study highlighting such a development is presented in the chapter.

- Production: Next is the production phase of the reservoir, which brings the fruits of the efforts of all the earlier phases. The reservoir is initially produced by primary drive mechanisms. The primary drive to produce oil and gas is based on the natural energy stored in the reservoir, and production is driven by one or more of the following: (i) expansion of oil and gas, (ii) liberation of vapor phase from liquid inside the reservoir, (iii) water encroachment from nearby aquifers, (iv) oil drainage due to gravity, and (v) compaction of unconsolidated formation. Primary drive mechanisms are discussed in Chapter 11.

- Secondary recovery efforts are mostly centered on waterflood operations or gas injection, or both. Under favorable conditions, significant quantities of oil are produced during this period in the production phase. Once secondary recovery operations run their course, EOR efforts are initiated to recover further amounts of oil. Common EOR methods include the injection of carbon dioxide in reservoirs subjected to waterflood earlier, and application of thermal energy to heavy oil reservoirs. Waterflooding of conventional oil reservoirs is discussed in Chapter 16. EOR operations are highlighted in Chapter 17. Secondary recovery and EOR are preferred parts of IOR efforts. The ultimate recovery during the production phase of a typical oil reservoir is 25–50% as worldwide statistics suggest. Petroleum reserves are discussed in Chapter 23.

- Abandonment: A reservoir approaches the final phase, namely, abandonment, when the declining production or certain operational issues lead to a production level that is no longer economical. The rate of return on the investment from the reservoir is below the acceptable level. The most common causes of abandonment are: (i) dwindling well rate, (ii) excessive water production, (iii) high gas–oil ratio, and (iv) frequent workover of wells, which are cost intensive.

- Role of professionals: The reservoir team is essentially multidisciplinary. Geologists, geophysicists, geochemists, reservoir engineers, well completion engineers, production engineers, facilities engineers, and others play important roles in various stages of the reservoir life cycle. Various studies of the reservoir must be integrated to achieve an accurate and detailed perspective, which leads to appropriate planning, development, production, and management of the reservoir. In the earlier phases, earth scientists make major contributions in exploration, including the depth and structure of the reservoir, rock types, lithology, source rock geochemistry, structures, stratigraphy, and traps. During appraisal, development and production phases, engineers from various disciplines play a leading role in effectively managing the reservoir. The studies may include material balance calculations and decline curve analysis to estimate reserves, analysis of primary drive mechanism, production and injection optimization, aquifer influx, reservoir simulation, design of secondary and tertiary recovery, and reservoir surveillance, among others.

Questions and assignments

1. What is the reservoir life cycle? How many phases does a typical reservoir have in its life cycle?
2. Describe at least five important factors that can influence the life cycle of a reservoir.
3. Based on a literature review, describe how a new well is appraised.
4. In what stage or stages of life cycle is reservoir simulation most effective?
5. How does an offshore field development differ from that of an onshore field?
6. What are the important points to recognize for the life cycle of an unconventional reservoir?
7. Which professionals are involved throughout the reservoir life cycle?
8. What are the contributions of the professionals?
9. What symptoms are evident when a reservoir is nearing the abandonment phase?
10. Describe the complete reservoir life cycle of a giant oil field that produced from hundreds of wells. Include any comments that could prolong the life of the reservoir.

References

[1] Satter A, Varnon JE, Hoang MT. Integrated reservoir management. J. Petrol. Technol. December 1994; p. 46–58.
[2] Gupta V, Grossmann IE. Offshore oilfield development planning under uncertainty and fiscal considerations. Department of Chemical Engineering, Carnegie Mellon University. Pittsburgh (PA); 2011.

Petroleum reservoir management processes

8

Introduction

The goal of reservoir management is to maximize reservoir assets within the framework of operational, technological, economic, regulatory, and other constraints. This is accomplished by optimizing production from a reservoir. Optimization in oil and gas production is attained by striking a balance between incremental revenue versus capital investment (Figure 8.1). Reservoir management is intimately involved in every phase of the reservoir life cycle, from exploration to production, and finally to abandonment.

In fact, since petroleum reservoirs have a definite life span going through various stages from beginning to end, the management process is aligned to the five phases of project management defined by the Project Management Institute. The framework is recognized worldwide regardless of the type of business activity and aims at leading the management process to success. The five phases of a project, presented in Figure 8.2, are as follows [1]:

- Project initiation
- Planning
- Execution
- Performance monitoring and control
- Project closure

The project initiation phase may include management approval and exploration for oil and gas followed by the discovery of a reservoir. The scope of the project is to optimize recovery of petroleum and maximize returns. In the planning phase, a field development plan is worked out based on all available data. As part of the plan, reservoir simulation is performed for locating future wells and scheduling the wells in order to optimize production. In the execution phase, the wells are drilled as per plan and production is commenced. Facilities that support the field operation on a daily basis are also built. In the monitoring and control phase, various performance indicators such as well rates, bottom hole pressure, recovery trends, water–oil ratio, gas–oil ratio, overall reservoir response, and others are monitored real-time as part of reservoir surveillance. Last but not least, the project comes to closure as the reservoir is abandoned due to the decline in well rates below an economic limit. Accomplishments and lessons learned during the entire project life are documented. The life cycle of a petroleum reservoir is discussed in Chapter 7.

Various phases of the project are dynamic and may interact with each other. For example, performance monitoring or reservoir surveillance may lead to a change in planning or execution of the project. Inherent in any management process are the

Reservoir Engineering. http://dx.doi.org/10.1016/B978-0-12-800219-3.00008-5

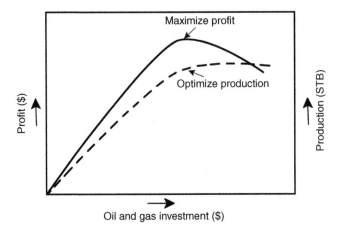

Figure 8.1 Maximization of reservoir assets by optimizing production against capital investments.

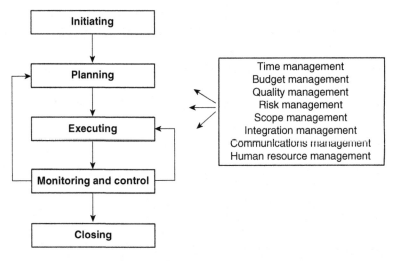

Figure 8.2 The five phases of project management. The management process is interactive. Monitoring and control may bring certain changes in planning and execution of a project. Major components of management are also shown at right.

various components, including time management, budget management, quality management, communications management, integration management, risk management, scope management, and human resource management.

The chapter highlights the various aspects of reservoir management and provides answers to the following:

- What are the objectives of reservoir management?
- What are the essential elements of reservoir management?

- How does the reservoir management process work?
- Would unconventional reservoirs be any different in the management process?
- Why is reservoir management characterized as integrated, dynamic, and ongoing?
- How is reservoir management strategy formulated?
- What disciplines, tools, and technologies are involved in reservoir management?

Elements in reservoir management

Reservoir management is based upon a specific need and a strategy to accomplish a realistic and achievable purpose. Reservoir management strives to add to reservoir assets by designing, engineering, and continuously evaluating primary, secondary, and tertiary recovery (Figure 8.3).

The important elements for setting an effective reservoir management strategy include, but are not limited to:

- Detailed knowledge of the reservoir, including all static and dynamic data gathered from the inception of the project
- Available and appropriate technology to produce the reservoir in an optimized manner
- Total environment that influences the development, production, and management of the reservoir

Knowledge of the reservoir is based upon geologic, seismic, petrophysical, and other studies that are conducted either periodically or regularly. Geological, geophysical, and geochemical data, which describe rock properties and reservoir structure, are static. The dynamic data are obtained from monitoring, collecting, and analyzing well production and related information. Wide-ranging technologies are involved in the management of a reservoir, including, but not limited to, geological, geophysical,

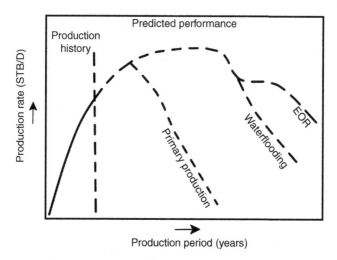

Figure 8.3 Plot depicts field production rate versus production period under strategic reservoir management. Primary, recovery, and tertiary recovery processes add to reservoir assets.

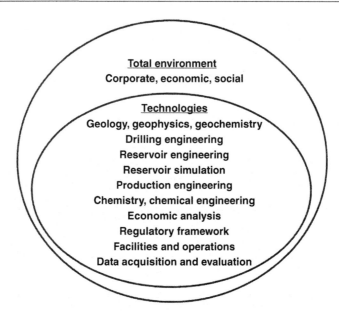

Figure 8.4 Major elements in reservoir management.

geochemical, drilling, completion, computer-aided simulation, well testing, well logging, and electronic surveillance to monitor reservoir processes (Figure 8.4). Total environment includes corporate, economic, and social as follows:

- Corporate – goal, financial strength, culture, and attitude
- Economic – business climate, oil and gas price, and inflation
- Social – conservation, safety, and environmental regulations

Reservoir management process

The reservoir management process can be characterized as integrated, dynamic, and ongoing. The process is integrated because various technical, economic, and other factors play important roles in managing a reservoir, all of which work in an integrated manner. For instance, the management may decide when to initiate an enhanced oil recovery (EOR) process on the basis of market conditions. Integration also takes place among human skills, available technologies, data, and tools. Reservoir management is dynamic because short- and long-term strategies to manage a reservoir are updated on a regular basis depending on reservoir performance. For example, the current reservoir behavior under waterflood may suggest that water injection should be ramped up in a certain part of the reservoir or infill wells need to be drilled that was not part of the original plan. The reservoir management plan needs to be revised and implemented as new information about the reservoir becomes available from reservoir surveillance and analyses. Reservoir management is an ongoing process as the reservoir is monitored in real-time and certain corrective actions are implemented by using automated tools and systems. Besides, reservoir performance is reviewed regularly throughout the life of the reservoir in order to manage it better. All the phases of the reservoir management process can be executed through appropriate planning and paying close attention.

In the words of Satter and Thakur [2], "Reservoir management involves making it happen and letting it happen. We can leave it to chance to generate some profit from a reservoir operation without ongoing deliberate planning, or we can enhance recovery and maximize profit from the same reservoir through sound management practices."

In a nutshell, the reservoir management process is accomplished by optimizing the recovery of oil and gas while minimizing the capital investments and operating expenses. Sound reservoir management practices are dependent on the use of available resources. Effective utilization of human, technological, and financial resources are needed to maximize the reservoir assets. Human skills needed in the overall management of the reservoir are multidisciplinary and cross-functional.

Management of unconventional reservoirs

In the broadest sense, management of conventional and unconventional reservoirs is aligned to the same principles. However, the development of unconventional reservoirs is often associated with immature technologies, lack of detailed information, unpredictable well performance, higher capital investment, and greater risks, which may require more focus on certain areas of reservoir management. For example, in shale gas reservoirs, drilling of large numbers of wells is required compared to conventional reservoirs in order to formulate a successful reservoir management strategy.

Reservoir management strategy

The reservoir management strategy focuses on how the reservoir will be developed, produced, and monitored for optimized production and smooth operation on a day-to-day basis, among others. The vital importance in setting management strategy is to understand the nature of the reservoir being managed. It requires the knowledge of geology, rock and fluid properties, fluid flow and recovery mechanisms, drilling and well completions, and past production performance. The proper uses of the various technologies make up the technological toolbox, which includes the technologies related to the exploration, drilling and completions, recovery processes, and production. These are essential elements to ensure the success of reservoir management. Available technologies are geology, geophysics, and reservoir and production engineering.

As indicated earlier, reservoir management strategy is influenced by the various facets of the "total environment." Business climate, market conditions, logistics, inflation, and other factors usually play a major role in setting a management strategy. Corporate goals, culture, and attitude set a direction in reservoir management strategy. Regulations and laws related to the environment, legal bindings, public opinion, and political stability play important roles in setting up a strategy. Last but not least, the availability and skill set of the reservoir personnel play a major role as well.

Developing a plan

Once the strategy for reservoir management is formulated, the next step is to develop a plan (Figure 8.5), which includes the following [2,3]:

- Data acquisition and information management
- Geological and numerical model studies for predicting reservoir performance
- Development of the reservoir scenarios including drilling of the wells
- Well production forecast
- Reservoir depletion strategy
- Estimation of reserves
- Facilities requirements
- Economic optimization
- Environmental and regulatory issues
- Management approval of the technical plan

A multidisciplinary, integrated team consisting of the following professionals is in charge of developing an economically viable plan for the reservoir.

- Earth scientists, responsible for the static description of the reservoir
- Reservoir engineers, responsible for providing production and reserves forecasts and economic evaluations
- Drilling and completion engineers, responsible for drilling and completing wells
- Equipment engineers, responsible for designing surface, sub-sea, and subsurface facilities
- Structural engineers, responsible for designing platforms and production decks for offshore projects
- Other professionals, including production and pipeline engineers, land managers, and others

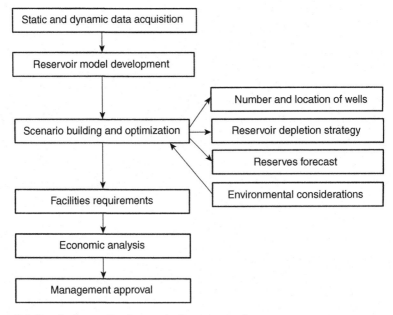

Figure 8.5 Developing a plan for a petroleum reservoir.

The makeup of the team and the number of professionals on it will depend on the size and goal of the project. The professional with overall knowledge of reservoir management would be the logical team leader.

Development and depletion strategies

The operational strategies for depleting the reservoir to recover petroleum by primary and applicable secondary and tertiary methods are the most important step in reservoir management.

The life cycle stage of the reservoir dictates development and depletion strategies.

In the case of a new discovery, how best to develop the field well spacing, well configuration, and recovery schemes needs to be addressed.

In the case of a depleted reservoir primary means, secondary, and even tertiary recovery schemes need to be investigated.

Data acquisition and information management

An enormous amount of data is collected and analyzed during the life of a reservoir. The sources of data are wide ranging. Some of the data is collected in real-time during production. Some of the reservoir data influencing the reservoir management strategy is presented in Figure 8.6.

The key steps in data acquisition and information management are planning, justification, timing, and prioritizing. Data needed before production include seismic, geologic, logging, coring, fluid properties, and results obtained from well testing. Data needed during production are obtained from well testing (Chapter 10), primary production (Chapter 11), waterflooding (Chapter 16), and EOR processes (Chapter 17). The collected data need to be analyzed, validated, and stored in a database for ongoing and future studies.

Geological and numerical model studies

The reservoir model is an integrated geoscience and engineering model to be built jointly by geoscientists and engineers. The model is based upon both static and dynamic data gathered over the life of the reservoir. Certain assumptions are also made in developing the models as all the pertinent information is not available in most cases.

The reservoir simulation model is concerned with rock and fluid properties, fluid flow and recovery mechanisms, drilling locations, completions intervals, production, and injection. The accuracy of the reservoir production performance analysis is dictated by the quality of the reservoir model. The geological model is derived by extending localized core and log measurements to the full reservoir using such technologies as geophysics, geochemistry, mineralogy, depositional environment, and diagenesis.

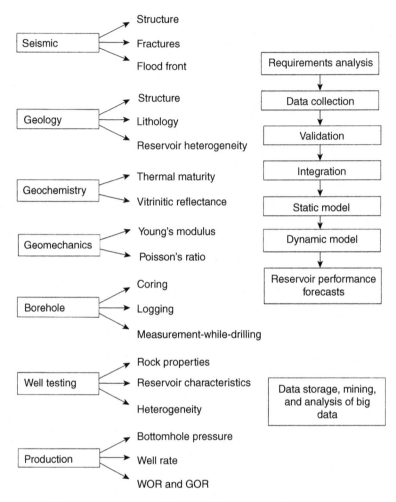

Figure 8.6 Reservoir data and sources; data acquisition and analysis serves as a primary tool in formulating a reservoir management strategy.

Well production forecasts and estimate of reserves

The reservoir production performance under the current and future operating conditions greatly influences the economic viability of a petroleum recovery project. Therefore, evaluation of the past and present reservoir performance and forecast of its future production, such as infill drilling and waterflooding, are essential in the reservoir management process. Volumetric analysis (Chapter 12), classical material balance (Chapter 14), decline curve analysis (Chapter 13), reservoir simulation (Chapter 15), and EOR processes (Chapter 17) are used for analyzing reservoir production performance and reserves forecasts (Chapter 23).

Facilities

Facilities are the actual physical connection to the reservoir. Various operations including drilling, completion, pumping, injecting, processing, and storing require the installation of surface facilities. Production performance results are used to estimate facilities requirements. Proper design and maintenance of facilities has a profound effect on profitability. The facilities should be designed to carry out effectively the reservoir management plan. Estimates of the capital and operating costs based on the facilities requirements are used for economic analyses.

Environmental issues

Environmental and ecological considerations are very sensitive and important aspects of the reservoir management process. In offshore field development, environmental issues can, and do, play a critical role. These considerations have to be included in developing and subsequently operating a field. In addition, there are regulatory constraints that must be taken into account in the reservoir management process.

Economic optimization

The ultimate goal of reservoir management is economic optimization. The project economics are based on the estimated production, investments, operating expenses, and financial data. Economic analysis required for the petroleum reservoirs is discussed in Chapter 24.

The steps in economic analysis include:

- Setting an economic objective based upon economic criteria such as payout period, discounted cash flow rate of return, and others (Chapter 24)
- Formulating a scenario for reservoir production and operation
- Data collection related to production, investments, operating expenses, and price of oil and gas
- Performing an economic analysis
- Performing a risk analysis
- Optimization of production and operation

Management approval

Management approval and support is the final step for developing a reservoir management plan. Field personnel commitment is also important for a successful project.

Implementation

After management approval of the project development plan, the next major assignment is the implementation of the plan to get oil and gas production on stream as soon as possible. A project manager with full authority is needed to manage the various activities as follows:

- Design, fabrication, and installation of surface and subsurface facilities. This critical path for the whole project requires tremendous efforts and experience to preplan, monitor, and complete the project on time
- Development of a drilling and completion program
- Acquisition and analysis of necessary logging, coring, and initial well test data from the development wells to characterize the reservoir
- Updating of the reservoir databases
- Revision of production and reserves forecasts

The key ingredients for successful implementation of a plan include the following:

- A flexible plan of action
- Management support
- Committed field personnel

It is critical to have periodic review meetings with all team members, mostly in the field offices.

Reservoir surveillance

Constant monitoring and surveillance of reservoir performance as a whole is essential to determine whether the performance is conforming to the management plan. Reservoir surveillance is discussed in Chapter 16.

In order to ensure a successful monitoring and surveillance program, coordinated efforts of the engineers, geologists, and operations personnel with management support and field personnel commitment are needed at the start of production from the field.

The surveillance program is dependent upon the nature of the project. Reservoir surveillance involves data acquisition and analysis, leading to effective reservoir management. Reservoir data that are collected and monitored include, but are not limited to:

- Oil, water, and gas production by wells
- Gas and water injection by wells
- Systematic and periodic static and flowing bottom hole pressures testing of selected wells
- Production and injection tests
- Injection and production profiles
- Recording of workovers and results, and any other data that aid in reservoir surveillance

Performance evaluation

The reservoir management plan must be reviewed periodically in order to ensure that the plan is being followed, that it is working, and it is still the best possible plan. The success of the plan should be evaluated by comparing the actual results against the expected reservoir performance.

It would not be realistic to expect the actual project performance to be the same as what has been planned. Therefore, the functional groups should establish certain technical and economic criteria to determine the project success. The criteria will depend on the nature of the project.

The answer to how the reservoir management plans working lies in a careful evaluation of project performance. The functional groups should routinely compare the actual performance (e.g., reservoir pressure, gas–oil ratio, water–oil ratio, and production) against the expected performance. In the final analysis, the economic yardsticks will determine the success or failure of the project.

Revision of plans and strategies

If reservoir performance does not conform to the management plan or the conditions change, plans and strategies should be revised. In order to ensure sound reservoir management, questions regarding reservoir performance must be asked and answered on an ongoing basis. The reservoir simulation model is updated accordingly to predict any changes in reservoir performance in the light of new information.

Abandonment

The reservoir management plan should include the final task of reservoir abandonment when all the depletion plans have been implemented and the reservoir can no longer be operated economically.

Case Study: Management of Means San Andres Unit, Texas

Means San Andres Unit, located near Midland, Texas, is a classic example of decades of reservoir management as the field went through primary, secondary, and tertiary recovery phases as well as an infill-drilling program [4]. Through reservoir management, changing economic and technical challenges were met to produce the reservoir successfully. The Means field was discovered in 1934. Reservoir management techniques were implemented within a year of discovery of the field. Reservoir management dealt with increasingly complex scenarios as the field went through various phases in production, i.e., from primary to secondary to tertiary.

Structurally, the Means field is a north–south trending anticline separated by a north and a south dome. It consists of Grayburg and San Andres formations. The depth of the formations ranges between 4200 ft. and 8000 ft. However, the upper 200–300 ft. of the San Andres formation is the most productive having good reservoir quality. The dolomite formation extends over 14,000 acres with a net thickness of 54 ft. However, the gross thickness is much larger, about 300 ft. Average porosity of San Andres is 9% and permeability is 20 mD, with their upper limits around 25% and 1 D, respectively. Average connate water saturation is 29%. In contrast, Grayburg formation is of poor reservoir quality.

The oil has a stock-tank gravity of 29°API and a viscosity of 6 cp. The original reservoir pressure is 1850 psi. The primary drive mechanism is a combination of fluid expansion and weak aquifer charge.

Reservoir management during primary and secondary operations

The first reservoir study was completed in 1935 concentrating on primary recovery. Later in1939, a reservoir study was conducted to evaluate the potential for secondary recovery. In order to conduct this study, additional data were accumulated including further logging, fluid sampling, and special core data.

In 1963, the field was unitized and water injection through peripheral wells was initiated. Twenty-four wells, distributed throughout the unit were permanently shut in and maintained as observation wells to monitor reservoir pressure during waterflood. In 1967, as a result of increased allowable, it was realized that the peripheral injection no longer provided sufficient pressure support.

In 1969, a reservoir engineering and geological study was conducted to determine a new depletion plan to offset the decline in reservoir pressure. In the north dome, pressure data were correlated with the geological data to identify three major San Andres intervals including Upper San Andres, Lower San Andres oil zone, and Lower San Andres aquifer. A permeability barrier was mapped between the Upper and Lower San Andres.

Analysis of the pressure data from the observation wells indicated that neither the north dome nor the south dome was receiving adequate pressure support. This study recommended a change in waterflood pattern. The new scheme was based on water injection through interior wells with a 3 to 1 line drive. Following implementation of the program, the daily unit production increased significantly.

After reaching a peak in 1972, oil production began to decline. A reservoir study conducted in 1975 indicated that all the pay zones were not being flooded effec tively by 3 to 1 line pattern. A detailed geological study showed a lack of lateral and vertical distributions of pay. Results of well logging conducted previously were correlated with core data to determine pore volumes. Original-oil-in-place (OOIP) was calculated in various zones based on data obtained from each well in the field. The study provided the basis for a secondary surveillance program and later the design and implementation of EOR by CO_2 injection.

A major infill-drilling program was undertaken based upon the potential of additional oil recovery. As the spacing between wells was reduced by infill drilling, recovery of about 15.4 million barrels of incremental oil was estimated. With the implementation of this pattern flooding, a detailed surveillance was developed, including:

- Monitoring of production of oil, gas, and water
- Monitoring of water injection
- Control of injection pressures with step-rate test
- Pattern balancing with computer balance program
- Injection profiles to ensure water flooding into all pay
- Specific production profiles
- Fluid level checks to ensure pump-off of producing wells

Reservoir management during tertiary operations

In 1981–1982, a reservoir study was conducted to recover oil by the injection of CO_2. The CO_2 project was implemented as part of an integrated reservoir management plan, which included CO_2 injection, infill drilling, pattern changes, and expansion of the Grayburg waterflood outside the project area.

Although Means was similar to other San Andres fields in the Permian Basin, some properties such as 6 cp oil viscosity, minimum miscibility pressure, and low formation parting pressure made Means Unit unique. A CO_2 pilot along with extensive laboratory and simulation works was initiated. A detailed reservoir program preceded this work and became the basis for planning the CO_2 tertiary project. Although the reservoir description was the building block for the project, it was continuously updated during the planning and implementation phase of the CO_2 project as more data became available.

Several 10-acre wells (generally injectors) were drilled as part of the CO_2–EOR project, which consisted of 167 patterns on 6700 acres. The wells covered 67% of the productive acres. It was further estimated that about 82% of the OOIP can be reached based upon the recovery efforts.

A comprehensive surveillance program had been present during the waterflood. Before developing a similar program for the CO_2 flooding, an operating philosophy was created by personnel from engineering, geology, and operations, and it was submitted to the management for approval and support.

Major operating objectives included:

- Completion of injectors and producers in all potential pay zones that can be flooded
- Maintenance of reservoir pressure near the minimum missile pressure of 2000 psi
- Maximizing injection of fluids below the fracture pressure of formation
- Pumping off the producers
- Obtaining good vertical distribution of injected fluids
- Maintaining balanced injection and withdrawals in each well pattern

Major areas of surveillance included:

- Areal flood balancing
- Vertical conformance motoring
- Production monitoring
- Injection monitoring
- Data acquisition and management
- Pattern performance monitoring
- Optimization

The objective of the surveillance program was to maximize oil recovery and flood efficiency. New opportunities were identified and evaluated. In addition, one of the key objectives was to obtain better reservoir description to understand the recovery process. This effort included the use of high-resolution seismic surveys to improve pay correlation between wells.

Summing up

The goal of reservoir management is to maximize reservoir assets within the framework of operational, technological, economic, regulatory, and other constraints. Reservoir management is intimately involved in every phase of the reservoir life cycle, from exploration of oil and gas to development and production, and finally to abandonment of the reservoir. Reservoir management is integrated, dynamic, and ongoing. It integrates human skills, experience, data, tools, and technologies. Furthermore, the reservoir management process deals with the total environment including economic, regulatory, social, and others in an integrated manner. The process is dynamic, as the plan for reservoir development and production may be updated as deemed necessary during the life cycle of the reservoir. The changes may involve infill drilling, well recompilation, enhanced recovery, and emphasis on producing from certain zones, to name a few. Reservoir management is an ongoing process as a wealth of data is collected in real-time and analyzed to make the process better and wells more efficient. The collected data often include well rates, water–oil ratio, bottom hole pressure, and composition of produced water, among others.

Reservoir management is based upon a specific need and a strategy to accomplish a realistic and achievable purpose. The important elements for setting an effective reservoir management strategy include, but not are limited to:

- Detailed knowledge of the reservoir, including all static and dynamic data gathered from the inception of the project
- Available and appropriate technology to produce the reservoir in an optimized manner
- Total environment that influences the development, production, and management of the reservoir

In a nutshell, the reservoir management process is accomplished by optimizing the recovery of oil and gas while minimizing the capital investments and operating expenses.

The reservoir management strategy focuses on how the reservoir will be developed, produced, and monitored for optimized production and smooth operation on a day-to-day basis. The essential elements to ensure the success of reservoir management from a technical perspective include the technologies related to exploration, drilling and completions, recovery processes, and production. The vital importance in setting management strategy is to understand the nature of the reservoir being managed. It requires knowledge of the following:

- Reservoir characteristics including geologic complexities
- Rock and fluid properties
- Fluid flow and recovery mechanisms
- Drilling and well completions
- Production history

In managing unconventional reservoirs, the role of innovative technologies as well as the high risk factor associated with those technologies must be recognized. The management process could be more dynamic than what is implemented in managing conventional reservoirs by established technologies.

A reservoir team is multidisciplinary and generally composed of highly skilled professionals including engineers, earth scientists, and others as follows:

- Earth scientists, responsible for the static description of the reservoir
- Reservoir engineers, responsible for providing production and reserves forecasts and economic evaluations
- Drilling and completion engineers, responsible for drilling and completing wells
- Equipment engineers, responsible for designing surface, sub-sea, and subsurface facilities
- Structural engineers responsible for designing platforms and production decks for offshore projects
- Other professionals, including, but not limited to, production and pipeline engineers and land managers

The stages and activities in reservoir management include, but are not limited to:

- Collection and analysis of static and dynamic reservoir data
- Formulation of field development strategies, including drilling of new wells and selection of EOR method
- Geological and numerical model studies
- Well production and reserves forecasting
- Environmental considerations
- Management approval of future plans
- Implementation of plans
- Best practices and quality control
- Reservoir monitoring and surveillance
- Evaluation of reservoir performance
- Required changes in plans and procedures depending of performance of the reservoir

The chapter concludes with a case study highlighting several decades of reservoir management of the Means San Andres Unit in Texas. Discovered in the first part of the twentieth century, the field has gone through various phases of recovery, including primary, secondary, and tertiary EOR. Infill wells were also drilled to augment recovery. The technical aspects of reservoir management demonstrate how the implementation of appropriate technologies can improve declining reservoir performance and add reserves. The highlights of the management of Means San Andres Unit are as follows.

The field was discovered in the 1930s and reservoir management techniques were applied within a year of production. In1939, a reservoir study was conducted to evaluate the potential for secondary recovery. In order to conduct this study, additional data were obtained including further logging, fluid sampling, and special core data.

In 1963, the field was unitized and water injection through peripheral wells was initiated. Twenty-four wells, distributed throughout the unit were permanently shut in and maintained as observation wells to monitor reservoir pressure during waterflood. In 1967, as a result of increased allowable withdrawal, it was realized that the peripheral injection no longer provided sufficient pressure support.

In 1969, a reservoir engineering and geological study was conducted to determine a new depletion plan to offset the decline in reservoir pressure. The study recommended a change in waterflood pattern. The new scheme was based on water injection through interior wells with a 3 to 1 line drive. Following implementation of the program, daily unit production increased significantly.

After reaching a peak in 1972, oil production began to decline. A reservoir study conducted in 1975 indicated that all the pay zones were not being flooded effectively by 3 to 1 line pattern. A detailed geological study showed a lack of lateral and vertical distributions of pay.

The study provided the basis for a secondary surveillance program and later the design and implementation of EOR by CO_2 injection.

A major infill-drilling program was undertaken based upon the potential of additional oil recovery. As the spacing between wells was reduced by infill-drilling, recovery of about 15.4 million barrels of incremental oil was estimated.

With the implementation of this pattern flooding, a detailed surveillance was developed, including, but not limited to:

- Monitoring of production of oil, gas, and water
- Monitoring of water injection

In 1981–1982, a reservoir study was conducted to recover oil by the injection of CO_2. The CO_2–EOR project was implemented as part of an integrated reservoir management plan, which included CO_2 injection, infill drilling, pattern changes, and expansion of the Grayburg waterflood outside the project area.

Before developing a similar reservoir surveillance program for the CO_2 flooding, an operating philosophy was created by personnel from engineering, geology, and operations, and it was submitted to the management for approval and support.

Major operating objectives included:

- Completion of injectors and producers in all potential pay zones that can be flooded
- Maintenance of reservoir pressure near the minimum missile pressure of 2000 psi
- Maximizing injection of fluids below the fracture pressure of formation
- Pumping off the producers
- Obtaining good vertical distribution of injected fluids
- Maintaining balanced injection and withdrawals in each well pattern

In conclusion, Means San Andres exemplifies the wide-ranging management strategies and techniques that were implemented to attain decades of successful oil recovery. Reservoir management tools and technologies included secondary and tertiary recovery (CO_2–EOR), well pattern balancing for injection and withdrawal, infill drilling, reservoir performance studies, and reservoir surveillance.

Questions and assignments

1. What is the objective of reservoir management and how it is implemented in the oil and gas industry? Name the essential ingredients of reservoir management.
2. Describe the challenges in implementing a reservoir management process. Why does a reservoir management process need to be dynamic? Why is integration of various disciplines necessary?
3. What role does reservoir simulation play in the management of oil and gas fields? What components need to be integrated to achieve success in reservoir management?

4. When does reservoir management start in the life of a field? What is the role of a project manager?
5. How can a reservoir management plan minimize costs and risks in a project?
6. Name the typical activities that are associated with reservoir management on short- and long-term bases.
7. Based on a literature review, compare the reservoir management processes between an onshore field and an offshore field.
8. What are the highlights of waterflooding in Means San Andres Unit from the reservoir management point of view?
9. Why and how was the EOR process implemented in Means San Andres Unit? Explain.
10. Your company has discovered a new oil field where rock permeability is only in fractions of a millidarcy. Two wells have been drilled so far, but only one has been successful. You have been asked to draw a detailed reservoir management plan. What efforts would you undertake to make the project a success?

References

[1] Project Management Institute. Avaiable from: http://www.pmi.org [accessed 30.09.14].
[2] Satter A, Thakur GC. Integrated petroleum reservoir management. Tulsa, OK: Pennwell; 1994.
[3] Satter A, Varnon JE, Hoang MT. Integrated reservoir management. J Petrol Technol December 1994; p. 46–58.
[4] Stiles LH. Reservoir management in the Means San Andres Unit, SPE 20751, presented at the ATCE. New Orleans; 1990.

Fundamentals of fluid flow through porous media

Introduction

Characterization of the flow of fluids through porous media serves as the foundation of reservoir engineering studies. Detailed evaluation of pressure, flow rate, and saturation of reservoir fluids as a function of location and time are essential in order to visualize the dynamics of fluid flow that relates to the past and future performance of the reservoir. Mathematical models, both analytical and numerical, are utilized to predict the effects of injection and production on reservoir pressure, and changes in fluid phase saturation, among others. In the case of a dry gas reservoir without any significant geologic complexities, simple analytic equations predicting reservoir pressure and well rate with time may be sufficient to evaluate reservoir performance. However, at the other end of the spectrum, study of a heterogeneous oil reservoir having many producers and injectors requires robust numerical models based on a suite of fluid flow equations that are solved simultaneously. This chapter presents an overview of processes involved in fluid flow through porous media, and introduces selected equations that are widely used in the industry. The chapter addresses the following:

- What are the mechanisms of fluid transport in conventional and unconventional reservoirs?
- What are the forces that affect the flow of fluid in porous media?
- How are analytical and numerical models developed based on the fundamental equations?
- What are the assumptions and limitations inherent in the development of fluid flow models?
- What are the major applications of fluid flow models in reservoir engineering?
- How do fluid and rock properties affect fluid flow behavior?
- How is fluid flow characterized as a reservoir produced? What are the effects of reservoir boundary and adjacent aquifers on fluid flow behavior?

Mechanism of fluid flow in porous media

In conventional and unconventional reservoirs, the types of fluid flow and related phenomena can be categorized as follows:

- Darcy flow
- Non-Darcy or turbulent flow
- Adsorption and desorption
- Diffusion

The flow of oil and gas in a permeable network of pores is largely laminar, which is referred to as Darcy flow. Darcy's law, which states that fluid flow rate is proportional to the difference in pressure between two points in a porous medium, is discussed in Chapter 3. However, in the vicinity of gas wells, turbulence may develop due to the

Reservoir Engineering. http://dx.doi.org/10.1016/B978-0-12-800219-3.00009-7

relatively high velocity of fluid. The net effect of turbulence is manifested as an additional pressure drop compared to what is predicted by Darcy's law.

In unconventional gas reservoirs, however, the phenomena of adsorption and diffusion play important roles during production. In shale gas and coalbed methane reservoirs, substantial quantities of gas are stored in an absorbed state onto the organic matter of rock. Once reservoir pressure decreases due to production, gas is desorbed and flows toward the wellbore through the microscopic porous channels and fractures present in rock. Furthermore, diffusion of the gas phase may occur through the micropores of rock contributing to production. Fluid flow simulation models for unconventional gas reservoirs take into account the phenomena of adsorption and diffusion.

Forces affecting fluid flow

The principal forces that affect fluid flow characteristics in conventional reservoirs include:

- Viscous forces
- Capillary imbibition
- Effects of gravity

The initial saturation distribution of oil, gas, and water phases in a reservoir prior to discovery and production is influenced by the effects of gravity and capillarity. The effects of capillary forces are evident in certain geologic formations having low porosity and permeability in the form of an oil–water transition zone. Movable water phase may rise to a significant height into the oil column resulting in the production of water along with oil. The capillary forces are counteracted by the effects of gravity. The length of the transition zone is also a function of density difference between water and hydrocarbons in the porous media.

Viscous forces are created as the reservoir is produced either by natural energy or by fluid injection. Viscous flow of fluid phases occurs due to the pressure differential that is created between various points in the reservoir as the reservoir is produced. In most cases, viscous forces largely dominate the fluid flow mechanism. However, dipping reservoirs may produce chiefly due to the effects of gravity.

Single and multiphase flow

Petroleum reservoirs have characteristically one of the following scenarios:

- Flow of single phase, either oil or gas: The primary production of oil from a reservoir above the bubble point pressure is a typical example of single-phase flow. Production from a dry gas reservoir is also single phase. Analytic models characterizing single-phase flow of fluid can be used to analyze reservoir performance as long as the reservoir is not highly heterogeneous.
- Two-phase flow: The simultaneous production of both oil and gas from a reservoir operating below the bubble point is an example of two-phase flow. A gas condensate reservoir produces natural gas as well as liquid hydrocarbons. Two-phase flow is encountered as water is injected into the reservoir to improve recovery. Water can also be produced naturally from the oil reservoir due to the encroachment of an adjacent aquifer into the reservoir.
- Three-phase flow: Certain reservoirs produce oil, gas, and water concurrently. Examples include the production from saturated oil reservoirs with a bottom aquifer drive, and reservoirs

where water and gas are injected alternately to enhance reservoir performance. Modeling of multiphase flow involves a suite of fluid flow equations that are solved simultaneously with the aid of reservoir simulators. Multiphase flow usually involves transfer of the fluid phase, i.e., dissolution of lighter hydrocarbons from heavier components, and condensation of heavier hydrocarbons from condensate-rich gas. Reservoir simulation of multiphase flow of fluids is discussed in Chapter 15.

Fluid flow model geometry

Reservoir fluid flow can be represented and modeled in various ways as the nature of reservoir study warrants. These are summarized in the following.

A fluid flow model can be linear (x, y, or z), radial (r), or spherical (r and z). Oil flows in the radial direction into the well. In the above, x, y, z, and r represent the axes along which the flow of fluid occurs as depicted in Figure 9.1.

Fluid flow in porous media can be treated as one-, two-, or three-dimensional depending on the complexities involved under a given circumstance. For example, a study may involve the linear displacement of oil by water from injector to nearby producer, which can be modeled and visualized in one dimension for gaining insight into the process. The study may indicate when an injected water flood front may break through the producing well and how the water cut will rise with time. A 2D radial model can be used to predict well performance where the well is producing from its own drainage area and the boundary conditions are known with reasonable certainty. However, the analysis of multiphase flow of fluids involving several wells in a layered reservoir would require a detailed three-dimensional study to evaluate and predict reservoir performance.

Fluid state and flow characteristics

Flow characteristics of oil and gas in porous media as well as reservoir performance largely depend on fluid compressibility. Reservoir fluids can be:

- Compressible
- Slightly compressible

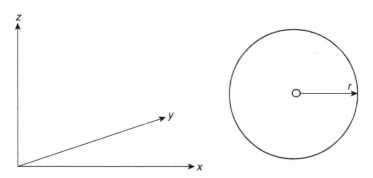

Figure 9.1 Flow in horizontal, vertical, and radial directions in porous media.

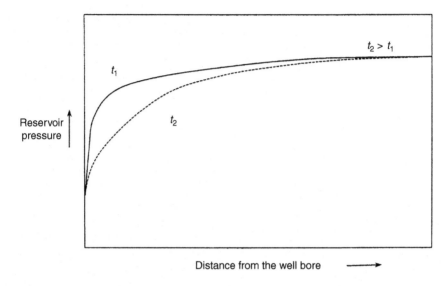

Figure 9.2 Unsteady-state flow. Fluid pressure and flow rate change with time.

Oil is considered to be slightly compressible while natural gas is a compressible fluid. However, transport equations for incompressible fluid are also provided in the petroleum literature highlighting the role of compressibility of reservoir fluids. Besides, fluid flow equations are developed where the reservoir fluid is considered to be incompressible.

Flow of fluids in porous media can be categorized as unsteady state, steady state and pseudosteady state. The classification of flow is based on how fluid pressure changes with time in porous media as in the following:

- Unsteady-state or transient flow is encountered when a well is put on production after shut-in for a certain period, and the reservoir pressure around the well changes with time. In most instances, flow within the reservoir, particularly around the production and injection wells, is unsteady state as pressure and flow rate change with time. Figure 9.2 is a time-lapse depiction of pressure changes in the reservoir from initial static condition as a well begins production.
- Steady-state flow occurs when pressure does not change with time in a given location of the reservoir and well flow rate is constant. Steady-state flow may develop when the pressure drop in a reservoir may readily be compensated by water encroachment from an adjacent aquifer.
- Yet another type of flow is characterized as pseudosteady-state flow, when pressure changes with time in a reservoir but the rate of change in pressure is constant. Pseudosteady-state flow may be encountered when a well produces from a bounded reservoir or from its own drainage area. A no-flow boundary can also be created at the edge of drainage area as nearby wells produce from their own drainage areas.

Equations describing transport of fluids in porous media

Fluid flow equations in porous media are primarily based on the following:

- Diffusivity equation: Predicts reservoir pressure as a function of location and time under dynamic conditions of production and injection

- Darcy's law: Correlates flow rate of fluids with pressure differential that exists in the reservoir
- Relative permeability data: Correlation between relative permeability and fluid phase saturation
- Capillary pressure data: Required to calculate individual fluid phase saturation

The diffusivity equation is formulated for various fluid phases (oil, gas, and water) and flow geometries as the case warrants. The equation is based on the law of conservation of mass, Darcy's law as discussed in Chapter 3, and equations of state related to fluid compressibility. The mass balance of flowing fluid recognizes the fact that the rate of input of fluid into an elemental volume of porous medium must be accounted for by the rate of output from the element and rate of accumulation within the element. Detailed derivation of the diffusivity equation is available in the literature [1,2].

Radial flow of single-phase fluid, oil, or gas

One of simplest fluid flow models in porous media is based on single-phase flow of fluid in a radial direction as encountered around a single vertical well producing from a formation where reservoir properties can be assumed to be homogeneous or nearly homogeneous. Based on mass balance, the equation of continuity is derived first, which can be expressed as follows:

$$\left(\frac{1}{r}\right)\left[\frac{\delta(r\rho u_r)}{\delta r}\right] = -\frac{\delta(\phi\rho)}{\delta t} \tag{9.1}$$

where r = distance in radial direction; ρ = density of fluid; u_r = fluid velocity in radial direction; ϕ = porosity of medium; t = time.

Next, Equation (9.1) is modified by applying Darcy's law. The fluid velocity term in the above equation can be written in terms of pressure differential that causes the fluid to flow, $\delta p/\delta r$, fluid viscosity, μ, and permeability of porous medium, k, by noting that:

$$u_r = -\left(\frac{k}{\mu}\right)\left(\frac{\delta p}{\delta r}\right) \tag{9.2}$$

Hence, Equation (9.1) can be recast as follows:

$$\left(\frac{1}{r}\right)\left[\frac{\delta(r\rho\delta p/\delta r)}{\delta r}\right] = \left(\frac{\mu}{k}\right)\left[\frac{\delta(\phi\rho)}{\delta t}\right] \tag{9.3}$$

Further modification of the above equation can be accomplished by noting the compressibility characteristic of the fluid. For slightly compressible fluid such as oil, the following approximation is valid:

$$\frac{\delta \rho}{\delta p} = c\rho \tag{9.4}$$

The above simply states that the rate of change of fluid density with pressure is a function of the product of fluid compressibility and density. Combining Equation (9.4) with Equation (9.3), the following form of diffusivity equation for the radial flow of slightly compressible fluids can be obtained:

$$\frac{\delta^2 p}{\delta r^2} + \left(\frac{1}{r}\right)\left(\frac{\delta p}{\delta r}\right) = \left(\frac{\phi \mu c_t}{k}\right)\left(\frac{\delta p}{\delta t}\right) \tag{9.5}$$

Finally, the diffusivity equation in oilfield units (fluid viscosity in centipoise (cp), k in millidarcy (mD), c_t in pounds per square inch (psi^{-1}), and t in hour (h)) can be written in the following form:

$$\frac{\delta^2 p}{\delta r^2} + \left(\frac{1}{r}\right)\left(\frac{\delta p}{\delta r}\right) = \left(\frac{1}{0.0002637}\right)\left(\frac{\phi \mu c_t}{k}\right)\left(\frac{\delta p}{\delta t}\right) \tag{9.6}$$

Furthermore, the diffusivity coefficient, which relates to the transmittal of pressure in porous medium, is defined as:

$$\eta = \frac{0.0002637\, k}{\phi \mu c_t} \tag{9.7}$$

As indicated above, the diffusivity coefficient is directly proportional to permeability of porous medium while rock porosity, fluid viscosity, and total compressibility are inversely proportional. Equation (9.5) can be solved analytically under appropriate initial and boundary conditions to compute reservoir pressure over time and radial distance.

The inherent assumptions in deriving the equation are as follows:

- A homogeneous and isotropic porous medium, which is seldom observed in reality
- Laminar flow of single-phase fluid
- Fluid is slightly compressible
- Fluid compressibility and viscosity are constant over the range of pressure

Note that the radial diffusivity equation can be readily extended to 1D, 2D, and 3D flow geometries in Cartesian coordinates. The diffusivity equation in 3D can be expressed as follows:

$$\left(\frac{\delta^2 p}{\delta x^2}\right) + \left(\frac{\delta^2 p}{\delta y^2}\right) + \left(\frac{\delta^2 p}{\delta z^2}\right) = \left(\frac{1}{\eta}\right)\left(\frac{\delta p}{\delta t}\right) \tag{9.8}$$

Since natural gas is a compressible fluid, the density term appearing in the diffusivity equation is a strong function of pressure and temperature as follows:

$$\rho = \frac{pM}{zRT} \tag{9.9}$$

where p = fluid pressure; T = temperature; Z = gas deviation factor; M = molecular weight; R = gas constant.

The equation of continuity, combined with Darcy's law and Equation (9.8), can be used as a basis for modeling compressible flow in porous media as follows:

$$\left(\frac{1}{r}\right)\left[\frac{\delta(rp / \mu z \, \delta p / \delta r)}{\delta r}\right] = \left(\frac{1}{k}\right)\left[\frac{\delta(\phi p / z)}{\delta t}\right] \tag{9.10}$$

The above equation is highly nonlinear and gas properties change markedly with changing pressure and temperature. However, at relatively high reservoir pressure over 3000 psia, the value of $p/\mu z$ does not change significantly. Furthermore, gas compressibility is a function of pressure and gas deviation factor as follows:

$$c_g = \left(\frac{z}{p}\right)\left(\frac{\delta(p / z)}{\delta p}\right) \tag{9.11}$$

As the total compressibility term is dominated by gas compressibility, the radial diffusivity equation for compressible fluid at higher pressure can be written as:

$$\left(\frac{1}{r}\right)\left[\frac{\delta(r\delta p / \delta r)}{\delta r}\right] = \left(\frac{\phi \mu c_t}{k}\right)\frac{\delta p}{\delta t} \tag{9.12}$$

At relatively low reservoir pressures, however, it is observed that μz does not vary significantly with pressure. Hence, the diffusivity equation takes the following form:

$$\left(\frac{1}{r}\right)\left[\frac{\delta(r\delta(p)^2 / \delta r)}{\delta r}\right] = \frac{(\phi \mu c_t / k)\delta(p)^2}{\delta t} \tag{9.13}$$

The above equation is considered valid for reservoir pressures under 2000 psia.

In order to deal with the nonlinearity of diffusivity equation for compressible fluids over all ranges of pressure, a pseudopressure is defined as follows:

$$m(p) = 2 \int_{p,\text{ref}}^{p} \left(\frac{p}{\mu z}\right) \delta p \tag{9.14}$$

where $m(p)$ = pseudopressure, psia^2/cp; p,ref = reference pressure, psi.

The above pseudopressure term replaces pressure terms in the diffusivity equation, which can be presented as follows:

$$\frac{\delta^2 m(p)}{\delta r^2} + \left(\frac{1}{r}\right)\left(\frac{\delta m(p)}{\delta r}\right) = \left(\frac{1}{0.0002637}\right)\left(\frac{\phi \mu c_t}{k}\right)\left(\frac{\delta p}{\delta t_{pseudo}}\right) \tag{9.15}$$

where

$$t_{pseudo} = \int_0^t \frac{1}{\mu z} \delta t \tag{9.16}$$

Steady-state flow

During steady-state flow, pressure does not change with time. Hence, Equation (9.6) can be simplified as follows:

$$\frac{\delta^2 p}{\delta r^2} + \left(\frac{1}{r}\right)\left(\frac{\delta p}{\delta r}\right) = 0 \tag{9.17}$$

Darcy's law, based on the assumption of steady-state flow, can be utilized to estimate the flow rate of a single phase when the reservoir pressure is known or estimated.

Slightly compressible fluids

For slightly compressible fluid, namely, oil, the following equation can be used (1):

$$q = \frac{7.08 \times 10^{-3} kh}{[\mu B_o c_o \ln(r_e / r_w)] \ln[1 + c_o(p_e - p_w)]} \tag{9.18}$$

where q = oil flow rate, STB/d; r_e = drainage radius, ft.; r_w = radius of wellbore, ft.; p_e = reservoir pressure at the external drainage boundary, psia; p_w = bottom-hole well pressure, psia; B_o = formation volume factor, rb/STB; c_o = compressibility of oil, psi^{-1}.

Equation (9.18) suggests that the transmissibility of the formation and pressure differential that exists between outer boundary and wellbore are directly proportional to the flow rate of oil.

Example 9.1

Estimate the production rate from an oil well based on the information given below. What would be the limiting drawdown below which the well will no longer be economic? Make necessary assumptions.

Average permeability of rock, mD: 18
Net thickness, ft.: 35
Oil viscosity, cp: 1.9
Oil formation volume factor, rb/STB: 1.2
Oil compressibility, psi^{-1}: 2.64×10^{-4}
External drainage radius, ft.: 5280
Radius of wellbore, ft.: 0.287
Pressure at external boundary, psi: 2300
Well bottom-hole pressure, psi: 1100
Steady-state flow rate for the oil well is estimated based on Equation (9.18) as follows:
$q = 7.08 \times 10^{-3}$ (18)(35)/[(1.9)(1.2)(2.64 $\times 10^{-4}$) ln(5280/0.287)] ln[1 + 2.64 $\times 10^{-4}$ (2300 − 1100)]
$q = 208$ STB/d
Assuming that the economic limit is reached at 20 STB/d, and considering all other parameters remain the same, the limiting drawdown is calculated as follows:
$20 = 7.08 \times 10^{-3}$(18)(35)/[(1.9)(1.2)(2.64 $\times 10^{-4}$) ln(5280/0.287)] ln[1 + 2.64 $\times 10^{-4}(\Delta p)$]
$\Delta p = 102$ psia

Compressible fluids

The equation for estimating flow rate of a gas well operating under 2000 psi is as follows:

$$q_{sc} = \frac{kh(p_e^2 - p_w^2)}{1422\,Tz\mu\ln(r_e / r_w)} \tag{9.19}$$

where q_{sc} = flow rate of gas in Mscf/d evaluated at 14.7 psia and 520°R.

Note that a similar equation involving pseudopressure function can be used to estimate the gas flow rate in all ranges of pressure.

Example 9.2

Compute the gas flow rate of a well located in a low permeability formation based on the following data. What are the inherent uncertainties in the calculation?
Permeability of gas zone, mD: 1.5
Net thickness of formation, ft.: 110
Pressure at external boundary, psia: 2000
Well bottom-hole pressure, psia: 1100
Reservoir temperature: 585°R
Gas compressibility factor: 0.82
Gas viscosity, cp: 0.014
Drainage radius, ft.: 660
Wellbore radius, ft.: 0.287
$q_{sc} = (1.5)(110)(2000^2 - 1100^2)/[1422(585)(0.82)(0.014) \ln(660/0.287)]$
$q_{sc} = 6.23$ MSCF/d

The uncertainties in the calculation of gas production include, but are not limited to, the following:

1. Formation permeability, net thickness, and other reservoir properties vary from point to point
2. The formation may be stratified with or without interlayer communication requiring further analysis
3. Drainage radius and pressure at the external boundary may not be known with certainty

Unsteady-state flow

A useful derivation of the diffusivity equation representing unsteady-state flow of fluid in porous media is as follows:

$$p_D = -\tfrac{1}{2} E_i \left[\frac{-948 \phi \mu c_t r^2}{kt} \right] \tag{9.20}$$

where

$$p_D = \frac{kh(p_i - p)}{141.2\, qB\mu} \tag{9.21}$$

E_i = exponential integral function; q = well flow rate, STB/d; p_i = initial reservoir pressure, psi; k = formation permeability, mD; h = net formation thickness, ft.; B = oil formation volume factor, rb/STB.

Unsteady-state flow is encountered during well testing. For example, during drawdown tests, well rate is held constant and the bottom-hole pressure varies with time. In Equation (9.20), the unknown is the permeability thickness product of the formation, kh, which is determined from the well response in terms of changing bottom-hole pressure, p. During the earlier stage of the test, pressure response is not affected by the outer boundary, and is referred to as infinite-acting flow period. During this period, the reservoir appears to be infinite in its response to flow. At the outer boundary, the pressure does not change with time during the infinite-acting flow period.

In deriving the unsteady-state solution, the well is approximated as a line source having zero radius.

The exponential integral function is defined as follows:

$$E_i(-x) = \ln(x) - \frac{x}{1!} + \frac{x^2}{2(2!)} - \frac{x^3}{3(3!)} \tag{9.22}$$

Pseudosteady-state flow

Under pseudosteady-state flow conditions, pressure changes with time; however, the rate of change of pressure is constant over time. The situation is encountered when multiple wells produce in a reservoir from individual drainage boundaries. The diffusivity equation takes the following form:

$$p_D = \frac{2t_D}{r_{eD}^2} + \ln r_{eD} - \frac{3}{4} \tag{9.23}$$

where

$$p_D = \frac{kh(p_i - p_{wf})}{141.2 q B \mu} \tag{9.24}$$

r_e = radius of outer boundary; p_{wf} = flowing bottom-hole pressure;

$$t_D = \frac{0.0002637 \, kt}{\phi \mu c_t r_w^2} \tag{9.26}$$

Under pseudosteady-state conditions, the extent of drainage area and average reservoir pressure can be estimated based on relatively simple equations. The rate of change of flowing bottom-hole pressure is an inverse function of reservoir or drainage pore volume as follows:

$$\frac{\delta p_{wf}}{\delta t} = -\frac{0.234 q B}{c_t V_p} \tag{9.27}$$

where $V_p = Ah\phi/5.615$; A = drainage area of the well, ft.^{2}.

The average reservoir pressure in the drainage area is related to the pore volume of drainage area as follows:

$$p_i - p_{av} = \frac{\Delta V}{c_t V_p} \tag{9.28}$$

where p_i = initial reservoir pressure, psia; p_{av} = average reservoir pressure, psia; ΔV = volume of fluid produced, bbl; V_p = pore volume, bbl.

It can be further shown that the average reservoir pressure can be obtained if bottom-hole flowing pressure, flow rate, and other parameters are known. An estimate of external drainage radius is also necessary for the calculation of the average pressure:

$$p_{av} = p_{wf} + \frac{141.2 q B \mu}{(kh)[\ln(r_e / r_w) - \frac{3}{4} - s)]} \tag{9.29}$$

where s = skin factor.

The above is valid for radial geometry, and is modified to apply for other geometries. This is accomplished by introducing various shape factors in the equation.

Multiphase flow: immiscible displacement of fluid

A common method of improving recovery from the oil reservoir is waterflooding as described in Chapter 16. Once oil can no longer be recovered efficiently by natural mechanisms, water is injected into the reservoir that displaces and drives *in situ* oil toward the wells. Displacement of oil from a porous medium by immiscible fluids, including water, can be described by the frontal advance theory presented by Buckley and Leverett [3]. Analysis of multiphase flow requires the knowledge of individual fluid properties and relative permeability of individual fluid phases, i.e., the ability of flow of one fluid when other fluid or fluids are present in the porous medium. In practical units, the fractional flow of water, defined as the ratio of flow rate of water over total flow rate, is as follows:

$$f_w = \left[1 + 0.001\,127\,k\left(\frac{k_{ro}}{\mu_o}\right) \times \left(\frac{A}{q_t}\right)\right] \times \frac{\partial p_c / \partial L - 0.433\Delta\rho\sin\alpha_d}{1 + (\mu_w / \mu_o) \times (k_{ro} / k_{rw})} \tag{9.30}$$

where A = area, ft.^{2}; f_w = fraction of water flowing; k = absolute permeability, mD; k_{ro} = relative permeability to oil; k_{rw} = relative permeability to water; μ_o = oil viscosity, cp; μ_w = water viscosity, cp; L = distance along the direction of flow, ft.; p_c = capillary pressure = $p_o - p_w$, psi; q_t = total flow rate = $q_o + q_w$, STB/d; $\partial\rho$ = water–oil density difference = $\rho_w - \rho_o$, g/cm^3; α_d = angle of formation dip to the horizon, degrees.

The fractional flow of water is a strong function of water saturation as the relative permeability of rock to water increases markedly with the increase in saturation. Neglecting the gravity and capillary effects, Equation (9.28) can be simplified as follows:

$$f_w = \frac{1}{1 + (\mu_w / \mu_o) \times (k_{ro} / k_{rw})} \tag{9.31}$$

Considering the conservation of mass and incompressible water, the linear frontal advance equation for water is given by:

$$\frac{\partial x}{\partial t} = \left(\frac{q_t}{A\phi}\right) \times \left(\frac{\partial f_w}{\partial S_w}\right)_t \tag{9.32}$$

The frontal advance equation can be used to derive the expressions for average water saturation as follows.

At breakthrough:

$$S_{wbt} - S_{wc} = \left(\frac{\partial S_w}{\partial f_w}\right)_f = \frac{S_{wf} - S_{wc}}{f_{wf}} \tag{9.33}$$

Following breakthrough:

$$S_w - S_{w2} = \frac{1 - f_{w2}}{(\partial f_w / \partial S_w)_{Sw2}}$$ (9.34)

where f_{wf} = fraction of water flowing at the flood front; f_{w2} = fraction of water at the production point; S_{wf} = water saturation at the flood front, fraction; S_{wbt} = average water saturation at breakthrough, fraction; S_{wc} = connate water saturation, fraction; S_{w2}= water saturation at the production point, fraction.

Average water saturation at and after water breakthrough can be determined graphically using the fractional water flow versus water saturation curve as shown in Figure 9.3.

For average water saturation at breakthrough, draw a tangent from $f_w = 0$ touching the fractional flow curve and extend the tangent line to $f_w = 1$.

For average water saturation after breakthrough, draw a tangent from a desired f_w after breakthrough to $f_w = 1$.

Multiphase and multidimensional flow

Detailed analysis of fluid flow in porous media often involves three phases, namely, oil, gas, and water in two or three dimensions. Multiphase flow is also associated with mass transfer between oil and gas, and vice versa, resulting in compositional changes. The above leads to the formulation of reservoir simulation models; the models are solved numerically with the aid of computers. Basic formulation of a multiphase, multidimensional fluid flow model is presented in Chapter 15.

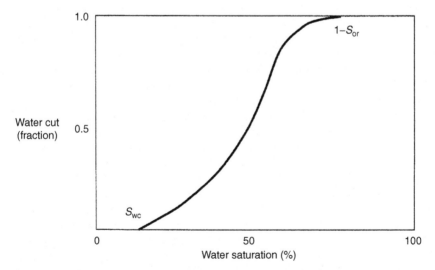

Figure 9.3 Fractional flow curve of water as it acts as a displacing fluid phase. Oil is the displaced fluid phase. Fractional flow curves indicate connate water satuaration, S_{wc}, and residual oil saturation, S_{or}.

Flow of water from aquifer into reservoir

Various mathematical models are available in the literature, ranging from simple to complex, which predict the effects of adjacent aquifer on reservoir performance. The models are broadly classified as steady state, pseudosteady state and unsteady state. A listing of the models is provided in the literature [1]. One of the simplest models pertaining to small aquifers is described in the following:

$$S(p,t) = p_i - p \tag{9.35}$$

$$U = (c_w + c_f)V_{aq} \tag{9.36}$$

$$W_e = U \times S(p,t) \tag{9.37}$$

where p_i = initial reservoir pressure, psi; p = current reservoir pressure, psi; c_w = water compressibility, psi^{-1}; c_f = formation compressibility, psi^{-1}; V_{aq} = pore volume of aquifer, bbl.

Besides analytic models, numerical models describing water encroachment are incorporated in reservoir simulation. Numerical aquifer modeling is described briefly in Chapter 15.

Summing up

Virtually all reservoir-engineering studies require a thorough understanding of fluid flow characteristics. Reservoir pressure, fluid flow rate, and the volume of individual fluid phases are affected by fluid flow behavior in porous media. It is described and analyzed by developing analytical equations and numerical models, which range from simple computation for single-phase radial flow to the complex simulation of multiphase, multidimensional models. Typical reservoir scenarios involve single- and multiphase flow. Fluid flow in porous media is caused by the viscous forces, effects of gravity, and capillary imbibition. Depending on the requirements of the study, fluid flow can be visualized in 1D, 2D, 3D, or radial geometry. Furthermore, fluid flow can be unsteady state, steady state, or pseudosteady state. During unsteady-state flow of fluid in porous medium, pressure and flow rate change with time, which is encountered as the well commences production or the well is shut-in following production. Steady-state flow of fluid occurs when pressure and flow rate do not change with time, a condition that may be encountered when any depletion of the reservoir is replenished by the effects of aquifer. A third type of flow behavior is referred to as pseudosteady state. During pseudosteady-state flow, pressure changes with time but the rate of change of pressure remains the same. This type of flow may be encountered when wells produce from their individual drainage areas in a reservoir.

Fluid flow models are based on the law of conservation of mass, which states that the rate of input of fluid into an element of porous medium must be accounted for by the rate of output from the element and rate of accumulation in the element. The above is combined with Darcy's law to derive the equation of continuity. The equation is modified for incompressible, slightly compressible, and compressible fluids, and for various flow geometries including radial, linear, and areal. For unconventional reservoirs, two other processes are of prime importance, namely, adsorption and diffusion.

There are large numbers of practical applications of the simple analytic equations describing fluid flow in porous media, some of which are illustrated in this chapter. However, numerical models must be employed to analyze multiphase, multidimensional flow systems.

Questions and assignments

1. Describe the diffusivity equation. What reservoir and fluid properties need to be known to solve the diffusivity equation?
2. Describe steady-state, unsteady-state, and pseudosteady-state flow. Why steady-state conditions are difficult to achieve in a reservoir? How do the boundary and well conditions affect the nature of flow?
3. What are the limitations of analytic solutions of fluid flow equations? How can these limitations be addressed in real-world situations?
4. Demonstrate how various reservoir and fluid properties would affect oil flow rate based on Equation (9.18).
5. Based on a literature review, describe how the diffusivity equation can be applied for stratified formations.
6. Describe the effects of gravity and capillary forces on fluid flow. How are the effects of these forces included in the formulation of equations? Provide an example.
7. How do the multiphase fluid flow models differ from single-phase models? What additional information is necessary to formulate multiphase models?

References

[1] Satter AS, Iqbal GM, Buchwalter JL. Practical enhanced reservoir engineering assisted with simulation software. Tulsa, Oklahoma Pennwell; 2008.
[2] Matthews CS, Russell DG. Pressure buildup and flow tests in wells. SPE Monograph, vol. 1. Dallas (TX); 1967.
[3] Buckley SE, Leverett MC. Mechanism of fluid displacement in sands. Trans AIME 1942;146:107–16.

Transient well pressure analysis **10**

Introduction

Well testing is one of the most valuable tools in the possession of reservoir engineers. It is routinely used to evaluate well and field performance, diagnose reservoir characteristics, integrate test results with other studies, plan for future development, and perform the overall management of the reservoir. The concept of pressure transient testing is rather straightforward. It is basically a three-step process as outlined below:

1. Design and create a pressure pulse in the reservoir by changing the flow of fluid through a well. Examples of rate change are planned stoppage in production or injection. Some pressure transient tests are based on incremental changes in well rates.
2. Monitor the resultant changes in pressure response over time by high resolution gauges downhole or at wellhead as the pressure pulse travels through various portions of the reservoir, from the immediate vicinity of the wellbore to the drainage boundary or physical limit of the reservoir
3. As the pressure pulse travels through the formation, the signature of response changes depending on the changing fluid flow characteristics and reservoir properties. Identify the signatures graphically by diagnostic and other plots. Evaluate well conditions and reservoir properties by applying appropriate fluid flow equations on the various signatures obtained from the plots.

This chapter provides a brief overview of well testing practices in conventional and unconventional petroleum reservoirs. The topics presented below attempt to address the following queries:

- What is well testing or pressure transient testing?
- What kind of information can a reservoir engineer obtain based on well testing results?
- What are the types of well testing conducted in conventional and unconventional reservoirs?
- What are the assumptions and limitations of well testing?
- How well are test results analyzed?
- How are the results integrated in reservoir studies?
- What are the design considerations in conducting a pressure transient test?

Role of well testing and pressure transient analysis

Well tests are performed throughout the life cycle of a reservoir to accomplish large numbers of important objectives and gather invaluable information, some of which are described below:

- Formation transmissibility and storativity
- Average reservoir pressure
- Evaluation of deliverability of gas producers

Reservoir Engineering. http://dx.doi.org/10.1016/B978-0-12-800219-3.00010-3

- Well condition including liquid holdup and segregation of liquid and gas
- Near wellbore phenomena including skin effects
- Presence of flow boundaries such as sealing fault
- Fracturing pressure for the formation
- Characteristics of hydraulic fractures
- Effect of induced and natural fractures on well performance
- Advance of flood front during fluid injection
- Interlayer communication in stratified reservoir
- Degree of communication between two wells

Types of well tests

The most common types of well tests are outlined below [1–5]:

- Drillstem test: As a well is newly drilled, the first test that is traditionally performed is the drillstem test. The test is performed to assess the potential of the well. It involves at least two sequences of flow period followed by periods of shut-in and pressure buildup, where the response in pressure is recorded and analyzed. The final shut-in period is longer in order to allow shut-in pressure to approach static reservoir pressure. In addition to reservoir pressure, formation permeability and skin around the well may be determined by this test. A fluid sample is collected as part of the drillstem test for analysis.
- Pressure buildup test: In order to conduct this test, the well is produced steady for a period of time, from a few days to months, followed by shut-in. The shut-in period is determined by the transmissibility of formation and test objectives. As a result, reservoir pressure builds up around the well, which can be monitored and recorded continuously until stabilization in pressure is obtained. Analysis of initial, middle, and late time response in pressure provides a wealth of data pertaining to the well and reservoir drainage area, including formation transmissibility, skin effects, drainage radius, fluid front location, and reservoir heterogeneity. Well rate and pressure response during the test is depicted in Figure 10.1.

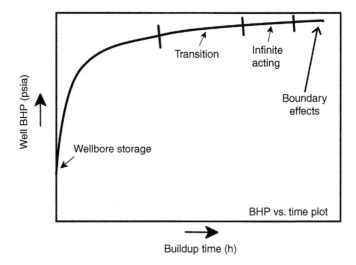

Figure 10.1 Pressure transient response during buildup.

- Practical considerations involve the length of time the well can be shut-in to reach stabilization and incur loss in production. Low permeability reservoirs require a longer duration of the test. For example, the test may not be applicable for shale gas formations where matrix permeability is in nanodarcies. Besides, the pressure buildup test can be complicated in the case of two-phase flow as encountered in water cut wells in conventional reservoirs and in coalbed methane reservoirs where both water and methane are produced.
- Drawdown test: The well is initially shut-in for a period of time till the pressure is stabilized. The well is then produced at a constant rate and decrease in pressure is recorded (Figure 10.2). The signature left by the decrease in pressure during the "drawdown" of fluid is analyzed to determine well and reservoir characteristics. In a practical setting, it could be difficult to produce the well at a constant rate for an extended period of time.
- Pressure fall-off test: Fluid is injected through the well at a constant rate for a period of time until the injection pressure is stabilized, followed by shut-in of the injection well. The bottom-hole pressure "falls off" in the absence of injection and reaches a stabilized value after a period of time.
- Minifrac test: A prefrac test is conducted before hydraulic fracturing in order to ascertain important reservoir characteristics related to fractures, including fracture gradient, fluid leakoff behavior, formation permeability, and fracture closure pressure, among others. A short fracture is induced in the formation by injecting fluid and monitoring pressure response during the fall-off period. Proppant is not used in the test. The information obtained from the minifrac test is used to design the actual hydraulic fracturing operation. In many instances, the duration of the test is relatively short to reduce waiting time for the field personnel to conduct the hydraulic fracturing operation. A postfrac is performed for hydraulically fractured wells to evaluate the effectiveness of fractures in producing the well. The tests are referred to as diagnostic fracture injection tests.

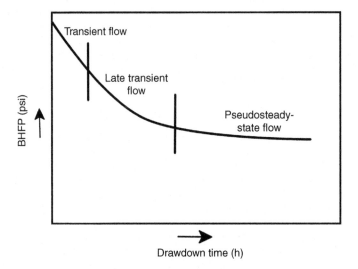

Figure 10.2 Pressure transient response during drawdown.

- Step rate test: A step rate test is conducted in an injection to a well determine the pressure at which the formation would be fractured. During waterflooding and enhanced recovery operations, fluid is injected below the fracture pressure to avoid any loss of fluids through fractures. During the test, fluid is injected into the formation at an increasing rate of steps, the duration of each step is about an hour or less, depending on formation permeability. The fracture pressure of the formation is readily identified by plotting the bottom-hole pressure against injection rate, which yields a straight line. At the fracture pressure of the formation, the slope of the line changes on the plot.
- Interference well test: The interference well test involves two or more wells, where one well is produced for a period of time and any pressure response is monitored in adjacent wells. This test provides valuable information regarding the degree of communication between two wells that is directly related to the effectiveness of waterflood or enhanced oil recovery operations in conventional reservoirs. In addition, the test is also conducted in unconventional coalbed methane reservoirs to determine the degree of heterogeneity, as well as face and butt cleat permeability. The test aids in the optimization of well location and spacing between the wells.
- Modular dynamic test: This test, developed in the late twentieth century, is run in a newly drilled borehole to determine the horizontal and vertical permeability of reservoir rock and provides information on the degree of interlayer communication in a stratified reservoir, for example. The test is conducted by drawdown of formation fluid followed by buildup. The test assembly includes a fluid collection chamber. The composition of fluid indicates whether a specific layer is watered out in a reservoir under waterflood.

Flow regimes

Depending on well geometry (vertical, horizontal), well characteristics (hydraulically fractured, partially completed), reservoir characteristics (low permeability, naturally fractured, faulted), and drainage boundaries (constant pressure, sealing), the observed pressure response creates a distinct signature at various times during the test. With the passage of time, one-flow regime transitions into the next and the signature changes. Diagnostic plots of pressure derivative against time on a log–log scale are generally used to identify the appearance and duration of the signatures. Other types of plots, including semilog plots and type curves, are also used to confirm the diagnosis. The flow regimes as typically observed in well tests are summarized below:

- Wellbore storage dominated flow: The first flow regime typically observed during a well test is wellbore storage dominated flow, as the fluid stored in the wellbore influences wellhead pressure gauge reading readings. The signature is indicated by a straight line of observed pressure plotted against time on a log–log scale.
- Linear or bilinear flow: In hydraulically fractured wells, flow through a fracture initially dominates the pressure response. Linear flow is indicated by a half slope on the log–log plot as well as on a diagnostic plot described later (Figure 10.3). The fracture is assumed to be infinitely conductive. Bilinear flow stems from relatively fewer conductive hydraulic fractures having a signature of quarter slope on the diagnostic plot.
- Infinite-acting flow regime: With the passage of time, the pressure transient moves past the region dominated by wellbore storage and skin effects into the infinite-acting flow regime. However, there is a transition period between the initial flow regime and the infinite-acting

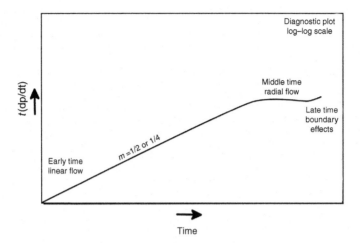

Figure 10.3 Linear or bilinear flow as encountered during pressure transient testing of a hydraulically fractured well. Linear flow is indicated by an ascending line having half-slope, while bilinear flow has a signature of quarter-slope. Well tests are customarily performed in hydraulically fractured wells in order to evaluate the success of fracturing and workover.

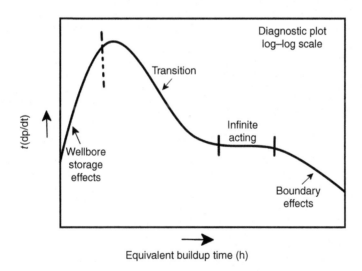

Figure 10.4 Diagnostic plot showing severe skin effects, infinite-acting flow regime, and boundary effects.

flow regime, lasting about one and half cycles in log scale of time. In case of severe skin damage, a "hump" is observed during the transition period (Figure 10.4). During middle time, an infinite-acting flow regime is observed, which leads to the calculation of formation transmissibility and other important properties. The pressure transient behaves in a manner as if the reservoir is infinite, since the boundary effects are not felt yet. The regime is identified by a horizontal line on the diagnostic plot, which is a log–log plot of pressure derivative over time. In a semilog plot, the region can be identified as a straight line having a slope, where the slope is proportional to formation transmissibility.

- Pseudosteady-state flow regime: This regime is observed as the effect of reservoir drainage boundary is felt by the pressure transient. A change in slope appears in the diagnostic plot beyond the horizontal segment of the plot indicating boundary effects. In a drawdown test, an ascending line of unit slope is produced by closed boundary. A line having half slope appears before the unit slope in case the closed drainage area is rectangular. However, if the well test is not run for a sufficient period of time, particularly in low to ultralow permeability reservoirs, the flow regime may not appear at all. A diagnostic plot showing various flow regimes is presented in Figure 10.4.
- Steady-state flow regime: At late times during the well test, a steady-state condition may be observed when any drop in pressure is readily compensated by water influx from an adjacent aquifer, for example. Replenishment of pressure can also be caused by gas cap. During a drawdown test, a continuously falling line indicates a constant pressure boundary. Again, the late time region is only observable when the test is run for a sufficiently long period.
- Horizontal well testing: During the testing of horizontal wells, the following flow regimes are likely to be evident (Figure 10.5):
  - Early time radial flow: The initial flow regime is radial centered around the axis of the lateral wellbore until the pressure transient reaches a boundary in the vertical direction.
  - Hemiradial flow: This flow regime is encountered when the vertical distance from the lateral to the top and bottom of the formation is not the same; pressure response is affected by either of the vertical limits.
  - Linear flow regime: This regime is observed as the fluid begins to flow through the lateral wellbore, which acts as a long fracture.
  - Late time radial flow regime: Once sufficient time has elapsed, a late time radial flow regime emerges where the lateral is in the center of the flow geometry. In low permeability formations, the late radial flow regime may require a month or more to emerge.

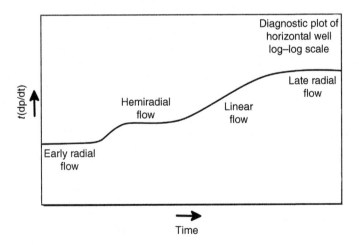

Figure 10.5 Depiction of various flow regimes in horizontal well testing.

Table 10.1 Flow regimes in well test

Flow regime	Signature on diagnostic plot	Pressure behavior	Notes
Early time	Ascending	Transient	Dominated by well-bore storage and skin
Linear	Half slope	Transient	Hydraulic fracture of infinite conductivity
Bilinear	Quarter slope	Transient	Hydraulic fracture of finite conductivity
Infinite-acting	Horizontal line	Transient	Flow unaffected by boundary
Pseudosteady state	Line with a slope	Rate of change in pressure is the same everywhere in drainage area	Bounded reservoir
Steady state	Line with a slope	Pressure is constant with time and location	Constant pressure boundary

The various flow regimes that may be observed in pressure transient testing are shown in Table 10.1.

Well test analysis equations

Pressure transient analysis is based on the mathematical models for unsteady-state, pseudosteady-state, and steady-state flow in porous media. Equations that are directly applied to analyze the pressure response from the well are deduced from the following:

- Law of conservation of mass
- Darcy's law
- Equation of state

The equations describing the flow of fluids in porous media are treated in Chapter 9.

Diagnostic plot

Diagnostic plot, as the name suggests, is used to identify various flow regimes and reservoir heterogeneities as these affect the pressure response during well tests. The y-axis of diagnostic plot is a function of the derivative of pressure. It is plotted against time or a function of time on a log–log scale. In the early days, traditional well test analysis involved the manual plotting of the well test results on semilog or log–log paper to identify the flow regimes required for analysis. The plot was typically based on limited data recorded by analog gauges. The importance of a diagnostic plot was not widely recognized. The advent of the digital age changed all that. Well test analysis is currently based on the collection of a large number of high quality data by electronic gauges and visual rendition of the derivative of pressure versus time (diagnostic) plots on the computer screen. A diagnostic may indicate a whole range of subtle changes in pressure response indicating boundary conditions and various reservoir heterogeneities located far from the well that could not be identified before. Some of the important features of the plot are listed below.

- Hydraulically fractured well: As described earlier, hydraulic fractures are indicated by the appearance of a half-slope or quarter-slope on the diagnostic plot as well as log–log plot (Figure 10.3). The line is apparent when the pressure transient travels beyond the time where the response is affected by wellbore storage. The half-slope is observed in the case of a finite conductivity fracture; a quarter-slope appears in the case of a finite conductivity fracture. The flow regime is called linear or bilinear, respectively.
- Dual porosity reservoirs: Naturally fractured reservoirs are characterized as either dual porosity or dual porosity dual permeability reservoirs. In a dual porosity system, only flow from the fracture network to the wellbore is considered, while in the dual porosity dual permeability system, flow from the fracture and matrix is assumed. A characteristic dip is observed in the diagnostic plot beyond the wellbore storage effects. Flow from a highly conductive fracture system leads to perceptibly less pressure change over time, causing the apparent dip (Figure 10.6).
- Reservoir with impermeable boundary: Depending on the type of test (drawdown or buildup), the late time region in the diagnostic indicates an ascending or descending line of unit slope as the effects of no flow boundary are felt on the pressure transient response. In this case, the rate of change in pressure is high as no external pressure support exists at the reservoir boundary.
- Constant pressure boundary: This type of boundary arises due to the strong water influx from adjacent aquifer or due to the presence of gas cap. A constant pressure boundary is identified by a continuously falling line on the diagnostic plot at late times during the drawdown test, as the rate of change in pressure is diminished due to the strong pressure support from aquifer or gas cap.
- Immiscible fluid front: Wells undergoing injection of water forms a radial or near radial bank of water around the well in relatively homogeneous formations. Two distinct horizontal lines are observed at different levels on the diagnostic plot, the first line represents the injected

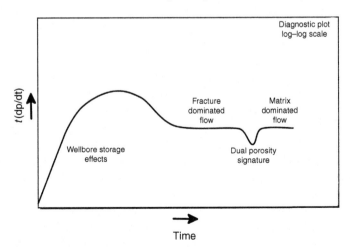

Figure 10.6 Signature of a dual porosity reservoir on a diagnostic plot. Since natural fractures are much more conductive than the rock matrix, pressure transient response is initially dominated by flow through fractures.

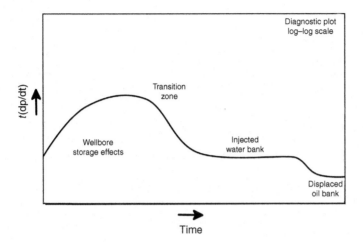

Figure 10.7 Diagnostic plot showing the injected water bank and displaced oil zone.

water bank, followed by the pressure response obtained from the displaced oil bank leading to the observance of a second line (Figure 10.7).

- Faulted reservoir: A sealing fault located on one side of a well is indicated by two distinct horizontal lines on the diagnostic plot separated by a transition region. The first line represents pressure response unaffected by the fault, while the second line appears as the influence of a fault becomes apparent. Sealing fault can be identified by the doubling of slope on a semilog plot.
- Pinchout: This type of reservoir boundary is indicated by a hump on the diagnostic plot at late times, followed by a continuously falling line having negative half slope (Figure 10.8).
- Stratified reservoir: A diagnostic plot obtained from layered formation having different rock characteristics exhibits a behavior that is very similar to a single layer system unless the transmissibility contrast between the two layers is quite significant. In such a case, use of a modular dynamic tester is recommended to identify and characterize the various layers.

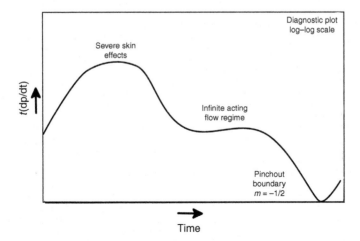

Figure 10.8 Pinchout boundary is identified by negative half-slope during late times.

Design of well test

The workflow for a well test begins with the design of the test. Clear objectives must be established before running a test, which may include the determination of formation permeability to be used in integrated studies, effectiveness of well workover, length and conductivity of a hydraulic fracture, confirmation of reservoir heterogeneity such as existence of a fault or fracture network, tracking of injected fluid front around an injector, and assessment of skin around the well, among others. The type of well test is largely determined by the objective of the test. The most widely practiced well tests in conventional reservoirs include pressure buildup and fall-off tests. Drawdown tests are also common. In evaluating the effectiveness of well workover, the test could be of relatively short duration. However, when the reservoir limit needs to be known, a well test can be quite long. In ultratight unconventional reservoirs, such as shale where production takes place through induced and natural fractures, well tests are mainly focused on the evaluation of the fracture characteristics.

The next factor to consider is the reservoir properties including transmissibility. Duration of a well test is an inverse function of rock permeability, meaning that reservoirs having very low permeability (5 mD or less) may require an unusually long period of time to reach middle time or late time flow regimes to evaluate the sought parameters like formation transmissibility and drainage boundary. In many cases, pressure buildup tests require shut-in of the well for days leading to a loss in production. Horizontal wells typically require a very long time to attain late time radial flow regime resulting in substantial loss in production. Drawdown tests require production at a stabilized rate for an extended period of time, which may not be operationally possible. However, well testing methodology is modified to account for the changes in rates requiring complex computations.

In designing the well test, certain information regarding the well is needed, such as the type of well (producer, injector, observer), whether the well is horizontal or vertical, whether it is completed partially or completely, whether well workover has been performed, what is the dominant fluid phase, and whether multiphase flow of fluids has been encountered. Pressure response of a partially completed well is dissimilar to a fully completed well. Again, the design process must include the review of bubble point pressure or dew point pressure. A drop in reservoir pressure during a test may result in the evolution of a new fluid phase. The presence of two-phase flow in wells may complicate well testing and unique results may not be achieved.

Interpretation of well test data

Before the advent of the digital age, most well test interpretations were performed manually where the pressure transient data were plotted on semilog and log–log paper. However, modern well test interpretations are performed with the aid of software applications available in the industry. This enables robust analysis of field data within a short period of time. Although well test interpretation is mostly based on the analytic equations of fluid flow in porous media, certain well tests can be analyzed by numerical methods in case of complex geologic heterogeneities.

A typical analysis of a pressure buildup or drawdown test conducted in an oil well leads to the estimation of the following parameters:

- Average effective oil permeability in the drainage area
- Wellbore storage coefficient
- Skin around wellbore
- Drainage area
- Oil in place in the drainage area
- Reservoir pressure in the drainage area

The steps that are involved in widely practiced pressure buildup tests are outlined in the following.

Inspect and validate well test data to identify any anomalies and unexpected trends. The reasons for anomalies can be many, including operational issues, instrument malfunction, failure to adhere to test procedure guidelines, and presence of unidentified heterogeneities in the reservoir. A data reduction scheme may also be needed as digital gauges may record thousands of data at very short intervals during the test.

Prepare a diagnostic plot to identify the various flow regimes. This is plotted as the derivative of pressure versus time on a log–log scale. The early time region is affected by wellbore storage and skin, following which a horizontal or flat line is identified as the middle time region. Formation transmissibility is determined from this region. In the late time region, the effects of drainage boundary are apparent as the slope of the diagnostic line changes its signature. A log–log plot of the changes in pressure buildup against time can also be drawn indicating the duration of wellbore storage and skin.

Prepare a semilog plot, also known as Horner plot, where the buildup pressure is plotted against the logarithm of Horner time. Horner time is defined as:

$$\text{Horner time} = \frac{t_p + \Delta t}{\Delta t}$$

During the middle time region, the value of formation transmissibility can be confirmed based on the slope of the straight line as indicated on the plot.

Other plots used in the pressure buildup analysis include Miller–Dyes–Hutchison plot, where the change in static bottom-hole pressure is plotted against time on a log–log scale.

Once the shape of the well drainage area (circular, rectangular) and the location of the well with respect to the drainage boundary are known or assumed (centered, off-centered), average reservoir pressure can be estimated from well test analysis.

Type curve analysis

The well test analysis described above can be performed with the aid of type curves. These type curves are available in the well test software applications. The pressure transient response obtained from the well test is matched against a family of type curves that are hypothetically generated for varying reservoir properties and wellbore conditions. Once a match is obtained between field data and one of the type curves by shifting the curves on the computer screen, well test results are calculated based on the match point.

Type curves are predetermined well responses in the form of dimensionless pressure versus dimensionless time. A family of type curves is generated for a variety of reservoir properties and wellbore conditions. The application of generic type curves for a specific well test analysis is made possible by the utilization of dimensionless pressure and time. The dimensionless variables eliminate the need for generating customized curves for a specific test. The methodology works by matching the pressure response of the test well with a member of an available family of curves, followed by the calculation of reservoir and well parameters based on the match point. In the earlier days, the procedure was tedious as it was performed manually by using tracing paper. However, with the advent of the digital age, type curve analysis has been made a lot easier as the match with well test data is made automatically on computer screen with an option for fine-tuning by the analyst.

Summary

Widely practiced well tests in conventional and unconventional reservoirs are shown in Table 10.2.

Table 10.2 Well test types and their applications

Well test type	Objective	Notes
Pressure buildup	Formation transmissibility, skin effects, average reservoir pressure, boundary conditions, formation heterogeneities including faults and fractures	Widely used in conventional reservoirs. May not be suitable in tight formations as test requires long shut-in time
Drawdown	Same as above	Requires stabilized flow before test, which may not be possible
Pressure fall-off	Hydraulic fracture characteristics, location of injected fluid front and various reservoir properties	Common test for injection wells. Fall-off test below fracture pressure is widely used in coalbed methane reservoirs
Minifrac test (prefrac)	Design and optimization of hydraulic fracturing operation	Typically implemented in ultratight shale reservoirs
Minifrac test (postfrac)	Fracture characteristics, formation permeability, and pore pressure	Also referred to as diagnostic fracture injection test
Step rate test	Formation fracture pressure	In waterflooding and enhanced oil recovery operations, fluid is injected below fracture pressure
Interference test	Characterization of reservoir in between wells, including directional permeability and existence of barriers	Provides valuable information for designing waterflood and enhanced recovery operations
Modular dynamic test	Formation permeability and degree of communication between layers	Tests are of short duration and conducted at various layers of the reservoir

Questions and assignments

1. What is well testing and what role does it play in managing a reservoir?
2. Describe the types of well tests widely used in the industry.
3. What are the assumptions and limitations of pressure buildup and drawdown tests?
4. What are the three distinct regions that are expected from a typical well test?
5. What is a diagnostic plot and what does it identify? Describe in detail.
6. How can a well test aid in characterizing a reservoir? Explain with examples.
7. How does well testing differ between conventional and unconventional reservoirs?
8. What are the design considerations for well testing?
9. What uncertainties can exist in the accuracy of well test data?
10. Does a well require testing on a regular basis? Why or why not?
11. Based on a literature review, describe a case study of well testing in a tight fractured reservoir. What information does it provide in managing the reservoir?

References

[1] Mathews CS, Russell DG. Pressure buildup and flow tests in wells, monograph, vol. 1. Dallas: SPE; 1967.
[2] Earlougher RC. Advances in well test analysis, monogram, vol. 5. Dallas: SPE; 1977.
[3] Lee J. Well testing. Dallas: Society of Petroleum Engineers of AIME; 1982.
[4] Gringarten AC, Bounder DP, Landen PA, Kniazeff VJ. A comparison between different skin and wellbore storage type curves for early time transient analysis. SPE 8205 Paper presented at the SPE Annual Fall Technical Conference and Exhibition; September 23–26, 1979, Dallas, Texas.
[5] Horner DR. In: Brill EJ, editor. Pressure Buildup in Wells. Leiden: Third World Petroluem Congress; 1951. p. 503. Proceedings, Chapter II.

Primary recovery mechanisms and recovery efficiencies

Introduction

Primary recovery of oil and gas is solely driven by the natural energy available to the reservoir. Internal energy in a reservoir is concentrated due to the intense pressurization of fluids and rock. The energy is released gradually as the reservoir is produced. Natural energy is also provided by adjacent aquifers in certain reservoirs. Oil may be recovered from a reservoir in multiple stages, which is commonly referred to as primary, secondary, and tertiary recovery. During primary recovery, petroleum reservoirs produce naturally. However, during secondary and tertiary recovery, energy is provided by injecting water or gas or chemicals into the reservoir to drive oil and gas. External energy is also provided by thermal methods.

For certain reservoirs, depending on fluid and rock characteristics, primary recovery from a reservoir may continue for a long period of time. Many small and older reservoirs have been produced by primary recovery mechanisms for the large part of their life. However, for large and complex reservoirs, where primary recovery is anticipated to be low, secondary recovery operations in the form of water injection and pressure maintenance are initiated quite early in the life of the reservoir.

This chapter outlines the primary drive mechanisms in oil and gas reservoirs and attempts to answer the following:

- What are the primary drive mechanisms for oil and gas reservoirs?
- Are the primary drive mechanisms effective in both conventional and unconventional reservoirs?
- How do the mechanisms affect reservoir performance and production characteristics?
- What oil and gas recovery can be expected during primary recovery?
- How do primary, secondary, and tertiary recoveries fit together in effectively managing a reservoir?

Primary drive mechanisms

In conventional oil and gas reservoirs, the natural sources of energy in oil and gas reservoirs are many, including the high initial reservoir pressure, volatilization characteristics of petroleum fluids, expansion of gas, and the effect of an aquifer, among others. The above sources control the primary reservoir performance to varying degrees. Multiple sources of energy from nature can act in combination during the primary recovery. Historical data have shown that the primary recovery can be as high as 80% or more for a gas reservoir having good porosity and permeability, and down to 10% or less for a heavy oil reservoir with unfavorable rock characteristics.

Reservoir Engineering. http://dx.doi.org/10.1016/B978-0-12-800219-3.00011-5

For most unconventional reservoirs, the forces of nature are inadequate to produce oil and gas. Innovative techniques such as multistage hydraulic fracturing in horizontal wells (as in shale gas) and applying thermal energy (as in oil sands) are needed to produce unconventional accumulations of petroleum. In fact, the unconventional reservoirs acquired their name as production of oil and gas in a conventional manner is not economically feasible.

The natural mechanisms for primary production from conventional reservoirs are discussed in the following paragraphs.

Oil reservoirs

The primary drive mechanisms in oil reservoirs are as follows [1,2]:

- Liquid and rock expansion drive
- Solution gas or depletion drive
- Gas cap drive
- Aquifer water drive
- Gravity segregation drive
- Compaction drive
- Combinations of the above

Liquid and rock expansion drive

When an oil reservoir is producing above the bubble point pressure, shown as line BB' in Figure 5.1, the primary mechanism for oil production is the expansion of liquid and rock. Oil is undersaturated, and all the volatile components are dissolved in oil as long as the reservoir operates above the bubble point pressure.

During primary production, natural gas evolves at the surface facilities due to the reduction in pressure and temperature, resulting in low and constant gas–oil ratio. No significant water production is anticipated during primary production except in reservoirs where water saturation is high.

The oil recovery mechanism is dominated by the volumetric expansion of the reservoir fluids and rock above the bubble point when no other external driving mechanism is present. Expected recovery efficiency is relatively low, and typically varies from 1% to 5%, with an average of 3%.

Solution gas or depletion drive

When the reservoir pressure declines below the bubble point into the two-phase region (refer to Figure 5.1) due to production, dissolved gas starts to come out of the solution, and a free gas phase is formed. Below the bubble point the gas phase increases rapidly in the reservoir. The dominant recovery mechanism is known as solution gas or depletion drive. The gas phase is significantly more mobile than the liquid phase in the reservoir because the viscosity of the gas is much lower than the oil.

Reservoir pressure declines rapidly from early stages of recovery. The gas–oil ratio is initially low, and then rises to a maximum, and finally drops as most of the liberated gas is produced. Again, no significant water production is anticipated from the reservoir except where the water saturation is high.

Typical oil recovery due to solution gas drive ranges between 10% and 30%, with an average of about 20%. Once secondary recovery operation is initiated, further recovery of oil is attained (Figure 11.1).

In the early years of the twentieth century, many small reservoirs were produced until abandonment based on a primary recovery mechanism. Common approaches to boost production during declining well performance involved the utilization of pump, gas lift, well recompletion, and workover, among others. Many older reservoirs are still produced in the above manner. However, as the reservoir characteristics and fluid flow behavior were better understood along with the introduction of technological innovations, large complex reservoirs were subjected to waterflooding and pressure maintenance early on following a relatively short period of primary recovery. The timing and strategy for improved oil recovery (IOR) operations are based on building various scenarios of drilling additional wells and fluid injection. The ultimate recovery of petroleum is maximized by introducing additional energy into the reservoir, and the added assets far outweigh the costs associated with drilling and IOR operations.

Gas cap drive

A gas cap present at the time of the discovery of the oil reservoir is known as a primary gas cap. Certain oil reservoirs are discovered with an initial reservoir pressure below the bubble point pressure where a primary gas cap forms long before the reservoir is discovered and produced.

In case an oil reservoir does not have a gas cap initially, but one is formed later by the dissolution of volatile components present in the liquid phase, the gas cap is referred to as secondary. Liberated gas forms a gas cap above the oil zone. At that point, the reservoir is located within the two-phase region (as in points B'' and V'' in Figure 5.1). Since gas is lighter than oil, it rises above the oil zone due to gravity segregation.

During production by gas cap drive, reservoir pressure falls slowly and continuously. The driving energy is predominantly provided by the expansion of the gas cap as the reservoir depletes. The gas–oil ratio rises continuously in updip wells. Water production is nonexistent or negligible where the water saturation is irreducible. Production from the gas cap drive reservoir is due to the driving energy imparted by both solution and free gases, resulting in higher oil recovery than the solution gas drive alone. Oil recovery due to gas cap drive is typically around 30% but could be as much as 40%.

Oil recovery under gas cap drive is improved by (i) completing the wells in the oil zone as deep as possible, (ii) reinjecting the produced gas in updip wells, and (iii) shutting off the wells as the gas–oil ratio becomes significant (Figure 11.2).

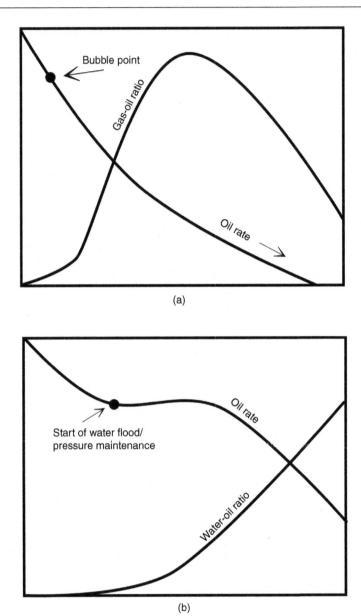

(a)

(b)

Figure 11.1 Comparison between reservoir performances based on two scenarios. (a) Solution gas drive alone and (b) solution gas drive followed by water injection (secondary recovery). Recovery is usually higher in the latter case.

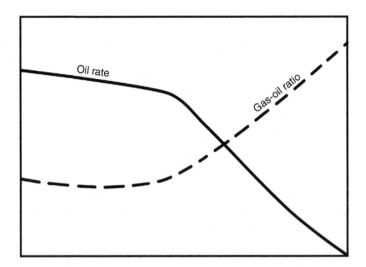

Figure 11.2 Reservoir performance under gas cap drive.

Aquifer water drive

Certain reservoirs are in communication with an aquifer, which may provide signifi-cant natural energy for production. Three types of water drive reservoirs are encoun-tered:

- Peripheral water drive: The aquifer is located at the periphery
- Edgewater drive: The aquifer is located at one edge
- Bottom water drive: The aquifer is located at the bottom of the oil or gas reservoir

As an oil reservoir is produced, water encroachment into the reservoir occurs due to high aquifer pressure. This leads to favorable oil recovery. Reservoir pressure remains high, and gas–oil ratio remains low during production. Early water production is encoun-tered at the downdip wells, and water production increases with time. Aquifer volume is quite large in comparison to reservoir volume, 10 times or larger than the reservoir.

Certain reservoirs experience bottom water drive where the aquifer is located below the reservoir. If the aquifer is below the oil reservoir, water coning into the oil reservoir results in lower oil recovery than what can be expected from a peripheral water drive.

Under favorable conditions, oil recovery efficiency under aquifer water drive could be as much as 50% or more. Hence, a strong influence by aquifer may be the most potent primary drive mechanism available in comparison to the others. Figure 11.3 presents the reservoir performance under aquifer drive.

Gravity segregation drive

Oil drainage due to gravity and subsequent production can be found in certain steeply dipping or fractured reservoirs located at shallow depths. The phenomenon may also

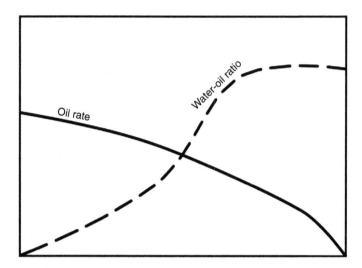

Figure 11.3 Reservoir performance under aquifer water drive.

occur where vertical permeability is more than horizontal permeability. Under gravity segregation drive, reservoir pressure declines continuously. Gas–oil ratio remains low in downdip wells, but a high value is observed in updip wells. Water production is either not observed or negligible at the wells.

Oil recovery due to gravity segregation drive could be 50% or more. Combined with gas cap drive, a recovery factor of 80% is achieved in certain cases. However, total recovery volume could be low in reservoirs that produce by the mechanism of gravity drainage.

Rock compaction drive

Certain reservoir rocks are unconsolidated and have very high compressibility much above the normal range of $3–8 \times 10^{-6}$ psi^{-1}. No significant decline is observed as the reservoir is produced and rock is compacted. Sizeable amounts of oil may be produced before the bubble point pressure is reached. Such a phenomenon is referred to as compaction drive. Certain North Sea and Gulf Coast fields are found to produce by compaction drive, although such reservoirs are not commonplace. Overpressured reservoirs may also produce by compaction drive.

Dry and wet gas reservoirs

The drive mechanisms in natural gas reservoirs include:

• Gas expansion or depletion drive
• Aquifer water drive
• A combination of the above

Dry and wet gas reservoirs exist in the single-phase region with the initial temperature exceeding the cricondentherm. The gas phase undergoes isothermal depletion inside the reservoir without any condensation. When a portion of produced gas condenses in the surface separators under reduced pressure and temperature, the produced gas is referred to as wet gas.

Dry gas reservoirs contain mostly lighter hydrocarbons with negligible condensate volume. Gas–condensate ratios could be 100,000 scf/STB or higher. Gas recoveries are observed to be 80% or more at relatively low separator pressures.

Gas reservoirs with aquifer drive

When a gas reservoir is in contact with an aquifer at the periphery, or below the reservoir, gas recovery efficiency is not as high. Due to water production in wells, recovery from aquifer-driven gas reservoirs can be as low as 50%. This is in contrast to oil reservoirs, where the presence of a strong aquifer usually augments oil recovery.

Summary

Conventional oil reservoirs are primarily produced by natural energy, including rock and fluid expansion, solution gas, gas cap, aquifer drive, and rock compaction. The recovery due to the various natural mechanisms is of varying degree, ranging from a few percent to 80%. In most modern-day reservoirs, primary recovery is followed by secondary and tertiary recovery to boost ultimate recovery.

In most unconventional reservoirs, primary recovery mechanisms are either inadequate or nonexistent to produce petroleum economically. Innovative techniques are required to produce from the reservoirs.

The natural drive mechanisms for conventional oil and gas reservoirs and expected recoveries are shown in Tables 11.1 and 11.2.

Questions and assignments

1. What are the primary drive mechanisms in petroleum reservoirs? What criteria distinguish the primary drive from the secondary drive?
2. What pressure and rate behavior would you expect from a solution gas drive reservoir?
3. Would the performance of a black oil reservoir be any different to a light oil reservoir under solution gas drive given all other factors remain the same?
4. Why do aquifer drive reservoirs generally perform better than solution gas drive and gas cap drive reservoirs? Explain.
5. How might well performance differ between updip and downdip wells in a gas cap reservoir? Why is gas reinjected in updip wells?
6. Based on a literature review, cite an example of combination drive and explain the mechanisms of primary recovery at work. Include the ultimate recovery efficiency from the reservoir.

Table 11.1 Primary drive mechanisms in oil reservoirs

Reservoir type	Primary production mechanism	Estimated recovery	Gas–oil ratio at well	Water–oil ratio at well	Notes
Oil	Liquid and rock expansion	1–5%	None	Insignificant	Predominant mechanism when reservoir pressure is above the bubble point
Oil	Solution gas drive	10–30%	Increases to a maximum, then decreases	Insignificant	Predominant mechanism when reservoir pressure declines below the bubble point
Oil	Gas cap drive	30–40%	High in updip wells. Low in downdip wells	Insignificant	Energy provided by both solution gas and free gas in gas cap
Oil	Aquifer drive	Around 50%	Low	High in downdip wells	Water coning could be an issue for bottom water drive
Oil	Gravity segregation	50–80%	Low in downdip wells, and high in updip wells	Insignificant	Observed in fractured or high vertical permeability rocks. Typical recovery volume is low
Oil and gas	Rock compaction	10% or more	Depends on circumstance	Depends on circumstance	Production declines significantly following rock compaction

Table 11.2 **Primary drive mechanisms in gas reservoirs**

Reservoir type	Primary production mechanism	Estimated recovery	Gas–oil ratio at well	Gas–water ratio at well	Notes
Dry and wet gas	Depletion	70–80%	Entirely gas, no oil	Insignificant	High recovery as gas is highly mobile
Dry and wet gas	Aquifer drive	50–60%	Entirely gas, no oil	Water production could be significant	Water production hampers recovery efficiency

7. Your company has made a new discovery of a volatile oil reservoir with a large gas cap and an aquifer at one edge. Reservoir permeability is quite high. Outline a reservoir development strategy in order to optimize oil recovery.

8. Why do gas reservoirs not perform favorably under aquifer drive like oil reservoirs do?

9. Are conventional and unconventional reservoirs produced by the same drive mechanisms? Compare the rock characteristics, drive mechanisms, development strategies, and reservoir performance between a conventional and unconventional gas reservoir.

10. Would you expect more recovery from a highly fractured reservoir with a water drive or a relatively homogeneous reservoir with depletion drive? Why or why not?

11. How would you determine the optimum timing for implementing secondary recovery? What are the deciding factors to implement secondary and tertiary recovery once the primary drive runs its course?

12. Is coalbed methane produced by a primary drive mechanism? Explain.

References

[1] Clark NJ. Elements of petroleum reservoirs, Henry L. Doherty series. Dallas, TX: SPE of AIME; 1969.

[2] Satter A, Iqbal GM, Buchwalter JA. Practical enhanced reservoir engineering: assisted with simulation software. Tulsa, OK: Pennwell; 2008.

Determination of oil and gas in place: conventional and unconventional reservoirs

12

Introduction

One of the first challenges for reservoir engineers is the estimation of oil and gas in place, chiefly due to the fact that the necessary information for performing such analysis is quite limited in the early stages of the reservoir life cycle. For conventional reservoirs, one of the simplest approaches is the volumetric method. The concept is quite simple and the method does not require any dynamic data from the reservoir. Once the reservoir dimensions are known with reasonable certainty leading to estimation of bulk volume, the pore volume of the reservoir can be determined based on the porosity of rock. The volume of hydrocarbon stored in a compressed state within the pores, referred to as hydrocarbon pore volume (HCPV), can be estimated based on the knowledge of fluid saturation. The above is influenced by the reservoir conditions, namely, pressure and temperature. In fractured formations, certain amounts of hydrocarbons are stored in the fractures that are produced initially.

Finally, the volumes of oil and gas in place under standard conditions at the surface, referred to as original oil in place (OOIP) and gas initially in place (GIIP), respectively, are estimated by using the formation volume factors, which convert reservoir volume to surface volume. Gas expands when produced at the surface while the volume of oil shrinks due to the dissolution of volatile components as pressure is reduced from reservoir conditions to surface conditions.

In conventional reservoirs, oil and gas remain in free and compressed states. However, in certain unconventional reservoirs such as shale gas and coalbed methane, significant quantities of gas are stored in an adsorbed state, in addition to free gas found in micropores and fractures. Hence, classical volumetric calculations outlined earlier lead to the underestimation of gas in place. For coalbed methane, the actual volume of gas stored in rock could be several hundred percent more than what can be determined by volumetric method. Shale gas reservoirs may also contain significant quantities of natural gas adsorbed onto the organic matter, as high as 80% of total gas in place. Hence, the adsorption characteristics of gas also need to be known in order to determine the total volume of unconventional gas.

This chapter deals with the estimation of hydrocarbon in place of conventional and unconventional reservoirs. The mechanisms of storage of hydrocarbon in porous media and volumetric estimation techniques are discussed. The chapter provides answers to the following questions:

- What are the fundamental concepts for estimating oil and gas in place?
- What are the data required to perform an estimate?

Reservoir Engineering. http://dx.doi.org/10.1016/B978-0-12-800219-3.00012-7

- What is the methodology for estimating OOIP and GIIP?
- How is the original hydrocarbon in place evaluated for unconventional reservoirs?
- How reliable are the volumetric estimates of oil and gas?
- How can the technique be used to identify the reservoir areas requiring further recovery efforts?
- Is the hydrocarbon storage mechanism in unconventional reservoirs any different to that of conventional reservoirs?
- What are the key factors in estimating gas in place in unconventional reservoirs?

Volumetric method

The volumetric method for conventional petroleum reservoir analysis is a simple but valuable technique for estimating the quantities of oil and gas in a petroleum reservoir. It is used to estimate the OOIP and GIIP. It is also used to estimate oil and gas reserves based on the knowledge of recovery factor or recovery efficiency from a reservoir.

The method is based on static data of the reservoir; it does not require the knowledge of dynamic reservoir data such as production history of oil and gas. Hence, it is most widely used in the industry to estimate hydrocarbon volumes, including reserves, particularly at the earliest stages of the reservoir life cycle. It must be emphasized that the calculation of OOIP and GIIP is only an estimate. Reservoir engineers and earth scientists strive to attach a high degree of confidence to the estimated oil and gas volumes by collecting and analyzing as much reservoir data as possible. The more accurate but involved techniques for estimating OOIP and GIIP are based on dynamic data of the reservoir. These include the classical material balance method, decline curve analysis, and reservoir simulation studies; the methods are described in Chapters 13, 14, and 15.

The analysis of original hydrocarbon in place, referring to both oil and gas, is the first step in calculating the estimated ultimate recovery or petroleum reserves. Based on the volumetric method, the ultimate recovery can be predicted at the early stages of the reservoir life cycle by analogy if the recovery efficiency is known from similar reservoirs with reasonable certainty. Typical ranges of recovery efficiency of various drive mechanisms in conventional reservoirs are provided in Chapter 11.

Original oil in place

The basis for the volumetric method for estimating the original hydrocarbon in place is the estimation of total HCPV of the reservoir. It is a product of the delineated area of the reservoir, net thickness of the pay zone, rock porosity, and the initial saturation of oil or gas or both. The following terminologies associated with the volumetric estimates are noteworthy:

$$\text{Bulk volume of the reservoir}, BV = \text{Reservoir area} \times \text{thickness} \qquad (12.1)$$

$$\text{Pore volume of the reservoir}, PV = \text{Reservoir area} \times \text{thickness} \times \text{porosity} \qquad (12.2)$$

$$\text{Hydrocarbon pore volume, HCPV} = \text{Reservoir area} \times \text{thickness} \times \text{porosity} \atop \times \text{saturation of hydrocarbon fluid} \tag{12.3}$$

Once HCPV is determined, knowledge of the formation volume factor of oil or gas is needed to convert the volume of oil and gas under reservoir pressure and temperature (reservoir barrels) to the surface volume at standard conditions (stock-tank barrels).

The general equation for estimating the OOIP in conventional oil reservoirs is as follows [1]:

$$\text{OOIP} = \frac{\begin{array}{c}\text{Areal extent of petroleum reservoir} \times \text{net thickness of pay zone} \\ \times \text{formation porosity} \times \text{oil saturation}\end{array}}{\text{Oil formation volume factor}} \tag{12.4}$$

Oil saturation in the reservoir can be calculated from water saturation; the latter is usually obtained from well logs and cores. In a two-phase reservoir where only oil and water are present, the following equation applies:

$$S_{oi} + S_{wi} = 1 \tag{12.5}$$

where S_{oi} = initial saturation of oil, fraction; S_{wi} = initial saturation of water, fraction.

The OOIP is reported in stock-tank barrels and the reservoir area is often referred to in acres. Noting that an acre is equal to 43,560 ft.^{2} and one barrel is equal to 5.615 ft.^{3}, Equation (12.1) can be written in oil field units as follows:

$$\text{OOIP, STB} = \frac{7758Ah\phi(1-S_{wi})}{B_{oi}} \tag{12.6}$$

where A = areal extent of the reservoir, acres; h = net thickness of the oil bearing formation, ft.; ϕ = average porosity, fraction; S_{wi} = initial water saturation, fraction; B_{oi} = oil formation volume factor, rb/STB; STB = stock-tank barrel.

In Equation (12.6), S_w and B_o are evaluated at initial conditions as we are estimating the OOIP at discovery. The thickness in Equation (12.4) refers to the net thickness of oil or gas zone. It is distinguished from the gross thickness of the formation, part of which may have little or no hydrocarbon accumulation. The bottom portion of a formation may be filled with connate water and excluded from the calculations. Oil, gas, and connate water zones across the formation are clearly identified by a suite of logs that are run at a new well.

In addition to stock-tank barrels, oil volumes are reported as follows [1,2]:

- bbls = barrels
- MMBO = millions of barrels of oil
- MMSTB = millions of barrels of oil under stock-tank conditions
- Metric tons = 7.333 bbls

Gas initially in place

For a gas reservoir, the GIIP is estimated as:

$$\text{GIIP, SCF} = \frac{7758 A h \phi (1 - S_{wi})}{B_{gi}} \tag{12.7}$$

Note that, for a dry gas reservoir, $S_{gi} = 1 - S_{wi}$, and the unit of gas formation volume factor is rb/SCF. In certain cases, B_{gi} is reported in ft.^{3}/SCF. Hence, Equation (12.7) is modified as follows:

$$\text{GIIP, SCF} = \frac{43,560 A h \phi (1 - S_{wi})}{B_{gi}} \tag{12.8}$$

B_{gi} can be computed if reservoir pressure, temperature, and gas compressibility factor are known.

$$B_{gi} = 0.02829 \left(\frac{z_i T}{p_i} \right) \tag{12.9}$$

GIIP is customarily reported in MMCF, BCF, or TCF. Note the following:

- MMCF = 10^6 ft.^{3} (million cubic feet)
- BCF = 10^9 ft.^{3} (billion cubic feet)
- TCF = 10^{12} ft.^{3} (trillion cubic feet)

The gas production rate of wells is usually reported in MCFD, i.e., 10^3 ft.^{3} per day.

Isopach, isovol, and isoHCPV maps

The simplest approach to determine the original hydrocarbon in place requires the knowledge of average or weighted values of the thickness of the oil or gas zone, porosity, saturation, and formation volume factor in the previous equations. However, better accuracy is obtained when relevant data from multiple wells are used to produce contour maps of h, $A \times h \times \Phi$ and $A \times h \times \Phi \times S$. These are referred to as isopach, isovol, and isoHCPV maps, respectively.

A typical isopach map showing gross thickness of the formation is presented in Figure 12.1. Isopach maps can also be based on the net thickness of the reservoir, where only oil is present.

Contouring the volume of oil per acre-ft. (area × thickness) of the reservoir is a useful technique for identifying future well locations and targeting improved oil recovery efforts (Figure 12.2).

The deterministic approach works well when the relevant data are obtained from a large number of wells and the reservoir is relatively homogeneous. Reservoirs with

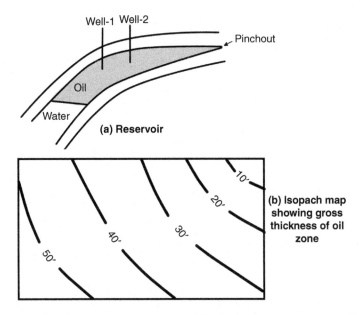

Figure 12.1 Isopach map of a pinchout reservoir. The contours represent gross thickness of the formation. Note the contour values decrease progressively toward the pinchout.

complex geology and limited information may require stochastic methods, among others, to determine the original hydrocarbon in place.

Data requirements

Reservoir data required for estimating the original oil and gas in place include the following.

Conventional reservoirs:

- Area × thickness from log studies, structure and isopach maps
- Porosity from core and log analysis
- Initial fluid saturation from log analysis
- Oil and gas formation volume factor at initial reservoir pressure, derived by laboratory analysis or correlation

Unconventional reservoirs – shale gas and coalbed methane:

- Quantity of adsorbed gas in micropores and rock matrix
- Quantity of free gas in pores and fractures

As noted earlier, the volumetric method requires static data, including fluid saturation and formation volume factor at initial reservoir pressure. Dynamic data such as well production rates, and changes in reservoir pressure and fluid properties, are not required in calculations. The estimation of the original hydrocarbon in place becomes more accurate as wells are drilled and valuable reservoir data are collected and compared against other methods of estimation.

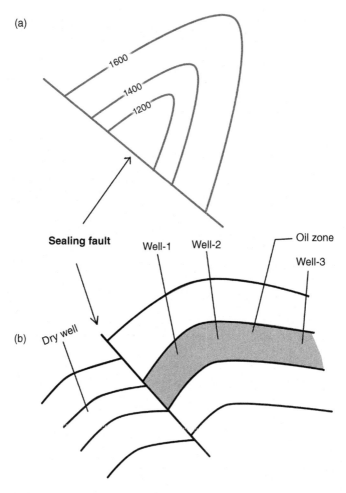

Figure 12.2 A faulted reservoir showing contours of oil volume per acre-ft. Note that no oil was discovered at the other side of the fault. Hence, the contours end abruptly at the fault location.

Estimation of original oil and gas in place

Oil reservoir with a gas cap:

For an oil reservoir with a gas cap, hydrocarbon in place can be estimated by considering the following three components:

- Oil in place
- Volume of gas dissolved in oil
- Volume of free gas overlying the oil zone

As shown earlier in Equation (12.6), the OOIP is estimated as follows:

$$\text{OOIP} = \frac{7758 A h \phi (1 - S_{wi})}{B_{oi}}$$

where A = reservoir area, acres; h = oil zone thickness above the transition zone, ft.; ϕ = reservoir porosity, faction; S_{wi} = irreducible or connate water saturation, fraction of pore volume; B_{oi} = initial oil formation volume factor, rb/STB.

Solution gas volume in the original oil:

$$G_{si} = NR_{si} \tag{12.10}$$

where G_{si} = solution gas in place, scf; N = initial volume of oil; R_{si} = initial solution gas–oil ratio, scf/STB.

Volume of gas in place in the gas cap:

$$G_{gc} = mNR_{si} \tag{12.11}$$

where G_{gc} = gas in gas cap, scf; m = volume of gas cap/volume of oil zone.

Petroleum reserves

Petroleum reserves are not the same as the volume of original oil or gas in place. A recovery factor is needed to estimate the reserves. It can be estimated as follows:

$$\text{Reserves} = \text{OOIP} \times \text{R.F.} \tag{12.12}$$

where R.F. = oil recovery factor, fraction.

Oil recovery factor can be estimated if the values of initial and residual oil saturation are known.

$$\text{R.F.} = \left(\frac{S_{oi} / B_{oi} - S_{or} / B_{or}}{S_{oi} / B_{oi}} \right) \tag{12.13}$$

Similarly, gas reserves can be calculated if the gas recovery factor is known.

The simplest approach to estimate the recovery factor is to use a regional trend from analogous reservoirs. More rigorous approaches include reservoir simulation to predict the estimated ultimate recovery that can be expected in a specific case. Petroleum reserves are discussed in detail in Chapter 23.

Applications of volumetric analysis

Oil reservoir with a gas cap:

Example 12.1

Using the data given below, calculate the pore volume, OOIP and gas in solution.
Reservoir drainage area, acres = 600
Average oil zone net thickness, ft. = 28
Average porosity, fraction = 0.205
Initial water saturation (fraction) = 0.20
Initial oil formation volume factor = 1.56 rb/STB
Solution gas/oil ratio at bubble point, scf/STB = 860
Size of gas cap, fraction of oil volume = 0.12

Solution

Using Equation (12.2), pore volume of the reservoir, rounded to two decimal places, is estimated as follows:

$$PV = 7758 \times 600 \times 28 \times 0.205 = 26.72 \text{MMbbl}$$

Note that 1 MMSTB = 10^6 STB of oil.
The volume of OOIP is estimated by using Equation (12.6):

$$OOIP = [7758 \times 600 \times 28 \times 0.205 \times (1 - 0.20)] / 1.56$$
$$= 13.7 \text{MMSTB}$$

Gas in solution is calculated as follows:

$$G_{si} = (13.7 \times 10^6 \text{ STB}) \times (860 \text{ scf} / \text{STB})$$
$$= 11.78 \times 10^9 \text{ ft.}^3$$
$$= 11.78 \text{BCF}$$

Finally, the volume of gas in the gas cap (G_{gc}) is determined by Equation (12.11):

$$G_{gc} = 0.12 \times 11.78 \text{BCF}$$
$$= 1.41 \text{BCF}$$

Volumetric gas reservoir:

Example 12.2

Using the reservoir data in Example 12.1, calculate the GIIP in a dry gas reservoir. Estimate the gas reserves if the recovery efficiency is known to be about 80% from similar gas fields in the region. The following data are available for connate water saturation and gas formation volume factor.
Connate water saturation, % = 22
Initial gas formation volume factor, rb/SCF = 0.00128

Solution

Using Equation (12.7), the volume of GIIP is estimated as follows:

Initial gas saturation, fraction $= 1.0 - 0.22 = 0.78$
GIIP $= (7758 \times 600 \times 28 \times 0.205 \times 0.78) / 0.00128$
$= 16.28 \times 10^9$ ft.^{3}
$= 16.28$ BCF

The recovery factor assumed in this example is used to calculate reserves.
Estimated gas reserves $= 16.28 \times 0.8 = 13.03$ BCF
Note that a range of values is usually reported for petroleum reserves rather than a single value.

Example 12.3

Consider the following properties of a gas reservoir. Gas is produced by volumetric depletion. Calculate the GIIP assuming a gas compressibility factor of 0.88.
Area of reservoir, acres $= 2000$
Thickness of reservoir, ft. $= 15$
Average porosity, % $= 17$
Connate water saturation, % $= 20$
Reservoir pressure, psia $= 3800$
Reservoir temperature, °F $= 160$

Solution
The first step is to compute the initial gas formation volume factor using Equation (12.9).

$$B_{gi} = 0.02829[0.88(160 + 460)] / 3800$$
$$= 4.062 \times 10^{-3} \text{ ft.}^3 / \text{SCF}$$

The volume of GIIP is calculated as follows:

$$\text{GIIP} = [43,560(2000)(15)(0.17)(1 - 0.2)] / 4.062 \times 10^{-3}$$
$$= 43.75 \text{BCF}$$

Oil reservoir under primary drive:

Example 12.4

Estimate the remaining oil volume and recovery factor of an oil reservoir following depletion. The reservoir does not have a gas cap. The following data are available:
Reservoir drainage area, acres $= 600$
Average oil zone net thickness, ft. $= 28$
Average porosity, fraction $= 0.205$
Connate water saturation, fraction $= 0.20$
Initial oil formation volume factor $= 1.56$ rb/STB
Reservoir pressure at abandonment $= 350$ psi
Oil formation volume factor at abandonment $= 1.07$ rb/STB
Gas saturation at abandonment, fraction $= 0.40$

Solution

The OOIP is calculated as in Example 12.1:

$$OOIP = (7758 \times 600 \times 28 \times 0.205 \times 0.80) / 1.56$$
$$= 13.702 MMSTB$$

At abandonment of a depleted reservoir, evolved gas fills the pore volume from which oil is produced. Oil formation volume factor is reduced to 1.07 as a consequence of gas evolution from the liquid phase.

Volume of oil remaining in the reservoir:

$$= [7758 A h \phi (1 - S_w - S_g)] / B_o$$
$$= [7758 \times 600 \times 28 \times 0.205 \times (1 - 0.2 - 0.4)] / 1.07$$
$$= 9.988 MMSTB$$
$$\text{Recovery factor} = (13.702 - 9.988) / 13.702 = 0.271 \text{ or } 27.1\%$$

Oil reservoir under waterflood:

Example 12.5

Consider the same reservoir dimensions, porosity, and initial water saturation as in Example 12.3. However, the reservoir is subjected to water injection and the reservoir pressure is maintained above the bubble point. Considering the residual oil saturation of 0.3, estimate the recovery factor.

$$OOIP = 13.702 MMSTB$$

Remaining oil volume following waterflood:

$$- [7758 \times 600 \times 28 \times 0.205 \times 0.3] / 1.56$$
$$= 5.138 MMSTB$$
$$\text{Recovery factor} = (13.702 - 5.138) / 13.702 = 0.625 \text{ or } 62.5\%$$

This example highlights the high recovery factor due to waterflooding the reservoir. Note that, due to reservoir pressure maintenance above the bubble point, the formation volume factor is assumed not to change significantly.

Recovery factor based on initial and residual oil saturation:

Example 12.6

Calculate the recovery factor of an oil reservoir based on the following limited information. List your assumptions.
Residual oil saturation = 30%
Connate water saturation = 24%
Initial oil formation volume factor = 2.0 rb/STB

Solution

Let us assume that the reservoir does not have any gas cap and produces under a primary drive mechanism. Furthermore, the oil formation volume factor at the end of recovery is 1.05.

The recovery factor can be estimated by using Equation (12.13). The initial oil saturation is 76%.

$$R.F. = (0.76 / 2 - 0.30 / 1.05) / (0.76 / 2)$$
$$= 0.248 \text{ or } 24.8\%$$

Unconventional gas reservoirs

In certain unconventional reservoirs such as shale gas and coalbed methane, gas may remain in an adsorbed state in the matrix and miniscule pores of rock. Calculation of GIIP requires the estimation of absorbed gas volume, in addition to free gas volume. The differences between conventional and unconventional gas reservoirs are shown in Table 12.1.

The adsorption capacity of shale can be modeled by Langmuir volume and Langmuir pressure [3,4]. The Langmuir isotherm model for shale is presented in Figure 12.3. The adsorbed gas content can be estimated by the following:

$$V_a = \frac{V_L P}{P + P_L} \tag{12.14}$$

where V_a = Volume of adsorbed gas, ft.^{3}/ton; V_L = Langmuir volume; the volume of gas adsorbed at infinite pressure; P = Pressure, psi; P_L = Langmuir pressure; the

Table 12.1 Conventional and unconventional reservoirs: Mechanisms of storage

Reservoir	Mechanism of storage of hydrocarbons	Location of gas	Determination of GIIP
Conventional gas	Gas in free state under compression	Rock pores and fractures in reservoir rock	Requires knowledge of HCPV and application of real gas law
Shale gas (unconventional)	Gas in free state under compression, 15–80%	Rock pores and fractures in source rock	
	Remaining gas in adsorbed state	Rock matrix and micropores in source rock	Requires the estimate of weight of rock and knowledge of Langmuir isotherm characteristic
Coalbed methane (unconventional)	Gas mostly in adsorbed state, up to 98%; adsorbed volume of gas in coalbed methane reservoir may be several times more than the pore volume	Rock matrix and micropores in source rock	

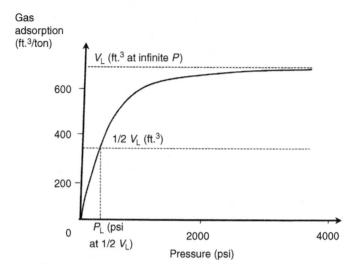

Figure 12.3 Typical plot of gas adsorption volume versus pressure in unconventional shale reservoir [5]. As reservoir pressure decreases, desorption of gas contributes to production.

pressure corresponding to half of Langmuir volume; The Langmuir isotherm for shale can be determined by the desorption test.

Free gas present in shale accounts for 15–80% of total gas; it is dependent on reservoir pressure, saturation, and rock properties including porosity. The amount of free gas, in cubic feet per ton, is calculated as follows:

$$V_f = C \phi_{\text{eff}} (1 - S_w) / \rho_b \tag{12.15}$$

where V_f = volume of free gas, ft.^{3}/ton; ϕ_{eff} = effective porosity, fraction; S_w = connate water saturation, fraction; ρ_b = bulk density, g/cm^3; C = conversion factor, 32.1052.

At relatively high reservoir pressure, free gas dominates the total gas content. However, as pressure declines, free gas is produced readily. Adsorbed gas is released from shale at a relatively slow rate and may hold a significant proportion of total gas at lower pressure. Change in the volume of free and adsorbed gas as a function of declining pressure is presented in Figure 12.4.

Furthermore, shale gas formations can be pervasive and may extend over several hundred miles. Only the "sweet spots" having favorable rock properties are of interest to the operators for drilling new wells. Hence, any attempt to estimate the original hydrocarbon in place is limited to the portion of the shale formation targeted for development.

Stimulated reservoir volume (SRV)

The concept of stimulated reservoir volume (SRV) is important in evaluating the performance potential of wells in ultratight formations such as shale. Following multistage

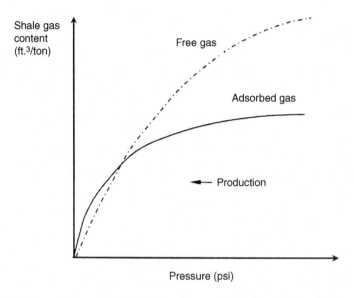

Figure 12.4 Release of free and adsorbed gas volume with decline in reservoir pressure [4].

fracturing of horizontal wells, large and complex network of fractures in multiple planes are created. The size of fracture network based on microseismic studies is represented by stimulated reservoir volume in 3D having enhanced formation permeability that facilitates production. SRV, along with fracture characteristics such as spacing and conductivity, is used to assess the ultimate recovery potential of a well in shale formations. A large stimulated reservoir volume may lead to better production performance of a well.

Case Study: Volumetric Estimation of Coalbed Methane in Place and Reserves

Typical data requirements in calculating gas in place and reserves include the following [6].

Reservoir data:

- Pressure and temperature
- Porosity
- Thickness
- Initial saturations of gas and water

Gas properties:

- Gas gravity and content, including CO_2 and other impurities

Water properties:

- Water formation volume factor

Gas adsorption properties:

- Methane content, Langmuir pressure and volume
- CO_2 content, Langmuir pressure and volume
- Initial gas content

Volumetrics:

- Reservoir drainage area
- Abandonment pressure

The steps involved in the volumetric determination of coalbed methane in place include the following:

- Determine free gas in place based on reservoir volume, porosity and fluid saturation. Convert volume of gas to standard conditions.
- Estimate the initial volume of adsorbed gas by calculating the total weight of rock and using the Langmuir isotherm.
- Calculate the total volume of methane by adding adsorbed and free gas volumes.
- Estimate the remaining volume of adsorbed gas at abandonment pressure from the Langmuir isotherm.
- Calculate the recovery factor based on total volume of gas in place and the remaining volume at abandonment.
- Account for any impurities present, including CO_2, N_2, and H_2S in methane. In case CO_2 is present in gas, use the Langmuir isotherm for CO_2 in estimating the adsorbed volume.

Role of volumetric analysis in integrated reservoir management

Results of volumetric analysis can be compared against the quantities of hydrocarbons recovered to date. The comparison, along with the analyses of well performance and other data, may point to the target areas in a large and complex reservoir where additional development efforts should be concentrated. These may include infill drilling, waterflooding, well recompletion, and enhanced oil recovery. Again, if the original hydrocarbon volumes estimated by other methods such as material balance and decline curve analysis differ substantially, further investigation is required to resolve the discrepancy as the reservoir may not be draining from all the areas due to the presence of unknown heterogeneities.

Summing up

The volumetric method for petroleum reservoir analysis is used to calculate OOIP and GIIP. Oil or gas reserves are also estimated when the recovery efficiency is known with reasonable certainty.

The basis for the volumetric method for estimating the original oil and gas in place is the bulk volume of the reservoir (product of the area and thickness), porosity (void space containing fluids), the initial saturation of oil or gas, and the formation volume factor, which is a factor to convert the reservoir volume of the fluid to standard conditions on the surface.

A deterministic method used to calculate both stock-tank barrels of oil initially in place and original gas in place is based upon a suitably averaged value of each parameter.

A very simple method is to use average or weighted values of thickness, porosity, saturation, and formation volume factor. However, better results can be obtained by using isopach, isovol, and isoHCPV maps.

For unconventional gas reservoirs, the phenomenon of gas adsorption in miniscule pores of rock is taken into account in estimating the volume of gas in place.

Data required for calculation are:

- Area × thickness from log analysis
- Structure and isopach maps
- Porosity from core and log studies
- Fluid saturation from log analysis
- Formation volume factor from lab analysis or correlation

Equations are presented for:

- Oil reservoir with a gas cap, solution gas in the original oil, and gas in place in the gas cap
- Reservoirs under primary recovery drive and waterflooding
- Volumetric gas reservoir

Questions and assignments

1. What are the assumptions and uses for the volumetric method for petroleum reservoir analysis?
2. What are the reservoir properties needed to estimate the original oil and gas in place? What reservoir maps are relevant to the volumetric method?
3. Derive Equation (12.6) by converting Equation (12.4) in oil field units.
4. Modify Equation (12.7) to estimate GIIP in a case where the initial gas formation volume factor is reported as gas expansion factor in cubic feet per scf.
5. List the uncertainties involved in volumetric analysis by including the sources of potential errors.
6. How might the volumetric analysis differ in an unconventional oil or gas accumulation as compared to conventional reservoirs?
7. Can the estimate of OOIP change over the life of the reservoir? Explain.
8. List the factors that are most sensitive to volumetric analysis and estimation of oil and gas reserves.
9. How would you estimate unconventional shale gas in place?
10. Why is information regarding oil–water contact needed to estimate OOIP? How would a long oil–water transition zone affect your estimates?
11. Your company has discovered a new gas condensate field offshore. The formation is highly inclined. One well has been drilled indicating quite high rock permeability and connate water saturation. Describe in detail a plan to estimate the original hydrocarbon in place and reserves. Make reasonable assumptions about the new reservoir and fluid properties.
12. List the various methods in reservoir engineering for estimating initial oil and gas volumes in a reservoir. Does well testing aid in estimating original hydrocarbon in place? Explain.

References

[1] Satter A, Iqbal GM, Buchwalter JA. Practical enhanced reservoir engineering: assisted with simulation software. Tulsa, OK: Pennwell; 2008.

[2] International Energy Statistics. Energy Information Administration; 2014. Available from: www.eia.gov.

[3] Langmuir I. Adsorption of gases on glass, mica, and platinum. J. Am. Chem. Soc. 1918;40:1361.

[4] Course notes: shale gas modeling, reservoir simulation of shale gas and tight oil reservoirs using IMEX, GEM and CMOST. Computer Modelling Group; 2014.

[5] Lewis R, Ingraham D, Pearcy M, Williamson J, Sawyer W, Frantz J. New evaluation techniques for shale gas reservoirs. Available from: http://www.sipeshouston.com/presentations/pickens%20shale%20gas.pdf [accessed 15.01.15].

[6] Available from: www.fekete.com [accessed 10.02.15].

Decline curve analysis for conventional and unconventional reservoirs

13

Introduction

Decline curve analysis, introduced in the 1940s, is one of the most popular methods to date for evaluating the future production potential of oil and gas wells [1,2]. Oil and gas reserves can be estimated by identifying and extrapolating the decline characteristics of wells in a field. The methodology is intuitive, and currently used to evaluate the future production potential of wells in both conventional and unconventional reservoirs based on current trends. As a reservoir is depleted during production, oil and gas wells exhibit an identifiable declining trend in rates that can be extrapolated for the future and analyzed to obtain valuable information.

This chapter describes decline curve analysis and provides answers to the following:

- What is decline curve analysis? What are the decline curve models?
- What are the advantages and limitations of decline curve analysis?
- What are the decline curve analysis methods for conventional and unconventional reservoirs?
- What valuable information can be obtained from decline curve analysis?
- How is a specific decline rate model identified graphically?

Decline curve analysis: advantages and limitations

The advantages of decline curve analysis are as follows:

- Decline curve analysis is a quick and intuitive method to predict future production rates and ultimate recovery. In certain cases, reservoir engineers perform analysis of hundreds of wells in a short period of time.
- The approach is based on empirical models that are simple yet powerful. Graphical techniques are used to match production rates and extrapolate in the future.
- Recent advances in decline curve analysis include the recognition of various flow regimes in a complex geological setting such as ultratight shale with induced and natural fractures.
- Decline curve analysis may involve implementation of multiple models in various stages of production in order to predict future performance accurately.
- The analysis may readily lead to the estimation of the cumulative well production until the economic limit for the well is reached.
- As monthly and annual production volumes are predicted, cash flow analysis for the well or the field can also be performed with ease.

Reservoir Engineering. http://dx.doi.org/10.1016/B978-0-12-800219-3.00013-9

- The method is applied not only for individual wells; in many cases, the aggregate declining trend of the entire field can be analyzed. The ultimate recovery from the entire oil or gas field, and petroleum reserves, can be estimated when all the producing wells are included in the analysis.
- Based on an identifiable trend, future water cut in a well can also be predicted.
- In the case where the well exhibits an unexpected trend, further analysis can be performed about the well and the reservoir. For example, oil production from a new reservoir may not show an appreciable decline. A strong water drive may be suspected, among other factors.
- Decline curve analysis is not resource intensive in comparison to reservoir simulation. The analysis can be conducted in a relatively short period of time, often with the aid of software applications available in the industry.

Assumptions of decline curve analysis

Traditional decline curve analysis, as applied to conventional reservoirs, is based upon a number of assumptions as follows:

- The well is produced by depletion drive alone. Water or gas injection, influx of water from an adjacent aquifer, or the presence of a gas cap usually influences production rate in a manner that a decline may not be identifiable.
- The well produces from its own drainage area without any interference from nearby wells. The flow regime is referred to as boundary dominated flow.
- The well produces at a constant bottom-hole pressure. In reality, such a condition may not be observed.

Limitations

The method, although straightforward and transparent, is applicable only when the well production rate is declining with an identifiable trend. The analysis requires sufficient well rate data ranging from several months to a year to predict future performance with confidence. In many cases, however, a definitive decline trend is not identifiable. This is due to the fact that management of oil reservoirs involves fluid injection as part of pressure maintenance operation. Other factors include two-phase flow of oil and gas, stimulation, hydraulic fracturing, operational issues, well recompletion, perforation to produce from a different layer, and water breakthrough.

Again, many wells produce under rate constraints where production rate remains the same without any decline for a long period of time. Hence, more robust methods, such as reservoir simulation are required to analyze well and reservoir performance.

With the development of unconventional resources including shale gas reservoirs, traditional decline curve analysis is found to be inadequate to estimate ultimate recovery or reserves. Fluid flow characteristics of shale gas can be quite different to that of conventional gas production. Shale has ultralow permeability and production takes place through an extensive and complex network of induced and natural fractures. Existence of various flow regimes (linear, transient, boundary dominated) during the productive life of the well is important to recognize as described later in the chapter.

As observed frequently, the decline trend of wells producing from shale formations changes significantly after the initial period of production. Extrapolation of initial decline characteristics to the economic limit of the well in the future may result in overestimation or underestimation in ultimate recovery.

Objectives

Decline curve analysis may provide the following information:

- A definitive trend that can be identified from well and field production data
- Future oil and gas rates
- Expected ultimate recovery (EUR) from the wells
- Economic life of the well and field
- Field reserves
- Predicted oil cuts and water cuts in a well
- Identification of flow regimes
- Analysis of reservoir characteristics based on production data

Decline curve models

Decline curve models are empirical, and predict the future well rates based upon past performance. In order to do so, the model equation requires a best fit to the existing production data by determining one or more unknown coefficients in the equation by graphical or mathematical techniques. The types of well rate decline models widely known in the petroleum industry are as follows [1,2]:

- Exponential
- Hyperbolic
- Harmonic

The above are collectively referred to as the Arps model. In recent times, other models have been proposed as the traditional models have been found to be inadequate in the analysis of unconventional reservoirs. For example, shale gas reservoirs having ultralow permeability and complex fracture network exhibit decline trends that are quite different than conventional reservoirs (Table 13.1). The following models have gained prominence in analyzing the decline of shale gas production:

- Stretched exponential decline model (SEDM) [3]
- Duong model [4]
- Multisegment model based on the implementation of more than one model at various stages of decline [5]

Theoretical background and working equations

During primary depletion, well production declines at different rates for oil and gas wells depending on reservoir characteristics, including storativity, transmissibility, and the presence of fractures and other types of heterogeneities, among others. Rock

Table 13.1 **Decline curve models for conventional and unconventional reservoirs**

Model	Number of unknown coefficients	Applicability
Exponential	1	Traditionally used for conventional reservoirs where boundary dominated flow is observed
Hyperbolic	2	As above
Harmonic	1	As above
SEDM	2	Works for transient flow encountered in ultralow permeability fractured formation such as a shale gas reservoir
Duong	3	As above
Multisegment model	>2	Decline history is analyzed by more than one model; typical in unconventional shale gas decline curve analysis

characteristics that influence reservoir performance are described in Chapter 3. The traditional decline curve models are frequently referred to as the Arps model based on his paper published in 1945 [1]. A generalized equation correlating the decline rate with production rate and time can be expressed as follows:

$$D = kq^b = -\frac{1}{q}\frac{dq}{dt} \qquad (13.1)$$

where D = instantaneous decline rate, 1/day; q = well production rate, MCF/day or bbl/day; t = time period of production, days; k, b = empirical constants depending on well decline characteristic.

Note that the consistent units of time are used in the above equation; for example, if t is in days then q is in MCF/day and D is in 1/day.

Based on the value of b, traditional decline curve analysis is classified into three types of rate decline as follows:

- Exponential decline: $b = 0$
- Hyperbolic decline: $0 < b < 1$
- Harmonic decline: $b = 1$

Note that a modified hyperbolic decline model is used in modern decline curve analyses to represent the linear flow at early times from fractured formations, where $b > 1$. The value of b is obtained by the best fit of the decline curve model with production data. Exponential and harmonic declines are viewed as special cases of hyperbolic decline. Note that the value of b can exceed unity in certain cases, including linear and transient flow through fractures. The phenomenon is typically observed in unconventional shale gas reservoirs for the entire productive life of a well due to the ultralow permeability of the formation. Use of the traditional hyperbolic model with $b > 1$ in modeling transient flow may lead to unrealistic reserve estimates. Hence, the flow regime requires a more rigorous approach for analysis as described later in the chapter.

Exponential decline

In the case of the exponential decline model, $b = 0$ in Equation (13.1), which can be integrated between initial and current time as follows:

$$q = q_i e^{-Dt} \qquad (13.2)$$

where q = oil or gas rate at time t; q_i = initial rate.

The exponential decline model is also known as constant rate decline; flow rates q_1, q_2, and q_3 recorded at equal time intervals t_1, t_2, and t_3 can be shown to hold the following relationship:

$$\frac{q_2}{q_1} = \frac{q_3}{q_2} = \ldots = \frac{q_n}{q_{n-1}} = e^{-D} \qquad (13.3)$$

where n = total number of time intervals.

Typically, the time interval is in months. Many oil wells are found to follow an exponential decline pattern although an enhanced decline rate may be observed at the initial stages of production. A semilog plot of production rate versus time would yield a straight line for wells undergoing an exponential decline in rate. The cumulative production of a well over time t under exponential decline is obtained by integrating Equation (13.2):

$$Q = \int_0^t q_i e^{-Dt} \, dt \qquad (13.4)$$

$$Q = \left(\frac{q_i}{D} \right)(1 - e^{-Dt}) \qquad (13.5)$$

where Q = cumulative production of oil or gas.

Combining the above with Equation (13.2), it can be shown that:

$$Q = \frac{q_i - q}{D} \qquad (13.6)$$

If two rates, q_1 and q_2, are known at time intervals t_1 and t_2, respectively, the value of D can be calculate as follows:

$$D = \left(\frac{1}{t_2 - t_1} \right) \ln \left(\frac{q_1}{q_2} \right) \qquad (13.7)$$

Equation (13.7) is obtained by taking the natural logarithm on both sides of Equation (13.2).

The value of D can also be obtained graphically. Equation (13.6) can be rewritten as follows:

$$q = q_i - (D \times Q) \tag{13.8}$$

Hence, a plot of q versus Q has a slope of $-D$ and an intercept of q_i.
The EUR can be expressed as follows:

$$EUR = \frac{q_i - q_f}{D} \tag{13.9}$$

where q_f = final rate at abandonment.

In case the exponential decline commences after a period of time (usually a few to several months), q_i should reflect the well rate where the decline is first identified. Equation (13.9) must be modified to account for the cumulative production Q_i that occurred prior to the onset of exponential decline as follows:

$$EUR = Q_i + \left(\frac{q_{i,exp} - q_f}{D} \right) \tag{13.10}$$

The decline characteristic of a large number of wells can be represented by the exponential decline model. It represents the behavior of an incompressible fluid producing from a bounded reservoir under a pseudosteady-state flow regime at constant bottom-hole pressure. However, note that hyperbolic and harmonic decline models are purely empirical and not supported by the physics of flow in porous media.

Hyperbolic decline

The decline of certain wells cannot be predicted by the exponential decline model. In such cases, a value of b between 0 and 1 is assumed, which leads to a better fit of well production history. Hence, Equation (13.1) is integrated to obtain the following:

$$q = q_i (1 + b D_i t)^{-1/b} \tag{13.11}$$

where D_i = initial decline rate.

In the exponential decline model, the value of D is constant; however, in the hyperbolic model, the value of D changes with time. An expression for cumulative production can be written as follows, following integration of rate of production over time in a similar manner shown earlier for exponential decline:

$$Q = \frac{q_i^b (q_i^{1-b} - q^{1-b})}{(1-b)D_i} \tag{13.12}$$

It can be further shown that the EUR is as follows:

$$\text{EUR} = \frac{q_i^b (q_i^{1-b} - q_f^{1-b})}{D_i(1-b)} \tag{13.13}$$

Harmonic decline

For certain wells, a harmonic decline model describes the production pattern more accurately. In this case, $b = 1$. Well rate, cumulative production, and EUR can be expressed as follows:

$$q = \frac{q_i}{1 + D_i t} \tag{13.14}$$

$$Q = \frac{q_i}{D_i} \ln\left(\frac{q_i}{q}\right) \tag{13.15}$$

$$\text{EUR} = \left[\frac{q_i}{D_i \ln(q_i / q_f)}\right] \tag{13.16}$$

Any production that occurred before must be added to the EUR calculations based on Equations (13.13) and (13.16).

Figure 13.1 compares the exponential decline patterns with two values of D. As Equation (13.2) suggests, higher value of D leads to a faster decline in rate. As a result,

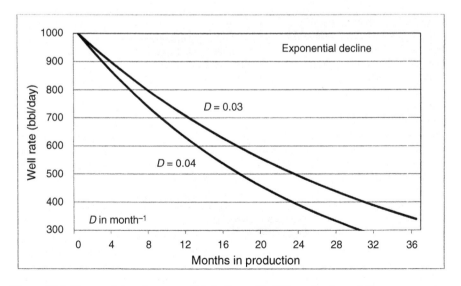

Figure 13.1 Comparison of exponential decline with different values of D.

the estimated EUR is less. In the case of hyperbolic decline, a higher value of b leads to slower decline and greater EUR given all other coefficients remain the same.

Method of identification

In the case of exponential decline at a constant rate, Equation (13.9) suggests that a plot of well rate versus cumulative production would be a straight line as shown in Figure 13.2. A semilog plot of rate versus time would also indicate a straight line. Since a straight line can be easily extrapolated, exponential decline curves are most commonly utilized wherever applicable. On the contrary, when the decline rate is not exponential, curvatures in rate versus cumulative plot are observed for hyperbolic and harmonic decline patterns. Rate versus cumulative production for a well under hyperbolic decline is presented in Figure 13.3. Finally, harmonic decline pattern is identified by plotting cumulative production against the log of well rate, which also yields a straight line (Figure 13.4). The above relationship can be deduced from Equation (13.15).

Table 13.2 summarizes the diagnostic plots to identify the Arps models.

Example 13.1

Decline curve analysis of oil production in a conventional reservoir.
The following production data are available from an oil well producing under depletion drive.

- Identify the model that best fits the data.
- Calculate future production rates and cumulative volumes produced.
- Estimate the reserves assuming an abandonment rate of 50 bbl/day.
- Calculate the life of the well.

Months in production	Well rate (bbl/d)
0	1000
1	978
2	958
3	937
4	915
5	894
6	876
7	857
8	838
9	820
10	803
11	784
12	768

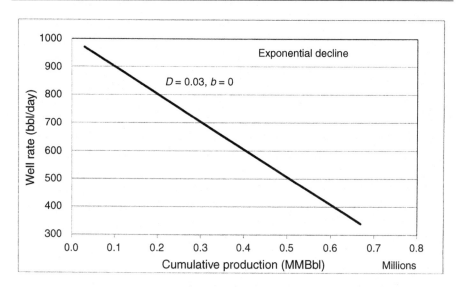

Figure 13.2 Well rate versus cumulative production in exponential decline.

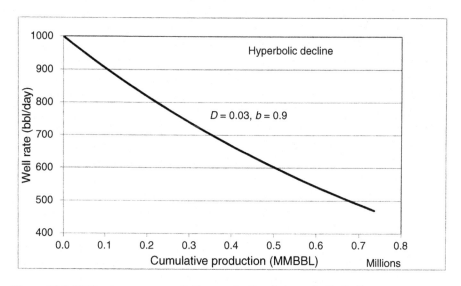

Figure 13.3 Well rate versus cumulative production in hyperbolic decline.

Step 1. The decline trend is identified by the procedure described in Table 13.2. First, a rate versus cumulative plot is drawn to determine whether the decline is exponential. In this case, a straight line is obtained. The plot is similar to Figure 13.2.

In case a straight line cannot be identified in step 1, cumulative production is plotted against the log of well rate. If a straight line is obtained, the decline is harmonic. If neither plot yields a straight line, the decline is hyperbolic. Curve fitting techniques can be used to obtain the values of coefficients b and D_i.

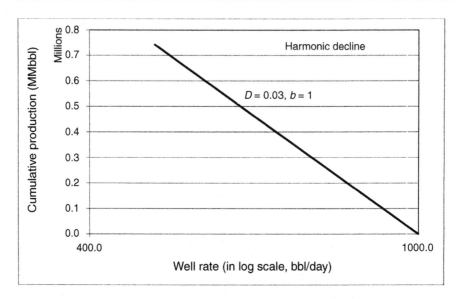

Figure 13.4 Well rate versus cumulative production in log scale for harmonic decline.

Table 13.2 Identification of Arps models

Model	Plot	Notes
Exponential	A plot of well rate versus cumulative production indicates a straight line (Figure 13.2). A semilog plot of rate versus time would also indicate a straight line.	
Hyperbolic	Curvature in rate versus cumulative plot is observed (Figure 13.3). Curve fitting methods based on regression analysis can be used to obtain best fit.	Curvature is also observed for the case of harmonic decline.
Harmonic	A plot of cumulative production against log of well rate yields a straight line (Figure 13.4).	

Step 2. Next, the value of D is calculated based on Equation (13.7):

$$D = [1/(1-0)]\ln(1000/978) = 0.022 \text{month}^{-1}$$

Step 3. The future production rates can be obtained by using Equation (13.2). For example, the rate at the end of 24 months is calculated as follows:

$$q = 1000 e^{-0.022*24} = 590 \text{bbl}/\text{day}$$

Step 4. The cumulative volumes can be forecast by converting D to per day and using Equation (13.6). At the end of 24 months, the cumulative volume produced is calculated as follows:

$$Q = (1000 - 590)/(0.022/30.4) = 566,545 \text{bbls}$$

Step 5. Oil reserves are calculated by using Equation (13.9):

$$EUR = (1000 - 50)/0.022 = 1,295,454\,bbls$$

Note that the reserves or the EUR can be obtained graphically by extrapolating the straight line in rate versus cumulative plot to the economic limit. The plot is prepared in step 1.

Step 6. Finally the economic life of the well is calculated by rearranging Equation (13.7) as follows:

$$
\begin{aligned}
t &= (-1/D)\ln(q/q_i) \\
&= (-1/0.022)\ln(50/1000) \\
&= 136\,months
\end{aligned}
$$

Modern software applications prepare the necessary plots and perform all the calculations for decline curve analysis, which allows the analyst to spend more time in focusing on the validity of data, applicability of a specific model to the production characteristic, and scenario building with multiple models.

Decline curve analysis for unconventional reservoirs

Modern decline curve analysis recognizes the influence of flow regimes on well rate decline that can be encountered at various stages of well life. In a hydraulically fractured well completed in ultralow permeability formation, the initial production is dominated by linear flow through fractures, while the transient flow can continue for years without observing the effects of drainage boundary. The traditional models, proposed by Arps [1] in the early part of twentieth century, were based on oil wells producing from conventional reservoirs where the "best fit" is obtained by overlaying the rates predicted by one of the three models on production data. The extrapolation of decline curve was based on the assumption that pseudosteady-state or boundary dominated flow is observed. However, well production characteristics from the unconventional reservoirs can be quite complex due to rock properties (ultralow permeability), formation characteristics (presence of natural fracture network), well geometry (long horizontal wells), and stimulation (multistage fracturing). For example, linear flow can be observed in hydraulically fractured wells. Furthermore, the decline pattern may change following the early stages of gas production. In unconventional shale gas reservoirs, wells exhibit a trend where initial rate of decline is high for a few weeks or months, followed by the attainment of exponential decline in later years of well life. The rapid decline in rates at the early stages of the well occurs as initial production is dominated by the limited quantity of gas stored in the network of fractures and stimulated portions of the reservoir in the vicinity of the well. The decline is rapid as the flow of gas from less permeable areas of formation cannot compensate for the relatively high rate of production from fractures and stimulated zones. Hence, initial production of shale gas occurring for a few months to about a year is followed by production that occurs at a much slower pace, which may last as long as 10–20 years or more. Hence, in determining the EUR from shale gas wells, only the decline trend identified after about a year of production is used.

Flow regimes

Decline curve analysis from unconventional shale gas reservoirs may identify one or more of the fluid flow regimes:

- Linear flow: This regime indicates the flow of gas through fractures during the initial stages of production. Linear flow regime exhibits the following relationship between flow rate and time:

$$q = q_i \, t^{-\frac{1}{2}} \tag{13.17}$$

  Hence, the linear flow regime is identified by plotting rate against time on a log–log plot where a negative half-slope is the characteristic slope. A match with production data can also be obtained by using the Arps model when the value of b is assumed to be 2. Note that a negative quarter-slope may also be observed in case of less conductive fractures; the above equation is modified accordingly.
- Transient flow: The flow regime can be observed when the effect of drainage boundary is not felt.
- Boundary dominated flow: As the name implies, the flow regime is indicative of the effects of boundary of stimulated reservoir volume on production. The regime typically occurs at the later part of well life. Some wells producing from very low permeability reservoirs such as shale may not exhibit this regime at all.

Recent models proposed to analyze gas production from hydraulically fractured shale gas production include the following.

Stretched exponential decline model

The empirical model is useful in ultralow permeability reservoirs where well production exhibits transient flow characteristics. The well can be horizontal with multistage fractures. In this approach, the value of b varies over the life of the well. It predicts well rate as a function of initial rate, time, and two other parameters, τ and n, as follows:

$$q = q_i \exp\left(\frac{-t}{\tau}\right)^n \tag{13.18}$$

where q_i = initial rate or peak rate, Mscf/month; τ = characteristic time, months; n = exponent, dimensionless.

Parameters τ and n can be obtained from a log–log plot of $\ln(q_i/q)$ versus t. Equation (13.18) is recast in the following form:

$$\ln\left(\frac{q_i}{q}\right) = \left(\frac{t}{\tau}\right)^n \tag{13.19}$$

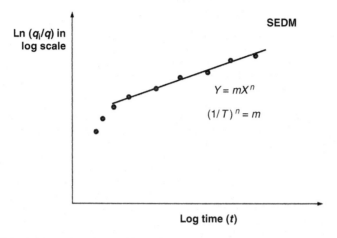

Figure 13.5 ln (q_i/q) against time t plotted in log-log scale.

Hence, a log–log plot of $\ln(q_i/q)$ against t yields a slope of $(1/\tau)^n$ as the straight line drawn through the points has the form (Figure 13.5)

$$y = mx^n \tag{13.20}$$

where

$$m = \left(\frac{1}{\tau}\right)^n \tag{13.21}$$

The characteristic time can be computed from the known values of m and n. Again, parameters n and τ can be obtained from a group of similar wells in the same reservoir.

The cumulative production can be calculated by integrating the rate over time as follows:

$$Q = \left(\frac{q_i^\tau}{t}\right)\left[t\left(\frac{1}{n}\right) - \tau\left\{\left(\frac{1}{n}\right) - \left(\frac{t}{\tau}\right)^n\right\}\right] \tag{13.22}$$

Duong model

As stated earlier, decline characteristic of transient fracture flow can be expressed as in the following:

$$q = q_i t^{-n} \tag{13.23}$$

where $n = \frac{1}{2}$ of linear flow. The volume of gas produced, G_p, can be obtained by integrating the above equation over the time period:

$$G_p = q_1 \left(\frac{t^{1-n}}{1-n} \right) \tag{13.24}$$

In the above equation, q_1 is the production rate on the first day. Finally, the following expression can be deduced to relate G_p and q with time:

$$\frac{q}{G_p} = \frac{1-n}{t}$$

It can be shown that a log–log plot of q/G_p against t yields a straight line of the following form (Figure 13.6):

$$\frac{q}{G_p} = at^{-m} \tag{13.25}$$

where a = intercept; m = slope.

Coefficients a and m can be obtained by regression analysis where the best fit of the straight line to field data is obtained. In unconventional shale gas reservoirs, the values of both the coefficients generally vary between 1.1 and 1.3.

The well rate is calculated based upon the initial rate, production period, and two coefficients, a and m, that are determined graphically. The equation can be written as:

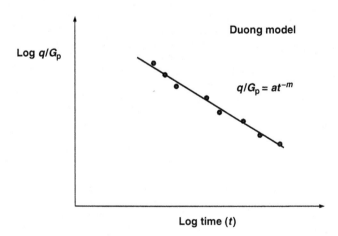

Figure 13.6 Log (q/G_p) versus log (t).

$$q = q_1 t^{-m} \exp\left[a \frac{t^{1-m}-1}{1-m}\right] \tag{13.26}$$

Cumulative production is calculated by the following:

$$G_p = \left(\frac{q_1}{a}\right) \exp\left[\frac{a(t^{1-m}-1)}{1-m}\right] \tag{13.27}$$

In the above equations, the value of q_1 can be obtained by plotting q versus $t(a,m)$, and drawing a best fit line through the data points (Figure 13.7). The expression of $t(a,m)$ has the following form:

$$t(a,m) = t^{-m} \exp\left[\frac{a(t^{1-m}-1)}{(1-m)}\right] \tag{13.28}$$

Duong proposes the following steps to perform decline curve analysis:

- Data validation, which includes screening for outliers due to skin effects and choked production, among others
- Determination of coefficients a and m by plotting log of q/G_p versus log time and obtaining the best fit by regression analysis
- Determination of q_1 by plotting well rate against $t(a,m)$ on log–log scale and obtaining best fit
- Production forecast, including the prediction of q and G_p over time (Figure 13.8)

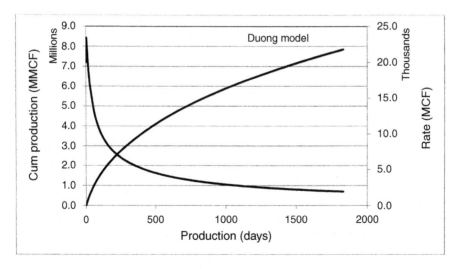

Figure 13.7 Cumulative production and rate versus time.

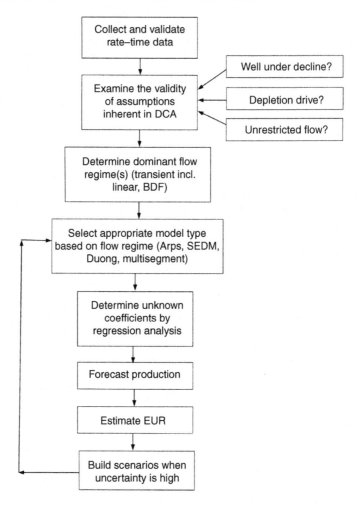

Figure 13.8 Decline curve analysis workflow.

Multisegment decline analysis model

Certain software applications allow the implementation of the Arps model in various stages of well production, including transient flow and boundary dominated flow where the values of coefficient b are varied to obtain the best match. Decline in production is analyzed in two to three segments to obtain a best fit. A common approach is to use the modified hyperbolic model initially and then switch to the exponential model when the decline rate in the hyperbolic model reaches a limiting value, for example, 5%. The exponential model is then extrapolated until the economic limit of the well in order to estimate the EUR (Figure 13.9).

An example of multisegment model application is presented in Table 13.3.

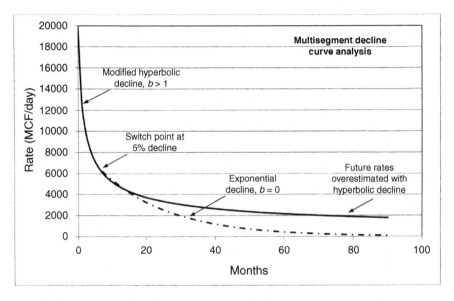

Figure 13.9 Multisegment decline curve analysis.

Table 13.3 **Multisegment decline curve analysis**

Segment	Model type	Notes
Initial	Modified hyperbolic decline	$b > 1$; fracture dominated transient flow
Final	Exponential decline	$b = 0$; exponential decline is extrapolated until the economic limit is reached

Estimation of EUR in shale gas reservoirs: a general guideline

A literature review on estimating EUR in shale gas reservoirs points to the following.

- Production rate characteristic from shale gas is likely to demonstrate more than one trend. Rate data obtained from the initial period of production lasting from a few months to a year or more may not be suitable for determining the EUR.
- Production from ultratight reservoirs having a large fracture network is dominated by transient flow regime. SEDM and Duong model provide more accurate predictions in case of transient flow.
- Use of modified hyperbolic decline with $b > 1$ to model transient flow regime is practiced widely; however, extrapolating the model to the economic limit may lead to inaccurate estimation of EUR.
- In cases where the boundary dominated flow regime is observed at later stages of well life, exponential or other classical models may be used.

- Studies have indicated that the SEDM may be more conservative in predicting EUR than the Duong model.
- In many cases, a limiting decline rate of 5% is implemented in order to avoid the overestimation of EUR. The time to switch models is a function of initial decline rate, limiting rate of decline, and coefficient b.

Decline curve analysis workflow

Decline curve analysis is based on a few simple steps as follows (Figure 3.8).

- Review production mechanism (pure depletion without any external injection or water influx) and well history (producing under full capacity, steady bottom-hole pressure, free of operational issues, and others).
- Determine whether the underlying assumptions of decline curve models are valid.
- Collect and validate well production data.
- Screen out any anomalous data; investigate the probable reasons for any anomaly
- In the case of conventional reservoirs, prepare diagnostic plots to identify the appropriate model (exponential, hyperbolic, harmonic). In the case of ultratight shale gas reservoirs, apply the models that are developed to represent transient flow through the fracture network (SEDM, Duong).
- Determine the need for using more than one model in various segments of production history, specifically for unconventional reservoirs having multiple flow regimes.
- Select the portion of data that is amenable to analysis and interpretation.
- Obtain the best fit with field data by regression analysis.
- In the case of the Arps model, obtain values of coefficients b, D, or D_i.
- In the case of other models, obtain values of appropriate coefficients.
- Forecast production rate over time until economic limit is reached.
- Calculate the EUR.

Decline curve analysis of coalbed methane

Production rate of coalbed methane (CBM) exhibits a trend that is quite different to the production behavior of both conventional gas and unconventional shale gas. Typical production characteristics of CBM are described in Chapter 22. Initially, the well produces a large amount of water that is stored in coal cleats. At this stage, the rate of CBM production is relatively low. The rate increases slowly over a few months during the dewatering phase. The reservoir pressure is reduced and CBM is desorbed from the matrix. The rate of production of CBM reaches a peak and finally declines. The declining portion is then analyzed to estimate EUR (Figure 13.10). A literature review indicates that both exponential and hyperbolic decline have been used to obtain best fit with the field data.

Type curve analysis: an overview

Type curve analysis is an advanced method of traditional decline curve analysis where field data are matched against a set of type curves to obtain the best fit. Various

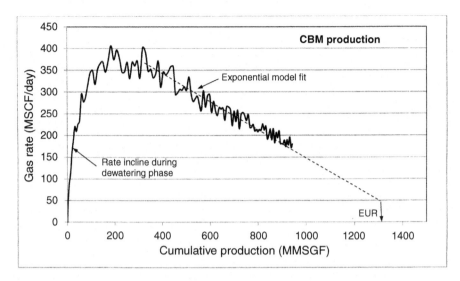

Figure 13.10 Analysis of CBM production data. Plot of rate versus cumulative can be extrapolated by a straight line if exponential decline is found to be best fit. With an abandonment rate of 50 MSCF/day, estimated EUR is about 13.2 MMSCF. Note that harmonic or hyperbolic decline leads to higher EUR.
Courtesy: IHS–Fekete.

software applications are available that perform the procedure quickly and transparently. The method was proposed by Fetkovich [6]. The analysis is based on a family of curves where dimensionless rate is plotted against dimensionless time. The type curves combine transient flow regime with boundary dominated flow regime that appears at later stages. Once a match is obtained with one of the curves, the coefficients b, D, or D_i are obtained to calculate the future production rates and reserves. Additionally, formation permeability and skin around the wellbore and drainage area can be obtained by the analysis. Various other type curves are also available, including Blasingame [7] and Agarwal–Gardner [8].

Summary

Oil and gas production rates decline with time during primary recovery as the natural energy to produce a reservoir depletes. In most cases, the declining trend can be represented by relatively simple empirical models. Based on the models, decline curve analysis techniques are used to predict future production trend and EUR from conventional and unconventional reservoirs. When the well has produced for a relatively long period of time, an analysis can be performed with a certain degree of confidence.

The decline curve analysis is applicable only when production is declining. In case the reservoir is produced under the influence of water influx or fluid injection, or when the well is producing under a predetermined rate, decline curve analysis is not applicable.

Traditionally, three types of rate decline were used to analyze well performance, namely, exponential, hyperbolic, and harmonic. The models are based on the assumptions of boundary dominated flow and pseudosteady-state flow regime. However, in ultralow permeability formations such as shale where horizontal wells are drilled combined with multistage hydraulic fracturing, transient flow period may dominate for the life of the well. Hence, various other decline curve models have been introduced in the industry, which represent the production characteristics more accurately. Notable are power law, SEDM, and the Duong model, among others.

The unknown coefficients that must be solved for each decline curve model along with the applicability of the model are summarized in Table 13.4.

Table 13.4 Summary of decline curve models

Model	Unknown coefficients	Applicability	Notes
Exponential	Constant decline rate D in Equation (13.1). Decline rate is the change of well rate with time divided by the rate.	Conventional and unconventional reservoirs. Observed in a large number of wells, particularly during the latter part of well production.	Valid for boundary dominated flow and assumes constant bottom-hole pressure.
Hyperbolic	Initial decline rate, D_i, and coefficient b in Equation (13.1), where $0 < b < 1$		A modified hyperbolic decline model is used to represent transient flow in the case of unconventional reservoirs where $b > 1$.
Harmonic	Initial decline rate, D_i, in Equation (13.1)		
SEDM	Characteristic time τ and exponent n in Equation (13.19)	Ultralow permeability reservoirs with fracture network. Fits well with shale gas production characteristics	Transient flow
Duong		As above	Transient flow. Need to switch model for boundary dominated flow
Multisegment model		Wide range of reservoirs. Various models can be used for initial and later portions of well production data.	

Questions and assignments

1. What is decline curve analysis and what are the objectives of analysis?
2. Why is decline curve analysis widely practiced in the industry?
3. What are the assumptions of decline curve analysis? Under what circumstances can decline curve analysis be applied?
4. List the information that can be obtained by conducting a decline curve analysis.
5. Distinguish among exponential, hyperbolic, and harmonic decline. How can the decline pattern of a well be identified?
6. The initial production from a well is 2000 bbls/day. The well is identified to be producing under hyperbolic decline. Perform a sensitivity analysis by assuming various values of the coefficients b and D_i in forecasting production.
7. Why are traditional models inadequate in modeling decline from unconventional reservoirs having low permeability and induced fractures?
8. What flow regime is typically encountered in shale gas reservoirs? Explain.
9. Why does well rate fall steeply in ultratight reservoirs during the initial period of production? Why might the rate data from the initial period lead to significant inaccuracies in estimating reserves?
10. What is type curve analysis and what advantages does it offer over traditional decline curve analysis?
11. Analyze the following rate data from an unconventional gas well and calculate EUR by assuming a suitable economic limit. Explain what model or models you have used to analyze the rate data and provide justification.

Days in production	Rate (MCF)
1	20,000
30	16,680
60	13,050
90	11,045
120	9,699
150	8,715
180	7,970
210	7,350
240	6,840
270	6,427

References

[1] Arps JJ. Estimation of decline curves. Trans. AIME 1945;160:228–47.
[2] Arps JJ. Estimation of primary oil reserves. Trans. AIME 1956;207:182–91.
[3] Freebom R, Russell B. How to apply stretched exponential equations to reserve evaluation. SPE 162631. Society of Petroleum Engineers: Dallas (TX); 2012.
[4] Duong AN. How to apply stretched exponential equations to reserve evaluation. CSUG/SPE 137748. Society of Petroleum Engineers: Dallas (TX); 2011.
[5] Decline Plus. Available from: http://www.fekete.com/SAN/WebHelp/FeketeHarmony/Harmony_WebHelp/Content/HTML_Files/Performing_an_Analysis/Decline_Analysis/Using_Segments.htm [accessed 28.09.14].

[6] Fetkovich MJ. Decline curve analysis using type curves. J. Petrol. Technol. 1980; June: 1065–77.

[7] Blasingame TA, McCray TL, Lee WJ. Decline curve analysis for variable pressure drop/ variable flowrate systems. SPE 21513. Paper presented at the Society of Petroleum Engineers Gas Technology Symposium. January 23–24, 1991.

[8] Agarwal RG, Gardner DC, Kleinsteiber SW, Fussell DD. Analyzing well production data using combined type curve and decline curve analysis concepts. SPE 57916. Paper presented at the 1998 Society of Petroleum Engineers Annual Technical Conference and Exhibition; September 1998. New Orleans. p. 27–30.

Reservoir performance analysis by the classical material balance method

14

Introduction

The material balance method is a valuable tool used by reservoir engineers to analyze and predict the performance of oil and gas reservoirs. The method is based on the fundamental concept that mass can be neither destroyed nor created. It is more detailed than decline curve analysis, yet simpler than full-fledged reservoir simulation requiring substantial resources.

This chapter describes the formulation and application of material balance models for various types of reservoirs and answers the following questions:

- What are the applications of the material balance method in reservoir engineering? How is reservoir performance predicted based on this method?
- How is the reservoir modeled in the material balance method?
- How is the material balance equation (MBE) formulated?
- How are the graphical techniques used in the analyses?
- Does the material balance method contribute to reservoir characterization?
- What types of reservoirs are suitable to use this approach?
- What are the assumptions and limitations of the method?
- What are the data requirements?

Applications of the classical material balance method

The classical material balance method is used to analyze various important aspects of oil and gas reservoirs as follows:

- Estimation of original oil and gas in place
- Assessment of natural producing mechanisms, including gas cap drive, solution gas drive, and water drive
- History matching of past performance of the reservoir
- Prediction of future reservoir performance

Basis for the material balance method

The material balance method, as the name implies, is based upon the law of conservation of mass or material in a given system, a reservoir in our case. The reservoir is modeled as a "tank" from which fluids are withdrawn or injected into (Figure 14.1). As fluids, namely, oil, gas, and water, are produced from a reservoir, the following effects are observed since the fluids and the porous rock are compressible:

Reservoir Engineering. http://dx.doi.org/10.1016/B978-0-12-800219-3.00014-0

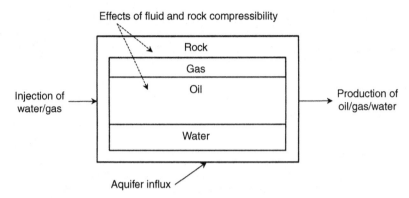

Figure 14.1 Depiction of reservoir "tank" model for the classical material balance method.

- Expansion of oil and dissolved gas
- Expansion of free gas in gas cap
- Expansion of formation water
- Reduction in pore volume of rock
- Water encroachment from adjacent aquifer

In essence, the net volume of fluids, namely, oil, gas, and water, produced from the reservoir within a specified period can be equated to the sum of the volume changes of fluids and rock within the reservoir as shown in Table 14.1.

In compact form, the MBE can be presented as follows [1,2]:

Table 14.1 Material balance of reservoir fluids and rock

Volume of fluids produced or injected (rb)			Changes in fluid and rock volume within reservoir (rb)	
Oil (produced)	$N_p B_o$		Oil and solution gas (expansion)	NE_o
Gas (produced)	$+ N_p(R_p - R_s)B_g$		Free gas in gas cap (expansion)	$+mNE_g$
Water (produced)	$+ W_p B_w$	$=$	Formation water (expansion)	$+NE_{fw}$
Water (injected)	$- W_i B_w$		Pore volume (reduction)	
Gas (injected)	$- G_i B_g$		Aquifer influx	$+W_e$

Notes: N_p = volume of oil produced, STB; B_o = oil formation volume factor, rb/STB; R_p = cumulative gas–oil ratio, scf/STB; R_s = dissolved gas in oil, scf/STB; B_g = gas formation volume factor, rb/STB; W_p = cumulative volume of water produced, STB; B_w = water formation volume factor, rb/STB; W_i = cumulative volume of water injected, STB; N = original oil in place, STB; E_o = expansion of oil and dissolved gas, rb/STB; m = fraction of initial gas cap volume over oil volume, rb/rb; E_g = expansion of free gas in gas cap, rb/STB; E_{fw} = expansion of free water and reduction in pore volume of rock, rb/stb; W_e = cumulative water influx, rb.

$$F = N(E_o + mE_g + Ef_w) + W_e \tag{14.1}$$

where

$$F = N_p B_o + N_p (R_p - R_s) B_g + W_p B_w - W_i B_w - G_i B_g \tag{14.2}$$

$$F = N_p [B_t + (R_p - R_{si})] B_g + W_p B_w - W_i B_w - G_i B_g \tag{14.3}$$

In Equation (14.3), R_{si} is the initial solution gas ratio and B_t is the two-phase formation volume factor defined as follows:

$$B_t = B_o + (R_{si} - R_s) \tag{14.4}$$

Furthermore, the terms related to fluid expansion and pore volume compaction in Equation (14.1) can be expanded as follows:

$$E_o = (B_o - B_{oi}) + (R_{si} - R_s) B_g \tag{14.5}$$

$$E_g = B_{oi} \left(\frac{B_g}{B_{gi}} - 1 \right) \tag{14.6}$$

$$E_{fw} = (1 + m) B_o \left(\frac{c_w + S_{wi} + c_f}{1 - S_{wi}} \right) \Delta p \tag{14.7}$$

where c_w and c_f are the water and formation compressibilities, respectively, and Δp is the observed pressure drop in the reservoir. Subscript i denotes the value of a parameter at initial reservoir conditions. Lastly, cumulative water encroachment from an adjacent aquifer can be modeled as:

$$W_e = US(p,t) \tag{14.8}$$

In the above equation, U is the aquifer constant in rb/psi and $S(p,t)$ is an aquifer function.

In the material balance method, appropriate aquifer models with varying degrees of complexity are used to predict the reservoir performance [3]. In the simplest case, the model assumes a small (pot) aquifer and uses an equation for influx, which is not dependent on time. More involved models assume steady-state flow and unsteady-state flow from the aquifer into the reservoir.

Assumptions and limitations

The important assumptions in formulating the MBE include the following:

- The oil or gas reservoir is modeled as a "homogeneous tank," i.e., rock and fluid properties are the same throughout the reservoir.
- Fluid production and injection occur at single production and single injection points, respectively.
- The analysis is independent of the direction of fluid flow in the reservoir.

However, petroleum reservoirs are inhomogeneous to varying degrees and reservoir fluids flow in definite directions. Furthermore, production and injection wells are drilled at different times and locations. Nevertheless, the material balance method can be used in reservoirs where the degree of reservoir complexity is not excessive. Result obtained from the material balance method can be compared with that obtained by other methods such as simulation, which would enhance the confidence in the analysis.

Requirements of data for analysis

The following are the data requirements in analyzing oil and gas reservoirs by the material balance method:

- Reservoir fluids data: Oil and gas formation volume factor, solubility of gas in oil, fluid compressibility
- Rock characteristics: Formation compressibility
- Reservoir production history: Cumulative volumes of oil, gas, and water produced, gas–oil ratio, and decline in reservoir pressure
- Reservoir drive mechanism: Solution gas, depletion, gas cap, aquifer influx, and combinations thereof.

Applications of the material balance method in oil and gas reservoirs

This section describes the techniques in using MBE to estimate the various reservoir properties by simple graphical techniques [4]. The MBE can be rewritten in the form of a straight line ($y = mx + c$) under certain simplifying assumptions, and can be solved for reservoir properties by noting the slope and intercept. Most of the present day analysis is based on material balance software applications available in the industry.

Oil reservoirs: estimation of the original oil in place, gas cap ratio, aquifer influx, and recovery factor

FE method

The FE method is used for solution gas drive reservoirs to estimate the values of N and m. We can rewrite Equation (14.1) in further compact form:

$$F = NE_t + W_e$$ (14.9)

where

$$E_t = N(E_o + mE_g + E_{fw})$$ (14.10)

If the effects of the aquifer are not significant, $W_e = 0$. Furthermore, $m = 0$ for reservoirs without a gas cap.

Hence, Equation (14.9) reduces to:

$$F = NE_t$$

A plot of F versus E_t yields a straight line with slope N, the original oil in place (Figure 14.2).

For a solution gas reservoir with a gas cap, the value of m is greater than zero. Graphical techniques can still be used by assuming reasonable values of m until a straight line is obtained (Figure 14.3).

Gas cap method

As the name suggests, the method is used for reservoirs with a gas cap. Additionally, the original oil in place can be estimated by the gas cap method. Neglecting the contributions of W_e and E_{fw}, Equation (14.1) can be simplified as:

$$\frac{F}{E_o} = N + mN\left(\frac{E_g}{E_o}\right)$$ (14.11)

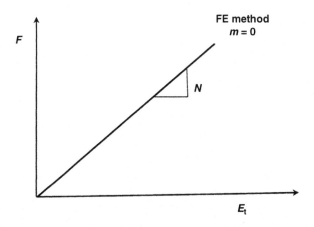

Figure 14.2 Plot illustrates the FE method to estimate the original oil in place.

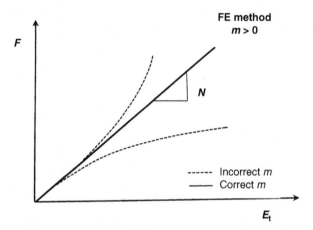

Figure 14.3 FE method to estimate the original oil in place and size of gas cap.

Again, the equation has the form of a straight line. Plotting F/E_o versus E_g/E_o yields a straight line with slope mN and intercept N.

Havlena and Odeh method

The method can be used to estimate the original oil in place for reservoirs with aquifer influx.

When $W_e > 0$, Equation (14.1) takes the following form:

$$\frac{F}{E_t} = N + \frac{W_e}{E_t} \tag{14.12}$$

$$\frac{F}{E_t} = N + U\left(\frac{S}{E_t}\right) \tag{14.13}$$

A plot of F/E_t versus S/E_t yields the original oil in place as the intercept N. The slope of the line is U, which is the aquifer constant (Figure 14.4).

Campbell method

This method can be applied by plotting F/E_t versus F, which yields a straight line. N is determined from the intercept of the straight line.

Besides the above methods, reservoir pressure may be plotted against oil produced by adjusting N, m, and W_e in Equation (14.1). The best match with the field data can be used to estimate the reservoir parameters.

The recovery factor of a solution gas drive reservoir with no gas cap or water influx can be estimated by simplifying the MBE when fluid properties and producing gas–oil

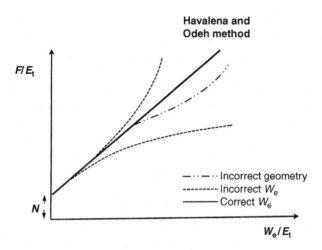

Figure 14.4 Illustration of the Havlena and Odeh method.

ratio are known. The following equations can be obtained for estimating the recovery factors above and below bubble points, respectively:

$$\frac{N_p}{N} = \left(\frac{B_{oi}}{B_o}\right) c_e \Delta p \tag{14.14}$$

$$\frac{N_p}{N} = \left[\frac{B_t - B_{ti}}{B_t + (R_p - R_{si})B_g}\right] \tag{14.15}$$

where B_t = two-phase formation volume factor, rb/STB; c_e = effective compressibility of oil, water, and rock, psi^{-1}.

$$\frac{N_p}{N} = \left(\frac{c_o S_o + c_w S_w + c_f}{1 - S_{wi}}\right) \tag{14.16}$$

Gas reservoirs: estimation of the gas initially in place and aquifer influx

In the absence of a liquid phase, Equation (14.1) reduces to the following in the case of dry gas reservoirs:

$$F = G(E_g + E_{fw}) + W_e \tag{14.17}$$

where

$$F = G_p B_g + W_p B_w$$

G_p = volume of gas produced, Mscf; B_g = gas formation volume factor, rb/Mscf; G = gas initially in place, Mscf; $E_g = B_g - B_{gi}$, rb/Mscf; $E_{fw} = B_{gi} C_e \Delta p$; C_e = effective compressibility that accounts for the compressibility of gas, water, and formation, psi^{-1}.

Plot of p/z versus G_p

The graphical method is quite simple and popular in estimating gas initially in place and gas reserves. For gas reservoirs under depletion drive, W_e and E_{fw} are assumed to be negligible, and Equation (14.17) can be simplified to the following form:

$$\frac{p}{z} = \left(1 - \frac{G_p}{G}\right)\left(\frac{p_i}{z_i}\right) \tag{14.18}$$

A plot of p/z versus G_p yields the estimate of gas reserves as depicted in Figure 14.5. Field data obtained in the early stages are extrapolated to abandonment. A value for the reservoir pressure at abandonment of the field is assumed in the analysis. The method has the limitation in that the static pressure can only be obtained by shutting down production for a long period. Furthermore, the method does not yield a straight line when aquifer influx influences static pressure data.

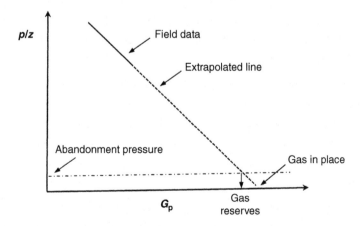

Figure 14.5 Plot of p/z versus G_p leading to the estimate of gas reserves and gas initially in place.

Havlena and Odeh method

This method estimates the gas initially in place for gas reservoirs under aquifer drive in a manner analogous to what is used for oil reservoirs. Equation (14.17) can be written in the following form:

$$\frac{F}{E_t} = G + U\left(\frac{S}{E_t}\right)$$

Values of F/E_t are plotted against W_e/E_t assuming various values of W_e until a straight line is obtained. The intercept of the plot is the estimate of G.

Pressure match method

The pressure match method applies to any type of gas reservoir. The methodology is analogous to that of oil reservoirs presented earlier.

Gas condensate reservoirs: estimation of wet gas in place

The same methodology can be applied to gas condensate reservoirs for the estimation of wet gas in place. In this case, Equation (14.1) can be modified as:

$$F = G_w(E_g + E_{fw}) + W_e \qquad (14.19)$$

where G_w = wet gas in place, scf.
   F can be evaluated as:

$$F = G_{wp}B_{gw} + W_pB_w \qquad (14.20)$$

where

$$G_{wp} = G_{dp} + N_{pc}F_c \qquad (14.21)$$

G_{wp} = cumulative wet gas production, Mcf; G_{dp} = cumulative dry gas production, Mcf; N_{pc} = cumulative condensate production, stb; F_c = condensate conversion factor, Mcf/stb; B_{gw} = formation volume factor of wet gas, rb/Mscf.

Role of material balance analysis in reservoir characterization

Reservoir drive mechanisms can be identified by the material balance method by various plotting techniques described earlier. Only the correct mechanism of reservoir drive should yield a straight line in an appropriate plot.

The material balance method to estimate the original hydrocarbon in place is based on dynamic reservoir and fluid data including cumulative volumes of oil and gas production, gas–oil ratio and formation volume factors. When the results are compared against that of a volumetric method, any significant inconsistency would warrant further investigation of the reservoir in terms of unknown reservoir heterogeneities and other uncertainties.

Summing up

The material balance method is a powerful technique in determining the important properties of a petroleum reservoir, including the estimates of original oil in place and gas initially in place, size of gas cap, and the strength of aquifer influx. It also provides an insight into the drive mechanisms at work, such as solution gas, water influx, and gas cap. The method is more involved than decline curve analysis and is applicable where decline curve analysis is inadequate, such as two-phase flow and a clear declining trend is not identifiable. However, the material balance method is significantly less resource intensive than that required for detailed reservoir simulation studies, which is an advantage when the reservoir is not overly complex and quick engineering and management decisions are sought.

The material balance method is based on the concept of the "tank model." The oil or gas reservoir is assumed to be a large tank where the volume of fluids withdrawn or produced from the reservoir is equal to the sum of all fluids injected plus the expansion of liquids and contraction of rock volume within the reservoir. Added to this is the effect of any water influx from adjacent aquifer. Expressed as an equation, the following describes the fundamental basis of the material balance method:

Volumes of reservoir fluids produced = Volumes of fluid injected (if any) + expansion of fluids within the reservoir + contraction of rock pore volume + water influx from aquifer (if any)

The material balance technique employs simple graphical techniques to estimate the reservoir parameters based on the following equation presented earlier:

$$F = N(E_o + mE_g + Ef_w) + W_e \tag{14.1}$$

In the above equation, N (original oil in place), m (ratio of gas cap size to oil volume), and W_e are the unknown parameters and the rest are known from reservoir production history, and rock and fluid properties. Equation (14.1) is simplified for various types of reservoir, and solved graphically as shown in Tables 14.2 and 14.3. It is interesting to note that reservoir porosity and permeability do not enter the calculation procedure directly.

The material balance method makes important assumptions regarding the reservoir characteristics (reservoir is homogeneous) and production method (production, as well as injection, occurs through a single point). In reality, reservoirs are heterogeneous and wells are drilled at different points in time and location. Nevertheless, the

Table 14.2 Classical material balance techniques for oil reservoirs

Reservoir type	Assumptions	Plot	Results	Notes
Solution gas drive without gas cap	$m = 0$, $W_e = 0$	F vs. E_t	N as slope	
Solution gas drive with gas cap	$m > 0$, $W_e = 0$	F vs. E_t	N as slope, m by trial and error	Values of m are assumed until a straight line is obtained.
Gas cap drive	$m > 0$, $W_e = 0$	F/E_o vs. E_g/E_o	N as intercept, mN as slope	
Aquifer water drive	$m = 0$, $W_e > 0$	F/E_t vs. W_e/E_t	N as intercept, W_e by trial and error	Values of W_e are assumed until a straight line is obtained.
Aquifer water drive	$m = 0$, $W_e >= 0$	F/E_t vs. F	N as intercept	A horizontal line is plotted when $W_e = 0$.
Any reservoir		P vs. N_p	N, m, and W_e are obtained by best match	Good knowledge of the reservoir is needed as three parameters are unknown.

Table 14.3 Classical material balance techniques for gas reservoirs

Reservoir type	Assumptions	Plot	Results	Notes
Depletion drive	$W_e = 0$	p/z vs. G_p	G by extrapolating the plot to x-axis	Gas reserves can be obtained when abandonment pressure is known.
Aquifer water drive	$W_e > 0$	F/E_t vs. W_e/E_t	G as intercept	Values of W_e are assumed until a straight line is obtained.
Aquifer water drive	$W_e >= 0$	F/E_t vs. F	G as intercept	Plot is linear when $W_e = 0$.

method is well recognized for estimating of oil and gas in place and providing useful insight into reservoir drive mechanisms under appropriate conditions.

The material balance method utilizes dynamic reservoir data, including production history and changes in fluid properties, to estimate the original oil in place and gas initially in place. The results can be compared to that obtained by the volumetric method, which relies on static data. If the two results are inconsistent, further studies of the reservoir are warranted to identify any unknown heterogeneities or processes involved.

Questions and assignments

1. What is the usefulness of the classical material balance method? Describe the reservoir parameters that can be estimated by the method.
2. Define the law of conservation of mass in the context of producing an oil and gas reservoir.
3. What is the contribution of Havlena and Odeh in the material balance method?
4. What are the limitations of the material balance method? Under what conditions might the application of material balance techniques not be appropriate? Cite an example.
5. What are the different methods used in graphical techniques for oil and gas reservoirs? How would you apply the technique in a gas condensate reservoir?
6. Deduce Equations (14.14) and (14.15) by making necessary assumptions.
7. Calculate the recovery factors of a solution gas drive reservoir based on the following data. Explain your methodology.

Initial reservoir pressure, psia	2640
Bubble point pressure, psia	1840
Initial oil formation volume factor, rb/STB	1.35
Oil formation volume factor at bubble point, rb/STB	1.365
Initial water saturation, fraction	0.26
Compressibility of water, psi^{-1}	3.08e-6
Formation compressibility, psi^{-1}	3.6e-6
Initial solution gas/oil ratio, scf/STB	1025
Oil formation volume factor*, rb/STB	1.028
Gas formation volume factor*, rb/STB	0.00128
Solution gas/oil ratio*, scf/STB	225
Producing gas/oil ratio*, scf/STB	1550

*At abandonment (900 psia)

8. How can reservoir characterization studies be performed with the aid of the material balance method?
9. You have recently performed a material balance analysis for a stratified oil reservoir that has been producing a moderate amount of oil with high water cut for a few years. Several wells are completed in various locations and layers based on reservoir quality and oil–water contact. The estimate of original oil in place provided by the method is about a third lower than what was estimated by the volumetric method earlier. Describe the sources of discrepancy between the two estimates. Propose your next course of action to resolve the discrepancies and improve your understanding of the reservoir.
10. Can the material balance method be applied to unconventional gas reservoirs? Why or why not?

References

[1] Havlena D, Odeh AS. The material balance as an equation of straight line. J. Petrol. Technol. 1963;896–900. August.

[2] Havlena D, Odeh AS. The material balance as an equation of straight line – part II, field cases. J. Petrol. Technol. 1964;815–22. July.

[3] Wang B, Teasdale TS. GASWAT–PC: a microcomputer program for gas material balance with water influx. SPE Paper #16484. Presented at the Petroleum Industry Applications of Microcomputers meeting. Society of Petroleum Engineers: Del Lago on Lake Conroe (TX); June 23–26, 1987.

[4] Satter A, Iqbal GM, Buchwalter JA. Practical enhanced reservoir engineering: assisted with simulation software. Tulsa, OK: Pennwell; 2008.

Petroleum reservoir simulation: a primer

15

Introduction

In the digital age, reservoir simulation holds the key for the formulation of the overall management strategy of petroleum reservoirs including development, production and technical analysis. Simulation is meant to replicate real-world processes or events based on physical or digital models. Reservoir simulation emerged as an invaluable tool for the petroleum industry in the later part of the twentieth century as powerful computers became available. Capital investments in exploration, drilling, and development of oil and gas fields can be staggering, specifically in cases of the development of offshore and unconventional reservoirs. Improved oil recovery (IOR) operations, including waterflood and enhanced oil recovery (EOR), require huge expenditure. Moreover, the risks and uncertainties involved in oil and gas ventures are quite significant, which require sound methodology in predicting the future performance of a reservoir. Reservoir simulation attempts to minimize the uncertainties by predicting reservoir performance based on best case, worst case, and most likely scenarios. Compared to the total investments required in oil and gas development, the resources needed for simulation are quite small. Hence, major engineering decisions and economic analyses are based on the outcome of reservoir simulation studies. In fact, reservoir simulation is required by law in some countries to prove the viability of oil and gas ventures and establish the assets of a company.

Advances made in computing technology in both hardware and software have enabled the analysts to utilize highly sophisticated and robust models, which combines state of the art computational techniques, dashboards, and virtual reality in 3D. This chapter provides a reservoir simulation primer for reservoir engineers and demonstrates how it can be used as a decision-making tool. The chapter attempts to answer the following:

- What is reservoir simulation and why it is a valuable tool for reservoir engineers?
- What is a reservoir model? How is it built?
- What is the mathematical basis for the reservoir models?
- How are the reservoir simulation models classified? What are the types of simulation models? What are the salient features?
- What are the data requirements for simulation? How are data collected?
- What is history matching in simulation studies?
- What is sensitivity analysis?
- How is reservoir simulation implemented in various types of reservoirs?
- What are the limitations of reservoir simulation?

Reservoir Engineering. http://dx.doi.org/10.1016/B978-0-12-800219-3.00015-2

Objectives of reservoir simulation

Reservoir simulation attempts to replicate real-world processes and events that take place in the reservoir based on available information. A typical simulation study forecasts well production rates, water–oil ratio, and gas–oil ratio with time. Reservoir pressure and fluid saturations at various locations and periods are predicted. Simulation studies also replicate the fluid phase changes (vaporization and condensation) that occur with changes in reservoir pressure and affect reservoir performance. The studies indicate optimum number of wells that need to be drilled and their locations. The best time to initiate secondary and tertiary recovery operations is studied by generating what-if scenarios.

Reservoir simulation efforts can be quite intensive depending on the complexity of the reservoir (stratified, fractured, faulted, and others), mode of recovery (primary, secondary, or tertiary) and the answers sought for the study (pressure, well rate, water breakthrough from thief zone, flow through fractures, etc.). Reservoir simulation predicts the performance of a well as well as the entire field. It provides a time frame for economically operating the field. Confidence in prediction is achieved by matching the results of simulation with production history of the reservoir that is available at the time of study.

Typically, a reservoir simulation study seeks one or more of the following:

- Oil and gas in place estimates, and periodic updates based on new information
- Recovery efficiency of a reservoir under various development scenarios, including infill drilling, waterflooding, and EOR operations
- Economic analysis and rate of return from the venture over the life of the reservoir
- Evaluation of marginal reservoirs, where small changes in reservoir performance can make or break a project
- Suitability of a particular well at a given location to attain maximum productivity
- Well completion in selective layers to avoid water production
- Design and location of well patterns, including injectors and producers
- Design and capacity planning of surface facilities
- Spacing between wells to optimize production and maximize economic value
- Horizontal well design in terms of length, trajectory, and number of laterals
- Optimization of well rates in terms of high oil rate, low water–oil ratio (WOR) and low gas–oil ratio (GOR)
- Effects of gas evolution or water influx on recovery
- Optimization of waterflood based on the pattern, location, and rate of injection wells
- Optimization of EOR operations, including thermal, chemical, and other methods
- Areas of high oil saturation left after primary or secondary production
- Deliverability of a gas reservoir
- Optimization of gas injection in a gas condensate reservoir
- Study of water coning in a well followed by corrective action
- Forecast of the potential of a well following workover or stimulation
- Effects of reservoir stratifications, fractures, faults, pinchouts, and other heterogeneities on reservoir performance
- Investigation of flow of oil and water across various regions within the reservoir as well as across geologic layers
- Identification of root causes of a problematic well performance

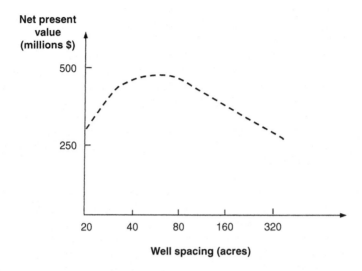

Figure 15.1 Optimization of well spacing based on reservoir simulation.

An example of reservoir simulation efforts to optimize well spacing in an oil reservoir is shown in Figure 15.1. Based on a reservoir model, various oil recovery scenarios are simulated by changing the well spacing and future waterflood pattern. Closely spaced wells result in an increase in production but require more capital investments. The outcome of simulation, coupled with economic analysis, is dependent on reservoir quality and estimate of reserves, among other factors. Poor reservoir quality, including low transmissibility and significant reservoir heterogeneities, would require the drilling of more wells for optimal performance.

There are many other uses of reservoir simulation, each serves a specific purpose and aids in formulating an overall reservoir management strategy.

Reservoir simulation at various phases of reservoir life cycle

It must be emphasized that reservoir simulation is a continuous process throughout the reservoir life cycle and serves as an important tool in managing the reservoir. At the early stages of field development, data needed for simulation are quite limited; hence, various scenarios are generated for the field by varying the range of input parameters, such as the extent of the reservoir, interwell permeability, and connectivity between adjacent strata and bottom water influx, within the realm of possibilities. For instance, Figure 15.2 depicts the high, low, and most likely range of pressure response during primary production of an oil reservoir when the effects of an aquifer are not known with certainty. Presence of a weak aquifer having negligible water influx may require an early implementation of water injection in order to maintain reservoir pressure and improve oil recovery.

As the field is produced over months or years, a definitive picture begins to emerge about the reservoir in terms of well performance, rock heterogeneities, and reservoir

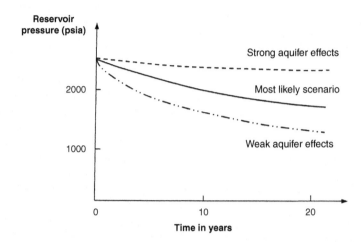

Figure 15.2 Simulation of likely scenarios of reservoir pressure over time due to varying degrees of water influx.

quality. The production history is then looped back to the reservoir model to adjust the various parameters used in the simulation study. The efforts of history matching go a long way in improving the accuracy and reliability of performance prediction. How any discrepancy between historical production data and the reservoir performance predicted by reservoir models would be resolved may also pose a challenge.

Reservoir modeling

A typical reservoir simulation model is built on cells or grid blocks that contain the local description of the reservoir pertaining to the grid block, and the properties of various fluid phases, namely, oil, gas, and water. Reservoir models predict fluid phase pressure, saturation, well production rate, and bottom-hole pressure (BHP), among others, over the life of the reservoir or for a specific period of time. The basis for simulation is a mathematical representation of the entire system, which integrates reservoir characteristics and fluid properties. Relevant equations for fluid flow in porous media are used for the purpose. The fluid flow and related equations used in the models are mostly nonlinear requiring iterative solutions.

Sound knowledge of the geological aspects of a reservoir and thorough understanding of fluid flow characteristics are requisites of meaningful simulation studies. Earliest attempts of reservoir modeling involved physical models such as a sand-filled tank for conducting various studies. Later on, pilot-scale projects based on a small portion of the reservoir gained popularity in designing EOR operations. As robust computers were developed to perform quite a large number of calculations at lightning speed, modeling of petroleum reservoirs entered the digital age.

However, it is neither possible nor practical to capture all the rock heterogeneities, specifically in microscopic scale, and their effects on fluid flow in a reservoir model. Most simulation studies are based on the approach that the model should be capable of matching reservoir production history on a well-by-well basis and predict future performance with a reasonable degree of certainty. In order to reduce uncertainties in prediction, modern simulation studies may involve multiple realizations of the reservoir based on geostatistical models.

The salient features of a typical reservoir model are as follows:

- A reservoir model is a representation of the reservoir based on grid cells of various shapes (regular, irregular) and sizes (large, small), and in 1D, 2D, 3D, or radial.
- Each cell is assigned static reservoir properties, such as thickness, elevation, porosity, and absolute permeability. Interblock fluid transmissibility, an important parameter in simulation, is calculated based on the grid block dimension, permeability of rock, fluid viscosity, and other properties.
- One or more regions may exist in the model; each region may be assigned a different set of relative permeability characteristics, oil–water contact, and other properties.
- During simulation, fluid phase pressure and saturation in each cell vary according to well operational characteristics (constant flow, constant bottom-hole pressure) and reservoir boundary conditions (constant pressure, noncommunicating).
- Changes in pressure and saturation, and phase transfer between liquid and gas, are influenced by the PVT properties of fluids.

Development of a reservoir model

Reservoir model development is based upon both static and dynamic data pertaining to the reservoir. A static component of the reservoir model is built from geological and geophysical studies, including the reservoir structure and rock characteristics. Changes in fluid saturation, pressure, and composition are handled by the dynamic component. Typical reservoir models include geologic description and known heterogeneities, existing and future locations of injectors and producers, reservoir pressure, oil–water and gas–water contacts, well completions and operating constraints, and fluid properties. Some models account for changes in fluid composition in detail. Depending on the maturity of the field, part of the reservoir simulation data is gathered from the field; the rest can be based on correlations, experience, regional trends, and valid assumptions.

Classification of reservoir simulation models

Reservoir simulation models can be categorized in various ways. Major classifications are outlined in the following:

- Model geometry (1D, 2D, 3D, radial)
- Number of fluid phases studied (oil, gas, water)
- Reservoir processes (phase and compositional changes of reservoir fluid, thermal, nonthermal)
- Target of study (individual wells, specific reservoir area or sector, entire field)

In each case, however, reservoir simulation is based upon an integrated model that includes both static (geology, geophysics, etc.) and dynamic (fluid pressure, saturation, etc.) aspects of the reservoir.

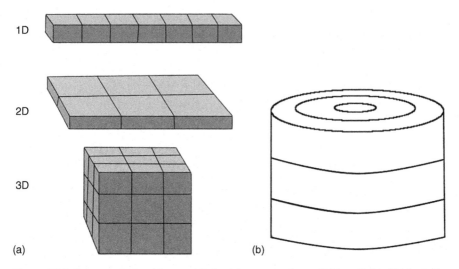

Figure 15.3 Representation of reservoir by (a) rectangular and (b) radial grid blocks in a model.

Model geometry

In contrast to the analytical methods that provide continuous solution of unknown variables, numerical methods provide discrete solutions at particular points in time and location in porous media. Hence, reservoir simulation models are typically based on hundreds of thousands of grid blocks where the solution for fluid flow is obtained in each block or cell in a discrete manner. Large reservoir models consist of a million cells or more, each cell containing a solution of pressure and saturation of fluid phase at a point in time. Based on geometry, reservoir simulation models can be classified as 1D, 2D, 3D, and radial models. In each case, the reservoir or its part is represented by grid blocks depicted in Figure 15.3a and b. The grid blocks are rectangular, cubic or radial; however, other shapes are also used that are thought to align better with flow geometry. With each added dimension, models become progressively large and complex. 1D and 2D models are usually much faster to run. 1D models may be used to understand a complex fluid displacement process. 2D models can be either areal, along x- and y-axes, or cross-sectional, along x- and z-axes. Certain laboratory studies can also be simulated using simpler 1D and 2D models. However, most field simulation studies are carried out in three dimensions, as the dynamic characteristics of fluid flow in lateral as well as vertical directions need to be simulated and analyzed correctly. However, a robust 3D model built on a very large number of regular or irregular cells is resource intensive in terms of computer memory and run time. Last but not least, well centric simulation efforts, such as the study involving water or gas coning, are performed by using a radial model as the flow around a well is predominantly radial.

In certain cases, fluid pressure and saturation need to be studied in fine detail in certain localized areas whereas larger grid sizes are adequate for the rest of the model.

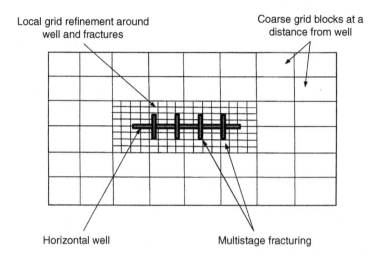

Figure 15.4 LGR to capture the finer details of horizontal well and multistage hydraulic fracture behavior in a shale gas reservoir.

Examples include areas around a horizontal well and in the vicinity of hydraulically created fractures. In order to accurately predict reservoir behavior, local grid refinement (LGR) is implemented in the model (Figure 15.4).

Since most reservoirs are structurally complex to varying degrees, regular grid blocks cannot represent the reservoir adequately. Irregular geologic features often include curved faults, pinchouts, compartments, dips, and odd-shaped boundaries. In cases where a more accurate description of the reservoir is needed, perpendicular bisector (PEBI) grids are used to simulate fluid flow more accurately. PEBI grid blocks are unstructured and may have a variety of shapes. The gridding scheme can be customized to fit any reservoir geometry. The grid system is also capable of representing flow of fluid around single- and multilateral horizontal wells in greater details. Grid orientation effects are also reduced, as the shapes of the blocks do not have the limitation of regular grids where the edges of blocks are perpendicular to each other. The PEBI grid scheme allows flow in more directions than what is accomplished by rectangular grid blocks. Although the time requirement for simulation of PEBI grids is more, the gain in accuracy of the results in simulation may be significant, specifically in cases where reservoir or well geometry does not conform to a regular grid pattern (Figure 15.5).

Number of fluid phases

The number of fluid phases treated in a reservoir simulation study may describe a simulation model. Reservoir simulation models can be single-phase, two-phase, and three-phase. In the simplest case, the model is based upon the properties of single-phase fluid that predict well rate and changes in pressure and saturation. A good example is a dry gas reservoir without any effects of aquifer influx; only gas is produced at

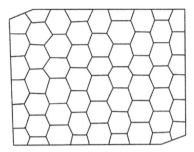

Figure 15.5 PEBI grids. Various other configurations of PEBI grid can be implemented in a reservoir model besides the honeycomb structure shown previously.

the well. A two-phase model may represent oil and water production from a reservoir where gas always remains in a dissolved state. The three-phase reservoir simulation model takes into account all the fluid phases present in the reservoir, namely, oil, gas, and water. Note that the latter may consist of both formation and injected water.

Reservoir fluid characteristics and recovery processes

Depending on fluid characteristics (volatile, nonvolatile) and recovery processes (thermal, nonthermal), reservoir simulation (Figure 15.6) models can be classified as follows:

- Black oil model: This model is most widely used in reservoir simulation. It considers two phases, namely, liquid and gas; the two phases have lumped hydrocarbon components. The solubility of vapor in liquid phase is assumed to be a function of pressure and temperature only. The gas phase can go in and come out of the liquid phase, but the liquid phase cannot vaporize into the gas phase. The third phase is water, which remains as a separate phase. The black oil model is widely used in the industry as it is relatively simple and provides acceptable results for the majority of petroleum reservoirs. For instance, primary production of less volatile oil and secondary production of oil by waterflooding or immiscible gas injection are good candidates for black oil simulation. The black oil model, or modified versions of the model, can be utilized for a large variety of reservoirs.
- Compositional model: Compositional simulation, although resource intensive, is necessary where the vaporization and condensation of hydrocarbon components play a pivotal role in the performance of a reservoir. In a typical compositional model, the PVT properties of lighter hydrocarbon components, C_1 through C_6, are represented individually. Intermediate to heavy components (e.g., C_7+) are lumped together so that the model does not become overly complex as compositional simulation requires a large number of runs to validate the model and generate various future scenarios. The model takes into consideration the changes in the composition of liquid and vapor phases under dynamic reservoir conditions. In a gas condensate reservoir, a reduction in reservoir pressure results in the condensation of heavier hydrocarbons from the gas phase. Similarly, during the injection of methane in an oil reservoir for enhanced recovery, lighter hydrocarbon components are vaporized from the liquid phase, resulting in significant changes in the composition of both liquid and gas. Applications of compositional simulation model include hydrocarbon gas injection for EOR, namely, condensing and vaporizing gas drives, and gas recycling in a gas condensate reservoir.

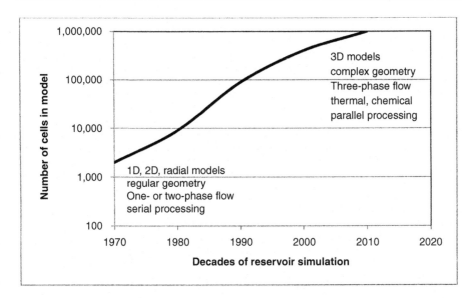

Figure 15.6 Evolution of reservoir models as more computing power became available over the decades [1].

In fact, the black oil model is viewed as a special case of the compositional model, where dozens or even hundreds of hydrocarbon components (C_1 through C_7+) present in the petroleum fluid are lumped simply into "oil" and "gas" phases instead of treating the components individually. In certain instances, wet gas and volatile oil reservoirs can be simulated by the black oil model [2]. This is accomplished by incorporating PVT data for any oil dissolved in the gas phase and gas dissolved in the oil phase.

- Simulation of unconventional gas reservoirs: Modeling of unconventional gas reservoirs, including shale gas and coalbed methane, incorporates the capabilities of simulating desorption and diffusion of gas. Significant quantities of gas may be stored in an adsorbed state in shale. As reservoir pressured is reduced, gas is desorbed and produced. Diffusion may occur as gas is liberated from the micropores of the rock.
- Thermal simulation model: As the name suggests, the model is used to simulate thermal EOR processes involving fluid and heat transport. Steam flooding, one of the most effective heavy oil recovery processes, is evaluated by thermal simulation.
- Chemical injection model: The model takes into account the transport of mass due to the phenomenon of dispersion, adsorption, and partitioning. Various chemicals are injected into the reservoir as part of the EOR process. For instance, evaluation of alkaline flood performance may be based on a chemical injection model.
- Streamline simulation: The model assumes that the flow of injected water in the reservoir can be represented by streamlines originating at the injection wells and moving towards the producing wells. Streamline models are generally used to study waterflood patterns and performance by envisioning that the injected water moves as streamlines toward the producers. These models, described briefly in Chapter 16, are relatively simple compared to compositional and black oil models. Streamline simulation has the advantage of visual representation of the path traced by injected fluid in the reservoir depending on pressure and reservoir characteristics.

Table 15.1 **Reservoir simulation models**

Simulation model	Application	Notes
Black oil	Reservoirs having oil of low to moderate volatility; the model is based on two phases, namely, oil and gas. Compositions of oil and gas phases are not considered.	Black oil simulation is very common and used in a variety of reservoir types.
Compositional	Highly volatile oil and gas condensate reservoirs where compositional changes in vapor or liquid phase may affect reservoir performance profoundly.	Compositional simulation is resource intensive compared to black oil simulation.
Unconventional gas	As unconventional reservoirs may store substantial quantities of gas in an adsorbed state and transport of gas may occur by diffusion, conventional simulators are modified to incorporate the features unique to unconventional reservoirs.	Shale gas and coalbed methane reservoirs store substantial quantities of natural gas in an adsorbed state; diffusion of gas takes place through micropores.
Thermal	Heavy oil reservoirs where thermal recovery methods are implemented	Models are based on equations of fluid flow and heat transfer.
Chemical	Oil reservoirs where polymer, surfactant, and other chemical substances are utilized in enhanced recovery	Models take into account the adsorption, dispersion and partitioning characteristics.
Streamline	Reservoirs under waterflood. Injected water is assumed to move in streamlines toward the producers.	Relatively simple model having the advantage of visually representing the flow characteristics of injected water.

Reservoir simulation models are summarized in Table 15.1.

Scope of simulation study

Reservoir models can also be classified according to their scope as follows:

- Well model: Certain simulation models focus on a single well. For example, the design of a horizontal well can be accomplished by a simulation study where the location, length, direction, and the number of laterals can be optimized to achieve the best possible results. A common implementation of a single well model involves water or gas coning issues. Certain wells are completed in a reservoir where oil production is affected by strong bottom water drive. Reservoir engineers need to know the completion interval and rate at which the well can be produced optimally. A 2D radial model is typically used to simulate the fluid flow characteristics around the well (Figure 15.7).

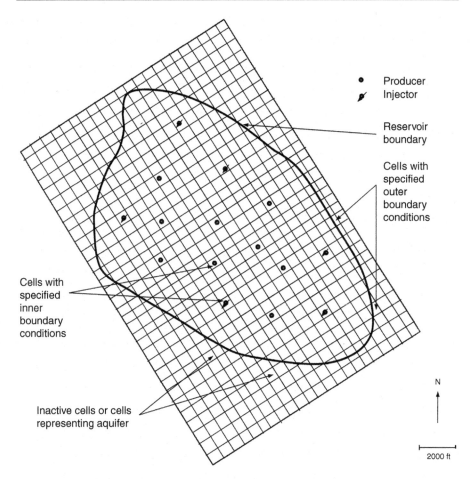

Figure 15.7 Representation of a reservoir by a 2D model including inner and outer boundaries. The inner boundary conditions are tied to injection and production wells. Cells outside reservoir boundaries are either inactive or used to represent aquifer effects. In practice, much smaller cells, including a LGR scheme at well locations, are often used to build a model. Fluid properties in each cell change with time under dynamic reservoir conditions.

- Vertical cross-sectional model: The model focuses on the vertical flow characteristics of fluids in a stratified or heterogeneous formation where vertical permeability is significant. Lateral flow of fluid in the reservoir is not represented in the model. A 2D cross-sectional model in the x and z directions is used.
- Sector model: The simulation model focuses on a specific sector or region of a reservoir to optimize the location, number, and rates of producers and injectors. The simulation study may determine the effectiveness of infill drilling or a balance between injection and production to achieve efficient displacement. Appropriate boundary conditions in terms of pressure and flow in and out of the model are required to connect the sector to the rest of the reservoir. The well and sector models can be very efficient and have a quick turnaround time. A 2D or 3D model can be built to represent a sector.

- Full field model: The comprehensive model is based on the entire reservoir and all the wells in order to provide a comprehensive picture of reservoir performance with time. The model is first history matched, followed by the generation of various scenarios related to the future well and production characteristics. The full field models are usually based on a large number of grid blocks; hence, the models are resource intensive, often requiring months to complete the study. The model is updated on a regular basis as more data on production and injection are available from reservoir monitoring. As a result, the prediction of expected ultimate recovery and other variables may significantly gain in accuracy. Full field models are based on 2D or 3D grid blocks. 2D models can be adequate in thin and relatively homogeneous reservoirs where the effect of vertical flow is not significant.

Applicability of simulation models

Depending on what information is sought, the simulation model may range between simple and complex. For example, a highly volatile oil reservoir in a complex geologic setting would require compositional model simulation in 3D, where the composition of fluids in both liquid and vapor phases are tracked over time and location. However, in cases of less volatile oil, a "black oil" model can be constructed where the changes in the composition of liquid and vapor phases need not be considered in the model. In order to analyze water-coning issues in a well, a relatively simple 2D radial model is appropriate for simulation. In any case, accurate representation of reservoir data holds the key to meaningful analysis. The reservoir team must use their expertise and experience in evaluating whether the results of simulation make sense.

Resource requirements

A compositional simulation model can be significantly resource intensive in comparison to a black oil model. As dimensions (2D to 3D) and fluid phases (two-phase to three-phase) are added to a reservoir model, the computational time may increase by one or several orders of magnitude. This is a major consideration as numerous runs may need to be performed during a reservoir study and important decisions about the reservoir must be made in a timely manner. Hence, black oil models are used wherever applicable. According to one estimate, over 80% of reservoir simulation efforts in the oil and gas industry are based on black oil models.

Mathematical basis for simulation

The following serves as a brief introduction to how reservoir models are built. The fluid flow equations that serve as the foundation for black oil, compositional, and other models include:

- Conservation of mass – mass can be neither created nor destroyed during flow of fluids in porous media
- Darcy's law – the correlation between pressure drop and resulting flow rate; however, non-Darcy flow is also incorporated in the reservoir models
- Fluid PVT properties – includes fluid viscosity, density, compressibility, solubility, formation volume factor, and others

It is noted that, for certain unconventional gas reservoirs such as shale gas and coalbed methane, fluid flow mechanisms may also include diffusion.

Let us consider one-dimensional flow of fluid along the x-axis through an elemental volume of length Δx representing the porous medium. The rate of inflow of mass is equal to the rate of outflow of mass and the rate of mass accumulation within the volume element.

Mass rate in − Mass rate out = Rate of accumulation of mass

For any fluid phase such as oil, the above can be written in equation form as follows:

$$\frac{\partial M_o}{\partial t} = \dot{M}_{oI} - \dot{M}_{oo} \tag{15.1}$$

where M_o = mass of oil; M_{oI} = mass rate of oil in; M_{oo} = mass rate of oil out.

The rate of accumulation of mass is expressed by the differential term on the left-hand side.

Further, note that the mass of oil can be expressed in terms of its flow rate, density, saturation, and formation volume factor as follows:

$$M_o = \rho_{os} A \Delta x \phi \frac{S_o}{B_o} \tag{15.2}$$

$$\dot{M}_{oI} = \rho_{os} \left(\frac{q_o}{B_o} \right)_x \tag{15.3}$$

$$\dot{M}_{oo} = \rho_{os} \left(\frac{q_o}{B_o} \right)_{x+\Delta x} \tag{15.4}$$

where A = cross-sectional area of the element; B_o = oil formation volume factor; q_o = flow rate of oil; S_o = oil saturation in the volume element, fraction of pore volume; t = time; Δx = length of the volume element, cm; ρ_{os} = density of oil at standard condition; ϕ = porosity of element representing porous medium.

Based upon Darcy's law, the flow rate of oil, q_o, can be expressed in terms of fluid viscosity, effective permeability, pressure drop along the length of the element, and its inclination. Hence, an expression for flow rate of oil can be obtained as follows:

$$q_o = -\frac{Akk_{ro}}{\mu_o} \left(\frac{\partial p_o}{\partial x} - \rho_o g \frac{\partial D}{\partial x} \right) \tag{15.5}$$

where D = depth of the element representing the porous medium; g = acceleration due to gravity; k = permeability; k_{ro} = relative permeability to oil, fraction; p_o = pressure in the oil phase, atm; ρ_o = oil density, gm/cm^3; x = direction of flow; μ_o = oil viscosity, cp.

Based on Equations (15.2), (15.3), (15.4), and (15.5), the differential equation for the flow of oil as presented in Equation (15.1) takes the following form:

$$\frac{\partial}{\partial x}\left[\frac{Akk_{ro}}{B_o\mu_o}\left(\frac{\partial p_o}{\partial x} - p_o g\frac{\partial D}{\partial x}\right)\right] = A\frac{\partial}{\partial t}\left(\frac{\phi S_o}{B_o}\right) \tag{15.6}$$

Similar expressions for gas and water flow can be obtained by following the same procedure. It must be noted that the expression for the flow of gas takes into account the presence of gas in the gas phase as well as in the dissolved state in the oil phase. The equations are presented in the following:

$$\frac{\partial}{\partial x}\left[\frac{Akk_{rw}}{B_w\mu_w}\left(\frac{\partial p_w}{\partial x} - p_w g\frac{\partial D}{\partial x}\right)\right] = A\frac{\partial}{\partial t}\left(\frac{\phi S_w}{B_w}\right) \tag{15.7}$$

$$\frac{\partial}{\partial x}\left[\frac{Akk_{rg}}{B_g\mu_g}\left(\frac{\partial p_g}{\partial x} - p_g g\frac{\partial D}{\partial x}\right)\right] + \frac{AR_s kk_{ro}}{B_o\mu_o}\left(\frac{\partial p_o}{\partial x} - p_o g\frac{\partial D}{\partial x}\right)$$
$$= A\frac{\partial}{\partial t}\left[\phi\left(\frac{S_g}{B_g} + \frac{S_o R_s}{B_o}\right)\right] \tag{15.8}$$

where R_s = gas in solution; g,o,w = subscripts referring to gas, oil, and water, respectively.

Note that Equation (15.8) incorporates gas solubility in oil to account for the gas phase in solution. In contrast, the oil and water phases in Equations (15.6) and (15.7) are standalone equations, and do not assume any phase transfer with gas. Such an assumption is valid in most reservoirs except for highly volatile and gas condensate reservoirs, where significant portions of the oil phase are likely to transfer into the gas phase. Any phase transfer between the oil and water phases is also deemed negligible.

In reservoir simulation, the above-mentioned partial differential equations are used to relate oil, gas, and water phase pressure values, namely, p_o, p_g, and p_w, with those of phase saturation, S_o, S_g, and S_w. In order to accomplish this, various fluid and rock properties such as viscosity, formation volume factor, porosity, and permeability need to be known. The reservoir model is not complete unless certain auxiliary equations are considered. It is noted that the sum of the saturations of three phases, namely, oil, gas, and water, must be equal to unity at any point in time and location. Hence:

$$S_o + S_g + S_w = 1 \tag{15.9}$$

Furthermore, it is noted that the capillary pressure of fluid phases are functions of phase saturations, and can be expressed as follows:

$$P_{cow} = P_o - P_w = P_{cow}(S_o, S_w) \tag{15.10}$$

and

$$P_{cgo} = P_g - P_o = P_{cgo}\,(S_o, S_g)\tag{15.11}$$

where P_c = capillary pressure; ow,go = subscripts referring to oil–water and gas–oil, respectively.

Hence, fluid flow in porous media can be modeled by Equations (15.6)–(15.11) as shown previously. The six equations have six unknown variables, namely, the values of pressure and saturation of oil, gas, and water phases. Assigning appropriate initial conditions in pressure and saturation, as well as boundary conditions at the wells and reservoir limits, the equations can be solved discretely in order to determine pressure and saturation at each grid block of the model described in an earlier section.

The outer boundary conditions are:

- Closed reservoirs – no flux at the grid blocks located at the outer boundary
- Aquifer influence – constant flux at the boundary

The inner boundary conditions at the well are:

- The Dirichlet condition – well operating under the constraint of specified BHP
- The Neumann condition – well operating under the constraint of specified rate

Discretization

The fluid flow equations presented previously provide continuous solutions of pressure and saturation in porous media in space and time. However, results of numerical solutions are not continuous; these are obtained in discrete intervals of both space and time. The derivatives appearing in fluid flow equations are discretized, as the reservoir model is built on a set of grid blocks depicted in Figure 15.3. Each grid block represents a small part of the reservoir, and reservoir simulation solves for pressure and saturation, among others, in each grid block or cell. The cells can be either regular or irregular, depending on the geologic structure of the reservoir and fluid flow geometry within the reservoir. For simplicity, let us consider a 1D reservoir model having n number of cells. Any three cells within the model are indexed as $i - 1$, i, and $i + 1$, at two consecutive time levels, t and $t + 1$, as shown in Figure 15.8a and b.

In formulating the model equations, the derivatives in space and time are replaced by a finite difference scheme. Note that other schemes are also implemented in numerical solution, including finite element and finite volume. For single-phase Darcy flow of fluid in porous media in one direction, the equation can be expressed in the following derivative form:

$$\frac{\delta P^2}{\delta x^2} = k\frac{\delta P}{\delta t}\tag{15.12}$$

where P = fluid pressure, which varies over distance and time; x = distance over which flow occurs; t = time; k = a coefficient.

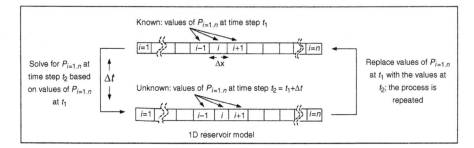

Figure 15.8 Illustration of a simple case in numerical solution of cell pressure.
Computation of new values of cell pressure at current time step is based on the values at
previous time step, cell dimension, and length of time step.

The finite difference approximations of the derivatives can be discretized as follows:

$$\frac{\delta P^2}{\delta x^2} = \frac{P_{i+1}^{(t)} - 2P_i^{(t)} + P_{i-1}^{(t)}}{2\Delta x} \tag{15.13}$$

$$\frac{\delta P}{\delta t} = \frac{P_i^{(t+1)} - P_i^{(t)}}{\Delta t} \tag{15.14}$$

Hence, Equation (15.12) can be cast in finite difference form:

$$k\frac{P_i^{(t+1)} - P_i^{(t)}}{\Delta t} = \frac{P_{i+1}^{(t)} - 2P_i^{(t)} + P_{i-1}^{(t)}}{2\Delta x} \tag{15.15}$$

In Equation (15.15), subscripts $i - 1$, i, and $i + 1$ denote the relative location of
grid blocks where the solution is sought. Similarly, superscripts t and $t + 1$ indicate the
current and future time steps, respectively. The variable in this case is fluid pressure,
which changes with time and location due to the dynamic conditions in the reservoir.

Numerical solution

At the first time step, the initial values of pressure are known at all the grid blocks,
which are typically calculated by simulation model based on user input of reference pressure, depth, oil–water contact and other pertinent information. As Equation
(15.15) suggests, reservoir pressure can be estimated at a future time step, $t + 1$, at
block i based on the known values of pressure in blocks $i - 1$, i, and $i + 1$ at the current time step t. Since we have more than three grid blocks in the model, we shall need
to solve an array of equations as shown in Figure 15.9. All the equations are solved
simultaneously for one time step.

In the next step ($t + 2$), the values of P_i are calculated based on the values of pressure obtained at time step $t + 1$. The solution marches on, as the values of pressure at
the current time step become the values at the previous time step as the calculation of

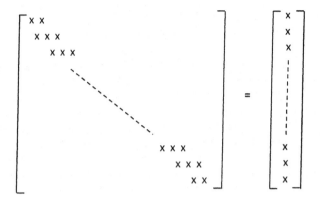

Figure 15.9 Array of model equations to be solved simultaneously for a time step. In the 1D formulation depicted here, the matrix has three bands. However, in the 2D and 3D models, the matrix is substantially larger and contains additional bands.

pressure for the next time step commences. In the case of multiphase flow represented in Equations (15.6) through (15.8), fluid phase pressure is coupled with saturation, which leads to the simultaneous solution of both pressure and saturation. The capillary pressure equations are shown in Equations (15.10) and (15.11).

As noted previously, the fluid flow equations are discretized for each grid block leading to a large number of finite difference equations forming a matrix. It is solved by a suitable computational algorithm in order to obtain the values of fluid saturation and pressure at a particular time step. Discrete solutions are quite sensitive to the number of grid blocks and length of time steps. Use of rather large numbers of grid blocks and smaller time steps increases the accuracy and stability of simulation; however, the computational process is resource intensive.

Explicit versus implicit solution

In Equation (15.15), the value of pressure at future time steps is based on the values obtained at the current step. Any coefficient appearing in the discretized equation is also evaluated at the current time step. The approach is referred to as explicit solution. However, in implicit solution, reservoir pressure is assumed for the next time step, and any coefficient appearing in the equation is also calculated at the assumed pressure. Implicit solution of cell pressure is obtained by iterating the values of assumed cell pressure until a convergence between assumed and calculated values of pressure are obtained. The topic is further discussed in the following.

Reservoir simulation equations are formulated in two broad categories according to the sought objectives and available computing resources as follows.

IMPES

Implicit pressure and explicit saturation (IMPES) methodology for simulating fluid flow in a reservoir has been widely used in the industry. During the early years of reservoir simulation when computing resources were limited, the method gained in popularity. IMPES derives its name from the fact that the mathematical formulation of the model is done in a manner where the value of pressure in a grid block for the current time step is computed implicitly, followed by explicit computation of fluid saturation for the grid block. Since fluid saturation in a grid block is computed explicitly without requiring a lengthy iterative procedure, simulation runs are highly efficient without any significant compromise in the accuracy of results. When used diligently, the method provides fairly accurate results in a variety of circumstances.

In the IMPES formulation, saturation terms are eliminated from the model equations so that the equations can be solved for only one variable, i.e., pressure for each grid block in the current time step. The sequence of calculation is provided below [3]:

- Step 1: For the current time step, assumption of pressure in each cell is made and the nonlinear coefficients of model equations, including fluid PVT properties, are read from a table at the assumed pressure.
- Step 2: Cell pressures are then computed by solving the matrix that is formed based upon all the grids in the model. Computed pressure in each cell is then compared against the assumed pressure in each cell.
- Step 3: If the compared values of pressure are sufficiently close and meet the convergence criterion, fluid saturations in cells are computed based on new cell pressure, length of time step, and previous saturation. Values of transmissibility and compressibility required to compute the current saturation in a grid block are also obtained from the previous time step. The solution then proceeds to the next time step, and the steps are repeated.
- Step 4: If the assumed and computed values of pressure are not close, further iterations are necessary for the current time step until the two values are sufficiently close to each other. If the time step is too large or the assumed pressure for the current time step is much higher or lower than that obtained in the last time step, convergence may not be achieved at all. Hence, the explicit approach requires that the time steps are sufficiently small and pressure changes are not too large between time steps.

As indicated earlier, the IMPES method is suitable where fluid phase saturation does not change significantly within a short period of time in the reservoir. For example, the method can be used in a full field model where the simulation model cells are relatively large and the main objective is to track reservoir pressure in various regions of the reservoir over months or years due to injection and production.

Fully implicit

In the fully implicit procedure, the coefficients that are used in both pressure and saturation equations are evaluated at the current time step in an implicit manner. The method is performed iteratively until a convergence is achieved for both pressure and saturation values in a cell. In contrast to the IMPES methodology, the coefficients of the nonlinear equations in the simulation model are updated in each iteration. The fully implicit procedure is outlined in the following:

- Step 1: For the current time step, both pressure and saturation values are assumed for each cell and the nonlinear coefficients of model equations, including fluid PVT properties, capillary pressure, and relative permeability data are read from relevant tables at the assumed pressure and saturation.

- Step 2: Pressure as well as saturation values in model cells are computed by solving the nonlinear equations simultaneously. Computed pressure and saturation values are then compared against the assumed values of pressure and saturation, respectively.
- Step 3: If compared values of pressure and saturations are sufficiently close and meet the convergence criteria set for pressure and saturation, the solution proceeds to the next time step.
- Step 4: In case convergence is not achieved following the prescribed number of iterations, the time step used previously needs to be reduced and the procedure is repeated.

The methodology is suitable where a rapid change in fluid saturation is anticipated, such as in the vicinity of a well. Cases where fully implicit solutions are required include water and gas coning that may be encountered around a producer, or where rapid evolution of the gas phase occurs from the liquid phase. Nevertheless, the fully implicit procedure is resource intensive and requires more computational time.

Treatment of wells

In reservoir simulation studies, wells are typically placed at or near the center of a grid block. For improving the accuracy of simulation results, a LGR scheme is implemented around the well.

Wells may be assigned to operate under limiting rate, pressure, and ratio of fluids to realistically simulate field-operating conditions. For example, a producing well may be constrained to produce at a lower rate when a specified minimum value of BHP is reached or the well may be shut when the WOR exceeds a maximum value. Similarly, an injector may be assigned to operate under a maximum injection pressure in order to ensure that the formation is not fractured inadvertently. The constraints that are used for wells in simulation are presented in Table 15.2.

A common practice in reservoir simulation is the control of wells by group, meaning that multiple wells are grouped together to impose the same set of operating constraints. Simulators also allow any adjustments to reflect the maintenance of downtime of wells.

Partially completed wells are represented accordingly in the model, where the well is connected to the reservoir through the perforated section or layers only.

Treatment of reservoir boundaries

Specification of two types of reservoir boundaries is common, which include:

- Reservoir with no flow boundary
- Constant flux boundary

Noncommunicating reservoir boundaries can be simulated in a variety of ways, including assignment of zero porosity for the grid blocks at the boundary. Bounded reservoirs can also be simulated by setting transmissibility values to zero that eliminates

Table 15.2 **Well constraints in simulation**

Well type	Constraint	Notes
Producer	Minimum BHP	As a well is produced, BHP decreases with time. As a specified minimum value of BHP is reached, the well produces at a diminishing rate.
Producer	Minimum tubing head pressure (THP)	BHP is correlated to THP by vertical flow performance curves.
Producer	Maximum WOR or GOR	A well may be shut-in or rate be reduced as WOR or GOR exceeds a limiting value. Wells with high WOR or GOR may not be economical to operate.
Producer	Constant BHP	Depending on the reservoir pressure, well rate varies as BHP is held constant.
Producer	Maximum liquid volume	Liquid volume is the sum of oil and water volumes.
Producer	Maximum drawdown pressure	Drawdown pressure is the difference between reservoir pressure and BHP of the producing well.
Producer	Constant production rate	Depending on reservoir pressure, BHP varies as the rate of production is held constant. The specified condition is prevalent in gas wells operating under deliverability constraints.
Producer	Constant voidage ratio	Well production is manipulated to match the injection volume of fluid in order to maintain reservoir pressure at the same level.
Injector	Maximum injection rate	The constraint is meant to avoid inadvertent fracturing of the formation during injection.
Injector	Constant injection pressure	Well injection rate may vary to maintain a constant injection pressure.
Producer/ injector	Open/shut	Wells can be opened or shut depending on a limiting condition or part of development strategy.

the flow of fluids across the boundary. Certain cells can be declared inactive, which do not contain any fluid or allow any movement of fluids. Reservoirs receiving pressure support from an adjacent aquifer can be simulated by assigning aquifer characteristics, including porosity, permeability, and dimension of the aquifer.

Treatment of hydraulic fractures

Hydraulic fractures are simulated by implementing LGR in the reservoir, as rapid changes in fluid flow characteristics occur due to the very high conductivity of the fractures in comparison to the matrix. Simulation of multistage fracturing in shale gas reservoirs is described in Chapter 22.

Treatment of natural fractures

Simulation of naturally fractured reservoirs is generally accomplished by envisioning a dual porosity dual permeability system. Both fracture and matrix have their distinct porosity and permeability. Flow can occur between the matrix and the fracture. Flow is also allowed to occur from the fracture as well as the matrix into the wellbore. Incorporation of natural fractures into the model essentially leads to an increase in the number of simulation cells.

Treatment of cross-flow between geologic layers

Conventional oil and reservoirs are usually stratified having multiple layers of varying reservoir quality. Intervening shale is often encountered between sandstone and carbonates, which leads to limited or no communication between the layers in a vertical direction. A common approach for simulating the varying degree of cross-flow between the layers is to assign the vertical permeability of the adjacent layer accordingly. An impermeable shale layer can also be simulated by deactivating the grid blocks that represent the layer. In yet another approach, net to gross thickness of the productive layer is adjusted to reflect the presence of shale.

Treatment of complex geology: block-centered versus corner point geometry

In the case of relatively simple geometry, the reservoir model is built on regular grid blocks. Blocks have flat faces, and their elevations are assigned. Computation of fluid transmissibility between two adjacent blocks is required to solve the equations. The transmissibility term is a function of fluid mobility and dimension of grid blocks. However, in the case of irregular reservoir geometry and complex geologic features, block-centered grids are inadequate in simulating fluid flow. When adjacent blocks are located at different elevations across a fault, unrealistic connections between the blocks are established. In reality, flow of fluid would not occur between the two blocks, which is otherwise expected in the case of regular geometry. Reservoir dips, fault planes, pinchout boundaries, irregular surfaces, and other structural heterogeneities are more accurately represented when corner point geometry is utilized. All eight corners of a gird block are specified in this approach. The edges of blocks may not be vertical. Fluid transmissibility computations are modified to reflect the true nature of communication of fluid in complex geological settings. However, the dataset requiring more information regarding a block is larger than that based on block-centered grids.

Treatment of faults and pinchouts

Edges of grid blocks are aligned to a fault. Corner point geometry is better suited to model a fault or dipping formation. Sealing, nonsealing, and partially communicating faults are modeled by modifying transmissibility between the grid blocks. For sealing

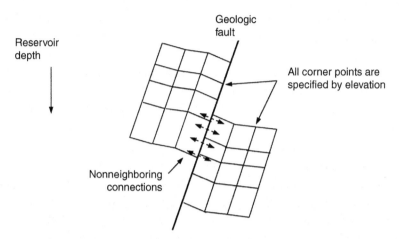

Figure 15.10 Modeling of a geologic fault based on the implementation of corner point geometry.

faults, the transmissibility modifier is 0. For partially communicating faults, the transmissibility modifier varies between 0 and 1. A corner-point grid system is better suited to represent geologic faults (Figure 15.10). In the case of curved faults, PEBI or other types of grids are utilized to represent the complex geometry more realistically.

Treatment of aquifers

Aquifers located at the edges or bottom of the reservoir are represented by grids that are much larger than reservoir grids as fine details regarding the changes in the aquifer are not needed in simulating oil and gas production. Since the reservoir model becomes quite large by including aquifer geometry explicitly, simulation requires a longer time to run. Other approaches are available to replicate the aquifer effects on the performance of a reservoir that require a single or a few grid blocks.

Reservoir simulation objectives

The first step in any simulation project is setting a clear objective and a practical timeline. The questions to be asked might include:

- What are the objectives of simulation (initiate waterflood, drill offset wells, combat high water cut, update reserves, understand reservoir complexities, etc.)?
- How the does the outcome of simulation studies aid in the short- and long-term management of the reservoir (e.g., optimization of drilling and production, and recompletion of problematic wells)?
- What kind of reservoir model needs to be built (2D, 3D, radial)?
- What would be the simulation methodology (black oil, streamline, compositional, chemical, etc.)?
- What are the data requirements? How much data are available? Where do the data reside (database, logs, surveys, reports, hardcopies, etc.)?

- What uncertainties may impact the reservoir performance most (e.g., presence of fractures or high permeability streaks)? Will further field tests and well monitoring be necessary to conduct a meaningful study?
- What is time frame to conduct the study? For example, how soon can an infill drilling campaign or waterflood operation be initiated based upon the results of simulation?

Once these questions are answered, the next focus is on reservoir data as discussed in the following section.

Data gathering and model building

The workflow starts with collection, validation, and integration of available reservoir data, usually from a large number of diverse sources, including but not limited to geological, seismic, geochemical, petrophysical, transient well testing, and production history. A literature review indicates that the following sources of data are typically utilized in reservoir simulation:

- Core analysis results, including porosity, permeability, compressibility, wettability, etc.
- Well logs showing formation thickness, porosity, fluid saturations, water–oil contact, gas–oil contact, stratification, presence of fractures, and all other valuable information
- Well completion data indicating perforated zones
- Hydraulic fracturing information, in the case of hydraulically fractured wells
- PVT analysis of fluids
- Relative permeability data of fluid phases based on lab studies or correlations
- Well production data, including rate, pressure, WOR, and GOR
- Reservoir surveillance data, including rate and pressure
- Transient pressure test reports
- Results of production logging tool, modular dynamic tester, and other surveys
- Reports on previous reservoir studies, including volumetric analysis, decline curve analysis, and material balance calculations
- Structural maps showing formation tops, net and gross thicknesses, water–oil contact, gas–oil contact, and other pertinent information
- Reservoir characterization reports related to the stratification of formation, flow units, high permeability streaks, natural fractures, faults, pinchouts, compartments, facies change, or any other feature that would influence reservoir performance
- Constraints on well rate and minimum BHP for the well to operate economically

A listing of reservoir and fluid data sources is provided in Table 15.3.

The reservoir engineering team must have a firm grasp of what data are available, what the available data represent, and how the data can be integrated to build a realistic model. At this stage, the necessity for further data gathering and subsequent integration may become apparent. For example, relative permeability data from a specific layer of the reservoir may not be available, which would require reviewing the existing data or even coring of future wells. Similarly, well pressure transient tests may be designed to gather average reservoir pressure from a specific region of a reservoir.

Challenges in data collection, validation, and integration

In many simulation studies, important reservoir data may not be available. Wells may be sparsely located in a field. Again, only the records of oil and gas production data

Table 15.3 **Data requirements in reservoir simulation [3,4]**

Source	Scope	Data	Notes
Seismic	Reservoir	Reservoir structure, anomalies, heterogeneities	
Microseismic	Stimulated reservoir volume around wells	Fracture network and characteristics	Extensively used in unconventional shale reservoirs
Geological	Reservoir	Lithology, structure, stratigraphy, boundary, reservoir heterogeneities	
Geochemical	Well/reservoir	Total organic content, maturity index	Unconventional shale reservoirs
Geomechanical	Well/reservoir	Young's modulus, Poission's ratio, fracture stress	Unconventional shale reservoirs
Well logs	Limited to the vicinity of wells	Porosity, saturation, formation thickness (net and gross), fluid contacts, water– and gas–oil transition zones	
Reservoir fluid samples	Reservoir/region	PVT properties	Fluid properties can be representative of the reservoir or a region within the reservoir, depending on degree of communication.
Cores	Limited to the wellbore locations; extremely localized	Porosity, absolute permeability, relative permeability, capillary pressure, wettability, stratigraphy	Cores represent a miniscule portion of the reservoir that is modeled.
Realizations	Reservoir	Interwell rock properties such as porosity and permeability	Based on geostatistical models. Multiple realizations of the same reservoir can be utilized in building simulation models.

Table 15.3 Data requirements in reservoir simulation [3,4] *(cont.)*

Source	Scope	Data	Notes
Transient well testing	Well drainage area, reservoir	Rock and fluid properties including effective permeability, near wellbore conditions, fracture characteristics, reservoir heterogeneities such as faults and pinchouts, reservoir boundary effects	Comprehensive well test program is implemented to better manage the wells and the reservoir as a whole. Interlayer communication can be characterized by modular dynamic tester.
Production history	Well, reservoir	Well rates, WOR, GOR, time to breakthrough, BHP, WHP	Available only when wells are in production
Correlations	Reservoir, region	Fluid PVT properties	Used in the absence of actual measurements
Reservoir surveillance	Reservoir, well	Well rates, pressure, water cut, GOR, time for breakthrough, individual layer performance	Surveillance includes production logging, spinner surveys, well tests, and tracer studies
Reservoir studies	Reservoir	Estimates of oil and gas in place and reserves by volumetric, decline curve, and material balance analyses	
Regional trends	Geographical region, basin	Reservoir attributes, production trends	

may be maintained; reservoir pressure or water production data are not being recorded. On the contrary, a reservoir may have a vast amount of data that may lead to building an overly complex model, which requires substantial time and resources.

During the early stages of the reservoir life cycle, a sizable portion of data required to build a model is not available. The reservoir model is generally built upon a number of assumptions and the experience of earth scientists and engineers. The static reservoir model is built on:

- Limited core data from a handful of wells
- Available correlations
- Analogy based on other reservoirs in the region

In order to reduce the uncertainties, multiple realizations of the reservoir can be used to conduct simulation study. The previous attempts to capture the possible variations in porosity, permeability, stratification, and various other geologic features based on geostatistical methods. However, the traditional approach in earlier decades was generally based on one reservoir model due to the limitations in computing resources.

An important issue in integrating the various sources of data is the scale of data. Permeability data obtained from small core samples are much smaller in scale compared to that obtained from a typical well test. Upscaling of rock properties obtained from a few small cores to a grid block, which may be several hundred feet in length and breadth, can be quite challenging.

Since the sources of data for reservoir simulation are large in number and diverse, resolution of any discrepancy between datasets can be a daunting task. Certain data could be outliers, meaning that these do not fall in the expected range. In such a case, data must be reviewed to ascertain their accuracy. Again, "inconsistent" data may point to any previously unknown reservoir heterogeneities.

Representation of reservoir by grid blocks

As noted earlier, a reservoir simulation is built on grid blocks or cells. The grid blocks can be 1D, 2D, 3D, or radial. The gridding scheme also uses LGR in order to capture the fine details of fluid flow behavior. Again, grids can be either regular or irregular. In each grid block, pressure and saturation values may change in each time step, depending on fluid properties and boundary conditions. Literature review indicates that reservoir simulation models are usually built on a large number of cells. Typical models range from tens to hundreds of thousands of cells; however, even larger models are also developed for simulation, such as the modeling of complex reservoirs and processes based on a million cells or more. Each grid block requires the following specifications.

- Dimension of grid block. The length, width, and height of each grid block depend on a number of factors, including, but not limited to:
  - Size and complexity of the reservoir – Large and heterogeneous reservoirs require a large number of grid blocks to improve the accuracy of simulation results.
  - Scope of simulation study – Sector models require fewer grid blocks than full field models.
  - Hydraulic fracturing – Vertical, horizontal, and hydraulically fractured wells require an appropriate LGR scheme to simulate flow of oil and gas in fine detail.
- Formation top. The source of data includes well logs and geological maps.
- Well locations. Certain grid blocks would include information related to production, injection, and observation wells present in the reservoir.
- Rock and fluid properties. Porosity, absolute permeability in x, y, and z directions for 3D models, pressure and fluid saturations. The traditional source of porosity and permeability data has been core and log data obtained from various wells. Rock properties between two wells were calculated by a suitable interpolation scheme. Present-day practices include very large geological models, comprising millions of cells, generated

by a geostatistical approach. Furthermore, a large number of realizations of the reservoir can be developed. Conversion of rock property values from geological modes comprising millions of cells to reservoir models requires the scaling of data. A simple approach includes the use of arithmetic averages of porosity and permeability values in lateral directions. However, harmonic averages are used for scaling the vertical permeability values.

Grid orientation effects

Due to certain limitations of finite difference approximations used in simulation models, the orientation of grids in relation to injectors and producers may affect the results of simulation in a significant manner. When an injector and producer are located parallel to the grid line in a reservoir model, injected water breakthrough in the producer occurs relatively early. When the injector and producer are located diagonally, the simulated time for breakthrough is longer (Figure 15.11). However, greater sweep efficiency and recovery are predicted in the latter case. The finite

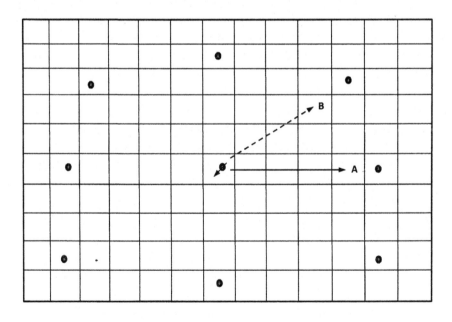

● Producer

✦ Injector

Figure 15.11 Demonstration of grid orientation effects on flow of fluids between injector and producer. Injected water apparently moves faster along path A than B, due to the location of producers with respect to the injector at the center. However, if the grids are rotated 45°, the effect will be reversed, and water will move faster along B than A.

difference scheme is modified to alleviate the grid orientation effects. A popular approach is based on a nine-point finite difference scheme instead of a five-point scheme for a 2D model. In the nine-point difference scheme, the corner points of a cell are taken into account. Modifications can also be made to calculate the transmissibility between adjacent grid blocks in order to reduce the grid orientation effects.

PVT data of fluids

As fluid pressure changes from cell to cell during simulation as a function of space, and within a cell as a function of time, various fluid properties change. Hence, various fluid properties, such as viscosity, compressibility, density, formation volume factor, and solution GOR are input into the simulation model. All the properties as a function of pressure mentioned previously may not be available. Established correlations are used in such cases.

Relative permeability data

In most reservoir simulation studies, one fluid phase, such as water, may displace another, such as oil. Consequently, fluid saturations change in a cell. Hence, relative permeability and capillary pressure data are required to characterize the flow of fluids as these are functions of saturation. When field data are unavailable, various correlations of fluid saturation and relative permeability are used in simulation.

Pseudorelative permeability

Many reservoirs are stratified due to the variations in depositional environment. Consequently, various rock characteristics, including porosity and permeability, vary from layer to layer. Incorporation of a large number of layers and rock properties including relative permeability and capillary pressure data for each layer in simulation models add to the size and complexity of the models, and the time required to run the models increases by orders of magnitude. Hence, pseudofunctions are generated that can reduce the number of layers used in a model, where the simplified model can still represent the complexity of the reservoir. An example to generate one set of pseudorelative permeability curves (oil, water, and gas) from corresponding relative permeability data obtained from individual layers in a reservoir involves the permeability–thickness weighted averaging technique as follows [2]:

$$k_{r,pseudo} = \sum \frac{(kh)_i \, k_{r,i}}{(kh)_i}$$

where k = absolute permeability; h = thickness; k_r = relative permeability; i = index of layer.

Pseudofunctions can also be generated dynamically. A single set of pseudorelative permeability data of fluid phases can be obtained by back calculation based on the results of a cross-sectional model having a number of layers. The pseudorelative permeability data are then used for a single layer in a 2D areal model, which is simpler and lot faster than a 3D model without significantly sacrificing the accuracy in simulation results.

PVT, relative permeability, and capillary pressure data by regions

Since reservoirs are inherently heterogeneous, it may not be adequate to build robust reservoir models based on a single set of relative permeability, capillary pressure, and PVT data. Rocks as well as fluid properties vary from one region to another within the reservoir. Consequently, a reservoir model may require input of multiple sets of relevant data by region in order to improve the accuracy in simulation results. A notable variation occurs in the values of end point saturation between two sets of relative permeability curves obtained from cores drilled at different well locations and at various layers in the reservoir. The end point saturations include irreducible water saturation and residual oil saturation. End point saturations are functions of various rock properties such as porosity, permeability, capillary pressure, and wettability, among others. In certain simulation models, relative permeability data are input by specifying end point saturations and employing a suitable correlation between fluid saturation and relative permeability.

Initialization of reservoir model

Prior to simulation, the reservoir model must be initialized successfully ensuring that all the fluid phases are in gravity–capillary equilibrium and the fluid contacts (gas–oil, oil–water) are at the correct depths in model cells. Failure to specify the fluid contacts in the model accurately results in incorrect estimation of hydrocarbon in place; it may also lead to unreliable prediction of well rate, pressure, and water or gas breakthrough. Reservoir simulation models allow multiple ways to provide necessary data for initialization as follows:

- Manual assignment of pressure and saturation of oil, gas, and water in individual grid blocks. The method is not recommended as it may lead to significant discrepancies.
- Assignment of oil–water, gas–oil, and gas–water contact depths, and reference pressure; a simulator calculates fluid phase saturations in grid blocks. Relevant data are obtained from well logs and pressure transient testing.
- Equilibrium of fluid phases based on capillary pressure characteristic between fluid phases. Long transition zones between oil and water may be encountered in certain reservoirs having low porosity and permeability, and where the contrast between the densities of oil and water are relatively small.

The location of gas–oil or oil–water contact is determined by the counteracting forces of gravity and capillarity of fluid phases. When initialization is performed by the simulator and a static equilibrium condition is attained throughout the reservoir model, movement of fluids from one grid cell to another should not occur as long as wells are not activated for production or injection.

In cases of complex reservoir geometry and noncommunication between various portions of the reservoir, more than one initialization would be required to represent the various regions in the reservoir.

Stability and accuracy in results: time step, grid size, and shape

Numerical solutions of reservoir model equations are obtained iteratively until convergence in pressure and saturation values is obtained for the grid blocks in the model. In general, simulation of rapidly changing fluid saturations and pressure in a reservoir requires the specification of smaller time steps and grid sizes to ensure that convergence is attained. Smaller grid blocks are required to enhance the resolution of a reservoir model, which requires the use of time steps of short duration. Models based on irregularly shaped cells would require more resources to run. The limits of simulation run parameters, including the length of time steps, can be deduced mathematically. Use of inappropriately large time steps often leads to unstable or unrealistic results in pressure and saturation in grid blocks that can be readily identified; in other cases, the inaccuracies can be subtle and misleading.

Production history matching

A reservoir model is not expected to capture all the inherent heterogeneities that exist in a reservoir at the micro-, macro-, and megascale. The only way some degree of confidence can be placed on simulation results is to verify whether the model is capable of replicating past production characteristics in terms of well rate, pressure, WOR, and GOR. Various reservoir properties can be varied within a valid range of values to match production history as closely as possible by conducting successive simulation runs. The first point to remember in the history-matching phase is that sufficient production history would be required to conduct a meaningful history matching. Second, a model with various combinations of rock properties and reservoir description (permeability, relative permeability, capillary pressure, porosity, stratification, faults, etc.) may lead to a match or near match with production history. Third, history-matching efforts could be resource intensive. Hence, modern simulators available in the industry are capable of automated history match.

Various approaches can be adopted to perform a production history match. A case study of shale gas simulation is presented in Chapter 22, where the simulated BHP of a well is matched with field data by using the observed production rate as an input parameter in the model. Various reservoir properties, including fracture characteristics, are adjusted to match well performance. In other cases, a production rate of one phase is input into the simulation model to match the rates of other fluid phases.

Table 15.4 provides a general guideline for history match of pressure, saturation, and well productivity index.

Computer applications are available in the industry that perform automated history match within a prescribed range of rock properties such as porosity, permeability,

Table 15.4 History match parameters in reservoir simulation

Variable to match	Parameters to adjust	Notes
Reservoir pressure	Well rates, permeability, porosity, thickness, total compressibility, water influx	High injection rates, strong water influx, reduced permeability, and thickness lead to higher pressure.
WOR, GOR (field and individual wells)	Relative permeability, pseudorelative permeability curves	Steep water relative permeability curve leads to early breakthrough of water; shifting of oil relative permeability curve to the left leads to poor recovery.
Well performance	Permeability, porosity, skin factor, BHP	Higher permeability around well increases well rate. Larger skin leads to increased pressure drop.

fluid saturation, length of fracture, and other properties. The automated procedure is capable of performing a large number of simulation runs within a short period of time and resulting in an increase in productivity. The procedure ranks the individual runs in the order of best to worst match with the field data. However, nonlinear equations with multiple variables may have more than one "best fit," and engineering judgment must be used in analyzing the results of the automated process. Furthermore, some of the parameters used in the matching process are correlated, including porosity and permeability, suggesting that any correlation between the variables must be taken into account when the parameters are varied to obtain a match.

Output of simulation study

Results from simulation studies are typically reported by the following:

- Individual well
- Type of wells (injector, producer)
- Group of wells
- Sectors in a reservoir based on development target
- Regions separated by varying fluid properties and fluid contact
- Formation layers highlighting geologic heterogeneities
- Entire field

The important results obtained from the simulation studies include, but are not limited to:

- Oil, gas, and water rate over time
- Cumulative production from wells, sectors, and field
- Percent recovery over time
- Expected ultimate recovery of oil and gas
- Economic life of wells and reservoir

- Water and gas breakthrough in wells
- WOR and GOR over time
- Changes in reservoir pressure over time
- Changes in saturation and distribution of fluids in the reservoir
- Changes in BHP
- Changes in well productivity index
- Movement of fluid from one area to another within the reservoir
- Identification of areas or layers where residual oil saturation is high

Restart file

Results of simulation, including pressure and saturation in model cells along with complete reservoir model description, can be written in a restart file. A restart file facilitates the simulation of the reservoir performance into the future without having to redo the original simulation study over and over again. For example, consider a reservoir that has been producing for 10 years. A restart file can be written at the end of 10 years if the history match is successful. The file contains up-to-date information regarding reservoir pressure and saturation. Simulation of future performance of the reservoir for another 15 or 20 years under various scenarios, including water injection, infill drilling, etc., can be based on the information contained in the restart file rather than starting with the original data that reflect the initial state of the reservoir. The approach can significantly reduce the computational time and enhance productivity.

Case Study: Simulation of Water Coning in Oil Well

In reservoirs with a bottom water drive, water coning poses a challenge in producing the well effectively. Water coning issues may be encountered in other situations where water is mobile in a reservoir and significant drawdown is encountered around a producer. As the well is produced, water begins to encroach from the bottom in significant proportions resulting in the decrease in oil production. Due to the high viscous forces created near the well, water having relatively less viscosity forms a cone around the perforated portion of the well and dominates flow through the wellbore. Water cut eventually rises to a level, 80% or more, when the well needs to be shut down due to operational and economic constraints. Water cut is defined as the rate of water produced over the total rate of water and oil produced. Solutions to water coning issues include recompleting the well in the upper part of the formation and producing the well at a lower rate to combat water coning. For wells plagued with water coning or gas coning issues, or where potential issues are anticipated, reservoir simulation can go a long way in optimizing the completion interval and well rate in order to maximize expected ultimate recovery.

A case study based on simulation of well performance under the adverse effects of water coning is presented later [4]. The black oil model simulates two-phase flow of oil and water with time. It is based on radial grid geometry, which is typical of single well models. The example highlights the data requirements for a water coning simulation study. It also presents well performance in terms of oil rate and water cut over time. Various scenarios for the well can be built based upon simulation, including selective perforation and oil production rate. A sensitivity analysis indicates that water cut is lower for the first few years of production when only the upper five layers are perforated as opposed to eight layers (Figure 15.12). Due to the presence of a bottom aquifer, lower layers produce more water that adversely affects oil recovery.

Oil production rate in the case of the well completed in five layers is also shown (Figure 15.13). Nevertheless, water cut is observed to increase significantly in both cases over the years as more water encroaches into all of the perforated layers. Simulation is terminated when a limiting water cut of 90% is reached.

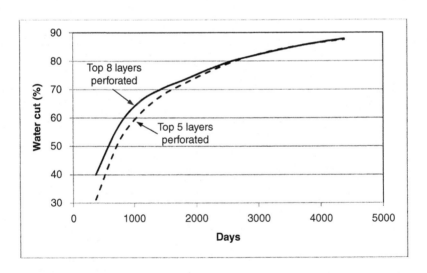

Figure 15.12 Sensitivity runs comparing the water cuts over time where the well is partially completed. In the base case, eight layers are perforated. A sensitivity analysis is performed where only five layers are perforated.
Courtesy: Computer Modelling Group.

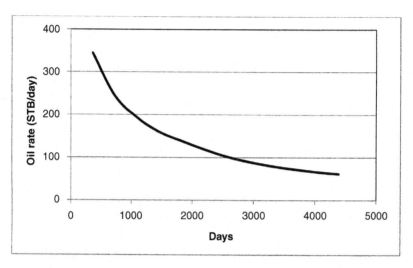

Figure 15.13 Oil production rate over time for the case where the well is completed in the top five layers.
Courtesy: Computer Modelling Group.

Data requirements for water coning study

A reservoir simulation dataset typically comprises several sections as shown in Table 15.5.

The values of model parameters are provided in Tables 15.6–15.11.

Case Study: Evaluation of Reservoir Performance Under Water Injection

This case study evaluates the performance of a reservoir under water injection. The reservoir model is built on 25 × 34 × 4 grid blocks [4]. The grid blocks are 360 × 410 ft. in length and width, respectively. Each of the four blocks in the vertical direction represents a geologic layer. A corner point gridding scheme is used, where the elevation data of all the corners of the grids are entered in the simulation dataset. The corner point elevations of each cell vary according to the structural map of the reservoir. Reservoir depth varies between 9,850 ft. and 10,500 ft. The reference pressure and depth are 4,000 psi at 10,000 ft. Oil–water contact is at 10,100 ft.

Range of formation porosity is between 1% and 19%, while permeability varies between 5 mD and 550 mD. Vertical permeability is assumed to be 10% of horizontal permeability.

Well locations, perforated layers, and maximum allowable production rates are shown in Table 15.12.

In the base case, seven producers are active. The reservoir produced under depletion drive. Cumulative oil production versus time is simulated. In the second case, four injection wells were drilled in specified locations and reservoir performance is simulated. The increase in oil recovery due to water injection is shown in Figure 15.14. The next steps in the overall management of the reservoir include facilities designed to handle oil, gas, and water production, followed by economic analysis.

Table 15.5 Description of dataset: Water coning study

Section	Notes
Title	A brief description of what the study accomplishes and relevant comments
Input/output control	Specification of units (field, SI, lab); type of information sought (well rate, field pressure, etc.)
Description of grids	Grid type (radial, 2D, 3D) and dimension. Porosity and permeability values in each grid. Reservoir depth
Aquifer properties	Type (bottom, edge), thickness, porosity, permeability, and radius of the aquifer
Fluid PVT properties	Table of formation volume factor, solution gas ratio, viscosity, compressibility, and density against pressure
Relative permeability	Table of oil–water and oil–gas relative permeability against saturation
Model initialization	Reference depth and pressure, oil–water and gas–oil contacts
Numerical specifications	Maximum time step; number of time step reductions allowed to maintain stability in results
Well information and constraints	Maximum liquid rate and water cut, minimum BHP and oil rate
Output specification	Type of output information related to well, grid, and field (oil, rate, water cut, BHP); interval at which output is printed

Table 15.6 Listing of model parameters

Parameter	Value	Notes
Grid type and dimension of model	2D radial (r, z): 9×12	Single well models are typically based on radial grid geometry.
Grid block size in radial (r) direction (ft.)	See Table 15.7	Inner grids are proportionally smaller to capture the details in pressure and saturation changes.

(Continued)

Table 15.6 Listing of model parameters *(cont.)*

Parameter	Value	Notes
Grid block size in vertical (z) direction (ft.)	See Table 15.8	Reservoir layers having distinct porosity and permeability may be represented by individual grid blocks in vertical direction.
Porosity in layers 1–12, fraction	See Table 15.8	Porosity ranges between 5% and 22%
Horizontal permeability in layers 1–12 (mD)	See Table 15.8	Permeability ranges between 3 mD and 92 mD
Ratio of vertical to horizontal permeability	0.1 in all layers	Relatively high values of vertical permeability result in adverse effects.
Aquifer properties	Location: At bottom Thickness: 500 ft. Radius: 35,000 ft. Porosity: 18% Permeability: 80 mD	
Fluid PVT properties	See Table 15.9	
Relative permeability data	See Tables 15.10–15.11	
Oil density (lb/ft.^{3})	45.0	
Gas density (lb/ft.^{3})	0.07	
Water density (lb/ft.^{3})	62.14	
Oil compressibility (psi^{-1})	1.0E-5	
Water compressibility (psi^{-1})	3.0E-6	
Water FVF (rb/STB)	1.014	
Water viscosity (cp)	0.95	
Model initialization parameters	Reference depth: 9,000 ft. Reference pressure: 3,600 psi Oil–water contact: 9,105 ft. Gas–oil contact: 9,049 ft. Bubble point: 2,000 psi	
Well parameters	Wellbore radius: 0.5 ft. Perforation: Layers 1–4 Skin factor: 0	
Well control	Maximum liquid rate: 1,000 stb/day Maximum water cut: 90% Minimum BHP: 1,500 psi Minimum oil rate: 50 stb/day	Well is to operate within the prescribed range of pressure, rate, and water cut

Table 15.7 Grid dimension in radial (*r*) direction

Grid block	1	2	3	4	5	6	7	8	9
Radius (ft.)	2.0	4.0	7.0	12.0	25.0	55.0	110.0	230.0	550.0

Table 15.8 Grid block properties in vertical (*z*) direction

Layer number	Thickness (ft.)	Porosity, fraction	Permeability (mD)	Notes
1	5.0	0.10	15.0	Layers are indexed from top
2	2.0	0.05	3.0	
3	4.0	0.12	10.0	
4	7.0	0.15	20.0	
5	3.0	0.10	27.0	
6	8.0	0.08	8.0	
7	5.0	0.16	18.0	
8	7.0	0.22	92.0	Highest porosity and permeability
9	5.0	0.12	10.0	
10	7.0	0.14	13.0	
11	4.0	0.17	20.0	
12	50.0	0.15	15.0	Bottom layer

Table 15.9 Fluid PVT data

Pressure (psi)	Solution GOR (scf/STB)	Oil FVF (rb/STB)	Gas expansion factor (scf/rb)	Oil viscosity (cp)	Gas viscosity (cp)
1200	100	1.038	510.2	1.11	0.014
1600	145	1.051	680.27	1.08	0.0145
2000	182	1.063	847.46	1.06	0.015
2400	218	1.075	1020.4	1.03	0.0155
2800	245	1.087	1190.5	1	0.016
3200	283	1.0985	1351.4	0.98	0.0165
3600	310	1.11	1538.5	0.95	0.017
4000	333	1.12	1694.9	0.94	0.0175
4500	355	1.13	1851.9	0.92	0.018

Table 15.10 Oil–water relative permeability

S_w	K_{rw}	K_{row}
0.25	0	1.0
0.4	0.19	0.52
0.5	0.352	0.3
0.6	0.551	0.108
0.7	0.795	0
0.8	0.96	0
1.0	1.0	0

Table 15.11 Gas–oil relative permeability

S_o	K_{rg}	K_{rog}
0.27	1	0
0.3	0.881	0
0.4	0.601	0
0.5	0.42	0
0.6	0.288	0
0.7	0.193	0.02
0.8	0.1	0.1
0.9	0.03	0.33
0.96	0	0.6
1.0	0	1.0

Table 15.12 Production and injection well data

Well type and number	Location (I,J)	Perforated interval (K)	Maximum allowable production rate (STB/day)	Maximum allowable BHIP (psi)
Producer-1	15,11	1,2	5430	
Producer-2	14,23	1,2	6132	
Producer-3	11,25	1,2,3	6338	
Producer-4	8,21	1,2,3	6132	
Producer-6	8,28	1,2,3	4633	
Injector-7	20,16	1,2,3		3625
Producer-8	10,7	1,2	5634	
Injector-9	7,12	1,2,3		3625
Producer-10	15,20	1,2,3	6132	
Injector-11	9,15	1,2,3,4		3625
Injector-12	13,23	1,2,3,4		3625

Summing up

Major decisions in reservoir engineering and management are based upon reservoir simulation studies. Simulation attempts to replicate and predict real-world processes and events by building physical and mathematical models. Reservoir simulation, developed in the latter half of twentieth century, attempts to reduce the uncertainties inherent in the prediction of reservoir performance over the life of the reservoir. The uncertainties are introduced by reservoir quality and geologic complexity, which are largely unknown at the time of developing the reservoir. Any reservoir development plan, including infill drilling and IOR, is evaluated and optimized based upon simulation. Development of reservoir models that predict future rates and pressure requires a large number of static and dynamic data from a

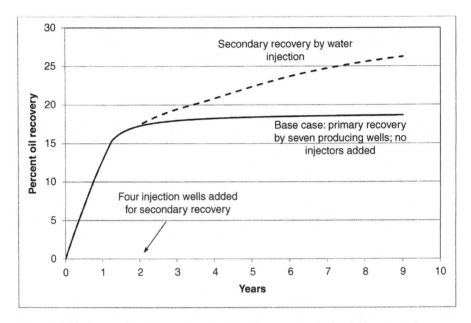

Figure 15.14 Comparison of two scenarios based on reservoir simulation. In the base case, the seven wells produce under depletion drive alone. In the water injection case, four injectors are added to the model. Oil recovery increases from about 18% to 26% over a period of 9 years.
Courtesy: Computer Modelling Group.

variety of sources, including geological, geostatistical, geophysical, petrophysical, well test, and other studies. Furthermore, a detailed analysis of production history is required.

Reservoir simulation can be categorized in a variety of ways, including 1D, 2D, 3D, two-phase, three-phase, black oil, compositional, thermal, chemical, and others. Configuration of reservoir models depends on the objectives of study, reservoir complexity, and the processes involved. Black oil simulation is the most popular, as it is efficient and appropriate in a variety of reservoir conditions (Figure 15.15).

Questions and assignments

1. What is reservoir simulation and why it is used extensively in the petroleum industry?
2. Describe a reservoir model, including mathematical basis and structural features.
3. How are reservoir models classified? What are the considerations in selecting a large reservoir model?
4. How does reservoir simulation integrate into the bigger picture, including the design of surface facilities and economic analysis?
5. Why do most reservoir simulators use numerical methods to obtain a solution?
6. What are the challenges in gathering and integrating data for reservoir models?

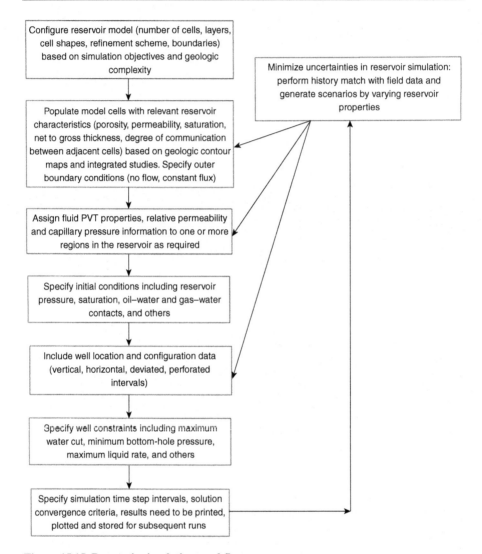

Figure 15.15 Reservoir simulation workflow.

7. Distinguish between black oil and compositional simulation. Describe the advantages and limitations of both. Why are black oil models used extensively in the industry?
8. Provide a detailed listing of data required in black oil simulation. Explain how each datum affects the outcome of simulation.
9. What is history matching? How and why is history matching performed in a simulation?
10. Based on a literature review, describe a simulation study that is used to develop a large field, enhance reservoir performance, and better manage the reservoir.

References

[1] Koederitz LF. Lecture notes on applied reservoir simulation. World Scientific Publishing Company, New Jersey; 2005.
[2] Satter A, Iqbal GM, Buchwalter JL. Practical enhanced reservoir engineering: assisted with simulation software. Tulsa, OK: Pennwell; 2008.
[3] Edwards DA, Gunasekera D, Morris J, Shaw G, Shaw K, Walsh D, et al. Reservoir simulation: keeping pace with oilfield complexity. Oilfield Review. Winter 2011/2012: 23, no. 4. Available from: http://www.slb.com/~/media/Files/resources/oilfield_review/ors11/win11/01_reservoir_sim.pdf.
[4] IMEX Manual. Computer Modeling Group: Calgary, Canada; 2015.

Waterflooding and waterflood surveillance

16

Introduction

The goal of waterflooding is to augment ultimate recovery from the reservoir by injecting water to increase oil production. Worldwide statistics indicate that only 10–30% of oil is produced by primary production, while waterflooding followed by primary recovery can recover up to 40% or more of the original oil in place (OOIP). As primary production declines in a reservoir due to reduction in reservoir pressure and other factors, additional energy is needed to produce the oil left behind in the formation. The basic concept is rather simple: inject water into the reservoir at high pressure to "push" or displace oil to the producing wells.

A typical waterflooding operation involves the injection of water through selected wells to provide additional energy to the reservoir to produce more. In most cases, injectors are converted from producers. A bank of water develops around an injector that is enlarged with continued injection and drives additional oil to the nearby producers. In certain reservoirs, water injection commences early on to maintain the reservoir pressure sufficiently high. The evolution of dissolved gas from the oil phase is avoided by maintaining the reservoir pressure above the bubble point pressure.

Waterflooding involves reservoir production, design of surface facilities, and operation engineering coupled with economic analysis. Sound reservoir management, including continuous monitoring of injection and production wells, is also required for the success of waterflood.

This chapter deals with the reservoir engineering aspects of waterflooding and waterflood surveillance. It attempts to answer the following questions:

- What is the process of waterflood?
- What are the design considerations for implementing waterflood?
- How did the practice of waterflooding evolve?
- What are the important concepts in waterflood?
- What reservoirs are suitable for waterflood?
- What are the injection and production well patterns in waterflood?
- What factors influence the performance of waterflood?
- What is reservoir surveillance?
- How does reservoir surveillance aid in managing waterflood operations and the reservoir as a whole?
- What diagnostic tools and methods are employed to monitor waterflood?
- What are the steps in workflow related to the design, implementation, and management of waterflooding?

Reservoir Engineering. http://dx.doi.org/10.1016/B978-0-12-800219-3.00016-4

History of waterflood

Waterflood has a long history [1]. It was discovered as far back as in the mid-nineteenth century and progressively developed since then. As early as 1865, the first waterflood occurred as a result of accidental water injection in the Pithole City area of Pennsylvania. Maintaining reservoir pressure, allowing wells to have a longer life than expected by primary depletion, was the primary function of waterflooding.

Earliest practices involved the reinjection of produced water back into the oil zone rather than disposing it into streams or rivers. Water was injected into a single well, then into many wells forming a circle drive, and a peripheral drive. In 1924, the first five-spot pattern flood was initiated in the Bradford Field in Pennsylvania.

In 1931, the application of waterflood grew from Pennsylvania to Oklahoma in the shallow Bartlesville sand, and in 1936 to Texas in the Fry Pool of Brown County. Waterflood became a common practice in the 1950s as the technology developed and matured. By the 1970s, most onshore reservoirs in the United States and many other oil producing countries were produced by waterflooding to tap additional oil.

Waterflood design

Waterflood design including the location, development schedule, and rate associated with the injection wells is a major area of interest to reservoir engineers. The overall design philosophy of waterflood operations is to optimize the ultimate oil recovery by targeting the areas and zones where relatively large volumes of oil are left behind following primary recovery. Waterflood design is also integrated with economic analysis, including payout period and cash flow as discussed in Chapter 24. Good reservoir quality, including high values of permeability and porosity, higher net thickness of formation, and relatively less heterogeneity leads to high cumulative volumes. On the other hand, relatively low production during primary recovery may indicate poor formation transmissibility, storativity, or isolated pay intervals. Poor mechanical condition at the wellbore and skin damage around the well are also causes of lesser than expected well performance and oil recovery.

In a typical reservoir, an ongoing waterflood operation is closely aligned with reservoir management on a daily basis. The effectiveness of waterflooding largely depends on rock and fluid characteristics, and how it is managed based on reservoir surveillance. Relatively homogeneous formations with favorable porosity and permeability, absence of highly permeable conduits and fractures, oil being light or medium gravity (20°API or greater), and relatively high oil saturation may lead to quite successful waterflood projects. Oil recovery as a result of waterflooding is referred to as secondary recovery. Industry experience indicates that about 15–30% of the OOIP is likely to be recovered in most cases. The performance of reservoirs under waterflood has improved notably in recent decades, especially in complex geologic settings, with detailed reservoir characterization, better well planning based on robust simulation models, better downhole equipment, deployment of "smart wells," and implementation of reservoir surveillance and analysis in real time.

The practice of waterflood

Waterflooding consists of injecting water into selected wells while producing from the surrounding wells. It displaces oil from the injector to the producer, while maintaining reservoir pressure. Water is an efficient agent for displacing light or medium gravity oil in a relatively homogeneous formation where high permeability channels are not encountered. As mentioned earlier, the success of waterflooding hinges on favorable economics, i.e., low capital investments and operating costs resulting in significant enhancements in oil production over a long period.

In earlier days, waterflooding had often been initiated in depleted or nearly depleted reservoirs with a free gas phase present. In the initial stage of the waterflooding process, injected water fills up the pores previously occupied by gas, which is redissolved in solution, and the reservoir pressure is restored. More efficient waterflooding practices, however, require water injection above the bubble point pressure of oil in order to avoid the evolution of gas in the reservoir. Liberation of dissolved gas leads to lower relative permeability to the oil phase, and lower production rates, as gas becomes mobile. There are instances in the past, however, where water has been injected slightly below the bubble point.

Applicability of waterflooding

Waterflood is widely used in the commercial recovery of oil for a variety of reasons, some of which are as follows:

- Water is generally available from wide-ranging sources, including subsurface formation, aquifers underlying or overlying the oil reservoirs, and surface streams and oceans.
- It is the least expensive of the fluids that are injected into the reservoir to enhance oil recovery.
- Water is an efficient agent for displacing light to medium gravity oil.
- Water is relatively easy to inject, and it spreads rather easily through the formation.
- Disposal of water at the surface is relative easy compared to certain other injection fluids.
- Waterflooding involves low capital investment and operating costs; hence, it leads to favorable economic returns.

Reservoir response due to waterflood

A typical waterflood response is characterized by an increase in oil rate, followed by a decline, and an eventual breakthrough of injected water at the producers. Figure 16.1 is a typical plot of the oil production rates versus waterflood life for a successful waterflood performance in a reservoir with a gas cap [2]. It presents the filling up of pore spaces initially occupied by free gas, and the incline and decline of the secondary oil saturation periods.

The water–oil ratio continues to rise with time, and the economic limit is reached when water production becomes excessive (Figure 16.2). The situation is exacerbated by the presence of highly conductive pathways or channels that exist in the formation in many cases.

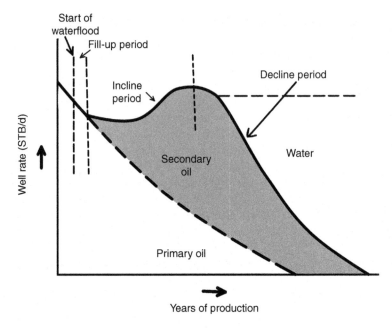

Figure 16.1 Example of a successful waterflood performance. A substantial amount of oil is recovered by waterflooding (secondary recovery), as indicated by shaded section.

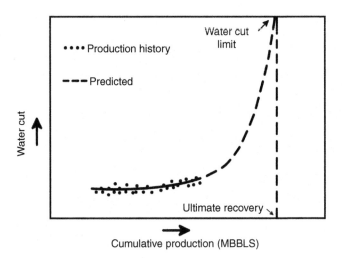

Figure 16.2 Rise of water cut in oil wells in a reservoir under waterflood. Finally an economic limit is reached when the well is abandoned.

Factors affecting waterflood performance

The response to waterflooding primarily depends upon the following:

- Well spacing, i.e., distance between producers and injectors
- Waterflood pattern, i.e., relative location of injectors and producers
- Schedule of conversion from injector to producer
- Fluid properties, including viscosity and gravity
- Rock properties, including the ratio of vertical to horizontal permeability, relative permeability characteristics, water–oil mobility ratio, capillary pressure, and wettability
- Multistage fracturing in tight formations to improve water injectivity
- Reservoir heterogeneities, including the presence of fractures and high permeability streaks, stratification or layering and lateral discontinuities
- Water injection rate and well injectivity, i.e., injection capacity
- Timing of waterflood operations

Waterflood pattern, well spacing, and conversion schedule

Historically, waterflooding has been implemented through injectors and producers where the location of the wells follows a definite pattern. Well patterns are selected after careful consideration given to rock characteristics such as permeability, presence of heterogeneities, and reservoir boundary. The most common patterns include line drive, peripheral drive, nine-spot, seven-spot, and five-spot, among others (Figures 16.3–16.6). A nine-spot pattern indicates that there are nine wells

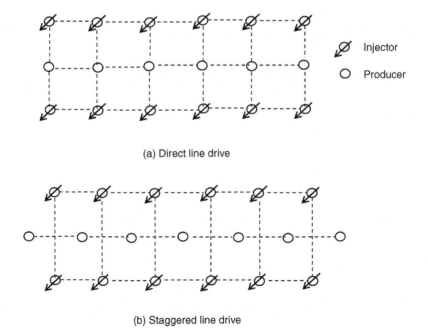

(a) Direct line drive

(b) Staggered line drive

Figure 16.3 (a) Direct line drive and (b) staggered line drive.

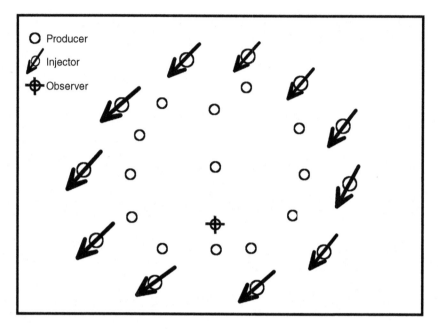

Figure 16.4 Peripheral water drive.

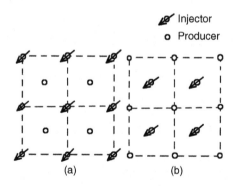

Figure 16.5 (a) Regular and (b) inverted five-spot pattern.

in a single section of the pattern. An oil field may have many sections. During the
early stages of waterflood, wells have larger spacings. As the production declines
and water–oil ratio increases at the producers, infill wells are drilled to produce
the remaining oil (Figure 16.7). As a result, the waterflood pattern may change.
For example, an inverted nine-spot pattern having eight producers and one injector
may change to a five-spot pattern later during waterflood. The latter comprises of
four injectors and one producer. The arrangement enables injection of more water
to drive oil to a producer. This is accomplished by drilling new producers and converting certain old producers to injectors.

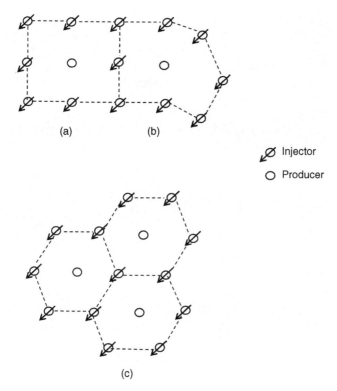

Figure 16.6 (a) Regular nine-spot, (b) irregular nine-spot, and (c) seven-spot patterns.

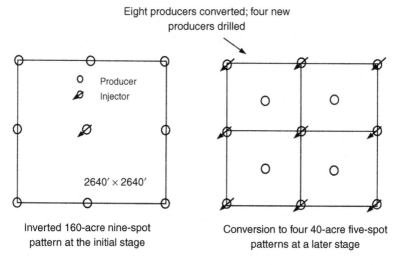

Figure 16.7 Conversion of pattern at later stages of waterflood. Also shows spacing.

Wells in a waterflood pattern are commonly found to have 40, 80, and 160 acre spacing. A literature review suggests that many reservoirs have wells drilled at 20 or 320 acre spacing as well. Tight or heterogeneous reservoirs require smaller well spacing and more injectors in a pattern. Directional permeability that may exist in the reservoir may influence the well conversion plans as water injection is implemented transverse to the direction to avoid early water breakthrough.

Production history also dictates well conversion. Wells showing high water–oil ratio or problematic wells may be considered for conversion early in the schedule. Some patterns are irregular, indicating that the distance between wells and their relative positions are not exactly alike in each pattern.

The selection and schedule of well conversion from producers to injectors are based on detailed reservoir simulation studies to optimize oil production and recovery.

Injection wells can be positioned around the periphery of a reservoir, which is referred to as peripheral injection. In contrast, crestal injection involves positioning of the wells along the crests of small reservoirs with sharp structural features.

In a dipping reservoir, water injection wells are located downdip to take advantage of gravity segregation. If a gas cap exists in the reservoir, produced gas may be reinjected through updip wells to maintain reservoir pressure.

Well drilling schedule, pattern selection, and well conversion are studied in detail by reservoir simulation and performance prediction before the actual implementation. Craig [3] suggests the following guidelines in designing a waterflood pattern in a reservoir:

- Optimization of oil recovery including maximum oil production based on minimum water injection
- Sufficient water injection rate to achieve target productivity
- Recognition of various reservoir heterogeneities including directional permeability, fractures and reservoir dip, and design waterflood accordingly
- Working with an existing well pattern to minimize drilling of new wells
- Knowledge of similar waterflood operations in adjacent leases

Existing reservoir heterogeneities must be taken into account in designing an efficient waterflood operation. For example, injectors can be placed transversely to the permeability trend or direction of natural fractures to maximize sweep, and water may be injected in downdip wells to avoid water slumping and adverse effects of gravity segregation. Reservoir heterogeneities affecting waterflood performance are described later in the chapter.

Oil gravity

Water is less efficient in displacing heavy oil having a high viscosity. Waterflooding can be uneconomic below oil gravity of 20°API. However, waterflood recovery increases in reservoirs where oil is of light to intermediate gravity (40–20°API).

Mobility ratio

The mobility ratio of water to oil is one of the most critical factors to influence waterflood efficiency. When mobility is greater than one, it is considered unfavorable as water is more mobile than oil in the porous medium; injected water tends to bypass oil and early breakthrough is experienced at the producers. At a mobility ratio of less than one, water is less mobile than oil leading to better displacement and recovery of oil.

$$M = \frac{\lambda_w \text{ in the water contacted portion}}{\lambda_o \text{ in the oil bank}} \tag{16.1}$$

$$= \frac{k_{rw}/\mu_w}{k_{ro}/\mu_o} \tag{16.2}$$

where M = mobility ratio; k_r = relative permeability; μ = viscosity; λ = mobility = k/μ; k = permeability; o,w = subscripts denoting oil and water, respectively.

The relative permeabilities are based on two different and separate regions in the reservoir during waterflood. Craig [3] suggested that the relative permeability to water should be obtained from the zone swept by water, while the relative permeability to oil should be based on an unswept region, which is located ahead of the displacement front.

During waterflooding, oil and water saturations change with time and distance from the wells, as the injected water displaces the oil toward the producer. Changes in fluid saturations are controlled by relative permeability characteristics along with other fluid and rock properties, including wettability.

Following secondary recovery by waterflood, a tertiary enhanced oil recovery process such as carbon dioxide flooding can be used to recover more oil.

Vertical permeability

In some reservoirs, a good vertical to horizontal permeability ratio is observed to have good vertical sweep efficiency and recovery of oil.

Reservoir heterogeneity

During waterflooding, the performance of the reservoir is greatly influenced by the inherent heterogeneities present in the geologic formation. Reservoirs are made of multiple strata or layers, which are not uniform in their properties such as lithology, porosity, permeability, pore size distributions, wettability, fluid properties, and interstitial water saturation. The properties vary in both areal and vertical directions. The heterogeneity of the reservoirs is attributed to the variations in depositional environments and geologic events, as well as the nature of the particles constituting the sediments. Waterflood recovery efficiency decreases with increasing reservoir heterogeneity.

Some of the common heterogeneities that affect the design, implementation, and management are listed as follows:

- Large number of reservoirs under waterflooding are identified with distinct stratification and layering, separated by impermeable or semipermeable shale. Variation in permeability in various geologic layers leads to reservoir heterogeneity. Water tends to flow through more permeable layers leaving a substantial quantity of oil in other layers. Consequently, premature water breakthrough, high water–oil ratio and less than expected well productivity are observed. Water does not sweep the entire formation and oil is not recovered efficiently from the less permeable layers.

- In certain reservoirs, large communication pathways exist between the geologic layers that may lead to water slumping leaving high residual oil saturation in upper layers.
- When microscopic fractures or fissures are present in rock, injected water moves quickly through these conduits as fracture permeability is much greater than the permeability of the rock matrix. The bulk of oil present in the rock matrix is not contacted by injected water leading to poor recovery.
- Rocks often exhibit directional permeability, which causes the injected water to flow in a certain preferential direction. For example, injected water may flow preferentially in a northwest–southeast direction in a reservoir rather than flow uniformly in all four directions. In such a case, water breakthrough occurs early in the producers that are located either in a northwest or southeast direction of the injection well. Areas around the other producers are not swept, leaving large pockets of oil behind.
- Sealing faults, compartmentalized formations, or a facies change in the formation may restrict the flow of injected water that impacts the ultimate recovery. Again, a nonsealing fault may lead the injected water to unexpected locations resulting in poor sweep in the intended sections of the reservoir.
- Reservoir dip also plays a significant role in waterflood performance. Water is heavier than oil but less viscous. In a dipping reservoir, water tends to underrun the oil leading to less than expected recovery.

Optimum well injection rate

The rate of oil recovery depends upon the water injection rate into a reservoir. For an injection well, the optimum injection rate ensures maximum contact with residual oil and recovers oil within the desired time frame.

The water injection rate, which can vary throughout the life of the project, is influenced by many factors. The variables affecting the injection rates are rock and fluid properties, fluid mobility values in the swept and unswept regions, and the well geometry, i.e., pattern, spacing, and wellbore radius.

Muskat [4] and Duppe [5] provided analytic injection rate equations for regular patterns with unit mobility and free gas saturation. With the advent of reservoir simulation, multiple waterflooding scenarios can be generated with various injection rates, well locations, and configuration of horizontal sections along with other parameters for an optimal waterflood design.

Injection pressure

During injection, care is taken to inject water below the threshold pressure that would fracture the formation. In case the formation is fractured, injected water is lost through the fracture and there is little buildup of reservoir pressure. As a result, water does not displace oil and the effectiveness of waterflood is minimal.

Water injectivity

Water injectivity is defined as the rate of water injection over the pressure differential between the injector and the producer. It has the unit of barrels per day per pounds per square inch (bbl/d/psi). Decline in water injectivity is observed during the early stages of injection into a reservoir depleted by solution gas drive. This occurs as pore spaces initially occupied by free gas are gradually filled up. Following fillup, the injectivity of water depends upon the mobility ratio. As shown in Figure 16.8, it remains constant in the case of unit mobility ratio and increases when the mobility ratio is greater than unity (unfavorable for displacing oil). It decreases when the ratio is less than unity (favorable for displacing oil).

Water injectivity is reduced due to the incompatibility between injected water with formation water.

Multistage fracturing

Tight reservoirs are unfavorable for water injection as the injectivity is quite low. Horizontal wells are drilled combined with multistage fracturing to enhance injectivity in low permeability formations.

Timing of waterflooding

Based on field experience and reservoir simulation studies, efficient waterflooding practices involve water injection above the bubble point pressure of oil. Oil remains as a single-phase fluid without any evolution of gas ensuring the attainment of maximum recovery. Hence, water injection and pressure maintenance operations are initiated early in the life cycle of the reservoir prior to the fall of reservoir pressure to the bubble point.

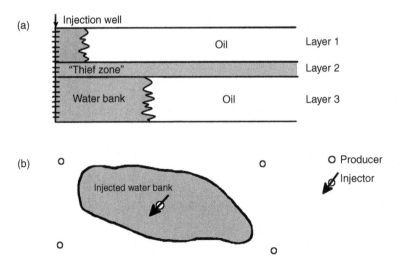

Figure 16.8 Waterflooding in (a) stratified reservoir and in (b) rock having permeability anisotropy. In stratified formation, presence of a high permeability layer, also referred to as thief zone, may lead to premature water breakthrough. In a reservoir with directional permeability trend, injected water may break through in certain wells bypassing large volumes of oil in other locations.

Analysis of waterflood performance

In modern times, waterflood performance is analyzed by collecting real-time data on well rates, bottom-hole pressure, water cut, and saturation profile of injected fluids, among others. Initially, waterflood performance is predicted by the simulation of an integrated, dynamic, and robust reservoir model [6]. Next, the model is updated on a regular basis as more information about the reservoir and well performance is obtained as the waterflood operation continues. However, in the early decades of waterflood, several empirical methods were developed, many of them based upon laboratory studies on scaled down models to predict waterflood performance. The analog solutions can provide an insight into the processes occurring during waterflood in case of small reservoirs and provide quick estimates of recovery.

Evaluation of waterflood performance is based upon several criteria, including the following:

- Waterflood recovery efficiency, which comprises areal and vertical sweep of the reservoir and fluid displacement efficiency
- Reservoir performance, including the volume of the recovered well
- Performance of individual layers in effectively producing oil
- Well performance, including water cut and injectivity

Waterflood recovery efficiency

The overall waterflood recovery efficiency is given by:

$$E_R = E_D \times E_A \times E_V \tag{16.3}$$

where E_D = displacement efficiency of waterflood to drive oil towards the wellbore, %; E_A = areal sweep efficiency of injected water, %; E_V = vertical sweep efficiency of injected water, %.

Only a portion of porous medium is swept by injected water due to the tortuosity of porous channels, miniscule pore throat opening, and various heterogeneities present in rock. Again, not all the oil is displaced that is contacted by water. Displacement efficiency is influenced by the rock and fluid properties and the volume of water injected.

The principal factors that determine the sweep efficiencies include the waterflood pattern, various reservoir heterogeneities, oil–water mobility ratio, and injected water volume.

Based on the assumption of homogeneous formation and unit mobility ratio, the areal sweep in various flood patterns are shown in Table 16.1.

Note that areal sweep efficiency at breakthrough in a staggered-line drive case is higher than the direct-line drive given the reservoir conditions are the same. Areal sweep efficiency at breakthrough in a nine-spot drive case is higher than all of the other cases, as more injectors are used to drive oil toward the producer.

Waterflood performance prediction methods

In earlier decades of waterflood, the availability of computing power was either non-existent or limited. Various empirical methods [7–12] were proposed for predicting

Table 16.1 **Areal sweep efficiency under ideal conditions**

Waterflood pattern	Description	Areal sweep efficiency (%)
Direct line drive	Injectors and producers are located on two adjacent lines	56
Staggered line drive	Injectors and producers are located on two adjacent lines but their relative positions are staggered	78
Regular five-spot	Four injectors and one producer	72
Regular nine-spot	Eight injectors and one producer	80

waterflood performance that were based on graphical methods or the solution of equations. Some of the well-known methods are as follows.

Dykstra–Parsons method [7]

In this method, recovery efficiency of waterflood operation is based upon reservoir stratification and consequent permeability variation of rock. It is well known that waterflood efficiency is adversely affected by reservoir heterogeneities in most cases. One common cause of rock heterogeneities is the significant variation in permeability from one location to another within the formation. The changes in permeability occur due to the variations in depositional environment in geologic times as well as due to certain postdepositional processes such as leaching. Dykstra and Parsons proposed a permeability variation factor based upon a large number of data collected from core samples. Based on a statistical approach, the permeability factor is defined as follows:

$$V = \frac{k_{50} - k_{84.1}}{k_{50}} \tag{16.4}$$

where k_{50} = log mean permeability of core sample; $k_{84.1}$ = permeability at 84.1% of the cumulative sample.

The log mean permeability is the value of permeability at 50% probability. When rock permeability is uniform as in an ideal case, $V = 0$. However, the upper limit of V is 1.0 in the case of a highly heterogeneous reservoir.

A series of charts is available in the literature that allows the prediction of waterflood recovery efficiency based on oil saturation, mobility ratio of oil and water, and water cut in the future. The assumptions in this method include piston-like displacement of oil by water.

Buckley–Leverett method [10]

This method is based on frontal advancement theory of injected water. It is described in Chapter 9.

Stiles method [9]

This method is based on piston-like displacement of oil by water, reservoir stratification, and mobility ratio = 1.

Craig–Geffen–Morse [11]

This method is based on the results of a series of five-spot model gas and water drives.

Prats [12]

This method is based on piston-like displacement of oil by water, and considers initial gas saturation, areal and vertical sweeps, mobility ratio and stratification, and five-spot pattern.

Surface facilities

A typical waterflood surface facility includes the following:

- Gathering and storage system
- Injection pumps
- Water distribution systems
- Flow metering
- Water treatment and filtering systems
- Oil–water separator
- Corrosion and scale inhibition systems

Surface facilities are monitored and managed by operational staff on a regular basis to ensure an effective waterflood.

Water quality management

Waterflooding operations require good water quality in order to avoid the plugging of rock pores resulting in higher injection pressure and poor injectivity. Good quality of water also ensures that well corrosion issues are not significant. Better quality water, although more expensive, is required for low permeability reservoirs as the tendency to clog the miniscule pores is quite high. Since the higher quality water does not require high injection pressure to inject into the formation, the possibility of fracturing the formation is also minimized.

Waterflood surveillance

Waterflood surveillance is all about monitoring the success for the entire operation, including well, facilities, and the reservoir as a whole. The ultimate goal is to optimize the secondary oil recovery. Modern waterflood surveillance efforts include the collection of production history, including oil rates, water cuts, bottom-hole pressures, and other valuable information in real time. The data are analyzed in detail by robust tools, including reservoir simulation.

Reservoir surveillance under waterflood or enhanced oil recovery is closely aligned to overall reservoir management. Continuous data collection and analysis enable the reservoir engineering team to manage the reservoir effectively and implement any changes in the long-term strategy or day-to-day operation if necessary. Examples of effective waterflood management are as follows:

- An adjustment in injection rate at an injector to balance the waterflood pattern
- Recompletion of a vertical producer located in a watered-out zone to a horizontal well targeting zones where the remaining oil saturation is high
- Selective shutoff of a watered-out zone at a producer while continuing production from other zones by using smart well technology
- Conversion of a producing well into an injection or observation well

Common surveillance efforts include, but are not limited to, deployment of automated downhole sensors to record pressure, rate and temperature in real time, waterflood pattern balancing, well testing, well logging, tracer studies, reservoir simulation including streamline simulation, ongoing reservoir characterization, flood front tracking, and periodic 4D seismic survey.

The following are important factors to consider in a comprehensive waterflood surveillance program [13]:

- Reservoir – average pressure, well rates, pore volume injected and cumulative production volumes; water cut trends, waterflood pattern balancing, and repositioning of injectors and producers
- Reservoir characterization – identification of hydraulic units, thief zones, presence of fractures, watered-out layers, and high permeability channels. Detailed information related to the common reservoir characteristics, including storativity, transmissibility, anisotropy, lateral continuity, oil saturations in various zones and locations, and oil–water contacts are required to manage waterflood operations
- Fluid flow analysis – identification of viscous fingering and gravity underriding by injected water
- Wells – production or injection rate, formation damage or skin around the wellbore, selection of target layers to optimize oil recovery and well integrity in terms of completion. Potential issues include formation damage, perforation plugging, and wellbore fractures
- Well testing – determination of pressure and pressure gradients for pattern balancing, which improves areal and vertical sweep
- Waterflood pattern – realignment or repositioning of injectors and producers for better waterflood efficiency
- Tracer studies – tracers are injected along with water and monitored at producers and observation wells to characterize water flow paths
- Observation wells – installation of observation wells to monitor the progress of waterflood
- 4D seismic surveys – periodic seismic studies to monitor the dynamic oil saturation profile and bypassed oil
- Facilities – water handling capacity, monitoring equipment, and others
- Water system – water quality for injection, presence of impurities, corrosive ingredients, dissolved gases, minerals, bacteria, and solids in dissolved and suspended state

The following points are important to note with respect to the monitoring and analysis of waterflood operations, and management of the reservoir.

- Duration of waterflood. Optimum waterflood performance can be achieved when all the producers reach their limiting values of water cut about the same period of time. Operating expenses are minimized in this approach.
- Streamline simulation. Streamline simulation, based on the solution of pressure and saturation in a numerical grid, allows the visualization of injected water movement in the formation as "stream lines" toward the injector. The method was introduced as early as the 1930s, and had been utilized in visualizing fluid flow, balancing waterflood pattern, comparison of waterflood scenarios, managing injection and production, and in the overall surveillance and management of the reservoir. The method is based on the solution of fluid flow equations in porous media and can take into account reservoir heterogeneity, fluid mobility ratio and changes in injection or production. The solution moves along the streamlines making the method more efficient than traditional reservoir simulation where the pressure and saturation values need updating at each grid block following a time step. Any imbalance in injection pattern can be readily identified from the results of streamline simulation [14] studies by visual inspection (Figure 16.9). As noted earlier, pattern balancing based on reservoir simulation, streamline simulation and other techniques minimize the probability of oil migrating out of the pattern, thus reducing recovery. Higher recovery is based upon effective pattern balancing. Fluid flow regulators are used to properly regulate the flow of fluid, at both surface and downhole.
- Flood front tracking. A flood front showing the edges of advancing water bank may pinpoint the reservoir areas and zones that are not swept. Identification of unswept areas allows the reservoir engineers to take appropriate actions, including any changes in well rates, workover, well conversion, drilling of new well, and others. Advance of waterflood can be monitored by various methods, including the analysis of fluid sample collection at wells, analysis of bottom-hole pressure data collected by downhole sensors, data obtained from observation wells, reservoir simulation based upon history matching, and period 4D seismic surveys.
- Injection well tests. Transient pressure tests are conducted to diagnose any issues with the injection. The objectives of testing include the reduction in skin, avoidance of formation

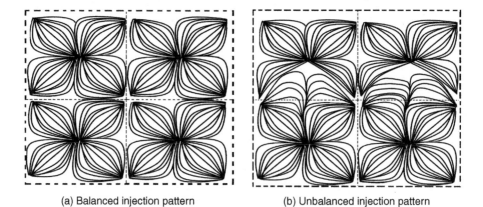

(a) Balanced injection pattern   (b) Unbalanced injection pattern

Figure 16.9 Balanced versus unbalanced waterflood pattern.

plugging, better injectivity, uniform injection profile across the formation thickness, and identification of any fractures or anomalies.

- Injection profile logging. The injection profile surveys are conducted periodically to identify "thief zones" having dominant water entry into the formation, formation plugging, water injection diverted from the intended zone, and zones that are not injected adequately, among others. One common method is to conduct a spinner survey, where the rotation of the tool indicates the relative rates at which injected water enters various sections of the formation (Figure 16.10). In an ideal case, the rate is the same across the formation. In thief zones, the rate is significantly higher than the rest. Uneven injection profile often results from the stratification of the reservoir.
- Management of water injection profile. Any anomaly in injection profile identified by surveys can be reduced by squeeze cementing and thief zone blockage through polymer treatments. It is a common practice to recomplete a well by selectively perforating the target zone that eliminates water entry in problematic zones. In recent decades, smart wells are deployed to selectively block the watered-out zones. During recompletion of wells, careful considerations are made with regard to the selection of perforation interval, workover fluid, and cleanup procedure.
- Well cleanout. Well cleanout on regular basis is necessary to improve injection profile and enhance well injectivity.
- Cased hole logging. Cased hole logs can detect any water flow behind casing due to the poor bonding between casing and/or cement.
- Analysis of produced fluid. Samples of produced water are analyzed for salinity. A sharp rise in chloride content would indicate the breakthrough of injected water at the well. A premature breakthrough results in poor recovery. The leading causes of premature breakthrough

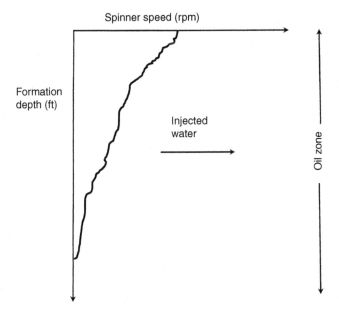

Figure 16.10 Spinner survey results indicate that the majority of injected water entering the top part of the formation results in poor vertical sweep.

include the presence of thief zones, fractures and other reservoir heterogeneities, and unbalanced injection in various patterns.

- Diagnostic plots. In waterflood surveillance, various diagnostic plots serve as effective visual tools; a few of them are discussed in the following.
  - Water cut versus cumulative production. Extrapolation of water cut versus cumulative production based on production history may lead to the estimation of ultimate oil recovery from a well. A representative plot of log of water cut versus cumulative production is presented in Chapter 13.
  - Bubble map. A bubble map is presented in Figure 6.3. The injected volume of water at each injection well is shown as a bubble or circle, the diameter of the circle being proportional to injected volume. Wells with larger bubbles indicate relatively large amounts of water injection. The bubble map of a field under waterflood readily indicates the areas of overinjection and underinjection. Underinjected portions of the reservoir may need more focus in recovering the large volume of oil that is left behind. Again, the producers may be labeled with water cut, which may pinpoint certain areas of the reservoirs where high permeability channels or fractures are suspected.
  - Hall plot [15]. This is a well-recognized graphical technique to analyze injection well performance, including well injectivity. A modified version of the plot is based on a plot of cumulative well-head pressure over time as ordinate versus cumulative injection as abscissa (Figure 16.11). The value of cumulative pressure over time have the unit of psi × days, and can be obtained by integrating the well-head injection pressure over time under certain simplifying assumptions. Hall plot is a continuous monitoring technique, which can provide a wealth of information regarding the characteristics of an injection well over the long term. Referring to Figure 16.11, the Hall plot may indicate the following:
    - Line concaving upward at the beginning of water flood – Expansion of water bank
    - Straight line – stable injection

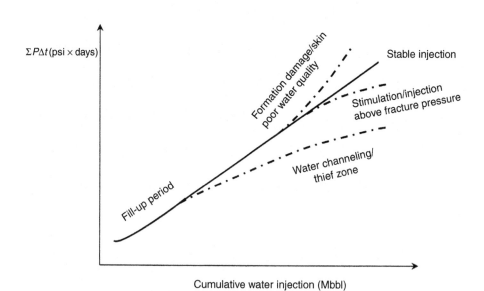

Figure 16.11 Water injection diagnostics based on Hall plot.

- Line concaving upward at later stages – Loss of injectivity due to skin damage or poor water quality
- Line with decreasing slope – Enhancement of injectivity due to negative skin or injection above formation fracturing pressure
- Line with abnormally low slope – Loss of injected water in high permeability channel or thief zone

- Waterflood operation. Waterflood operation on a day-to-day basis is managed by field operations staff, who are responsible for data collection and identifying potential problems related to the mechanical and electrical aspects of the equipment used. Chemical aspects of the fluids pertaining to corrosion and any changes in composition are also monitored.

Summing up

Introduction: Most oil reservoirs are subjected to waterflooding or water injection to maintain reservoir pressure and recover additional amounts of oil, thus adding to the reservoir assets. It has been the industry experience that generally 15−30% of OOIP can be recovered by waterflood. Light to intermediate oil reservoirs having relatively low viscosity are good candidates for waterflooding. Oil having an API gravity of 20° or more is amenable to efficient displacement and production in a reservoir with favorable rock characteristics.

History of waterflood and maturation of technology: The first waterflood occurred as a result of accidental water injection in the Pithole City area of Pennsylvania as early as 1865. In the absence of detailed reservoir characterization and robust dynamic models of the reservoir, achieving success in waterflooding was a challenge. Waterflooding technology progressively matured since then to the point where waterflood is managed in real time or near real time by a team of experts who have a vast knowledge of the reservoir, modern surveillance tools, and large-scale reservoir simulation.

Waterflood process: Waterflooding process consists of injecting water into a set of wells that provide energy to the reservoir in the form of high pressure. Under an ideal situation, a water bank is formed around the injector. The water bank expands as more water is injected, which pushes or displaces the oil toward the producer wells where the pressure is relatively low. Essentially, waterflood or injection maintains reservoir pressure and displaces oil from the injector to the producer.

Response of waterflood: Typical waterflood response is characterized by an increase in oil rate as oil trapped in rock pores is pushed toward the producing wells. Eventually, breakthrough of injected water is encountered at the producers. The water−oil ratio continues to rise with time, oil production rate declines, and the economic limit is reached when the water production becomes excessive.

Advantages of waterflood: Waterflood is a widely used process due to the abundance of water in rivers, streams, lakes, and oceans that can be used economically in the oil recovery process. Waterflood operation is the least expensive method in comparison to other enhanced oil recovery processes such as surfactant or chemical injection. Moreover, water is relatively easy to inject and contacts large areas in the reservoir by spreading rather freely.

Waterflood pattern: This refers to the relative location of injectors and producers in a reservoir. A literature review suggests that common waterflood patterns are line drives (direct and staggered), and five-spot, seven-spot, and nine-spot patterns. A nine-spot pattern indicates the presence of eight injectors at the corners of the pattern while one producer is located centrally. On the other hand, an inverted nine-spot pattern indicates eight producers and one injector; the latter is located in the middle.

Well spacing: A literature review also suggests that common well spacings are 40, 80, and 160 acres. However, 20 and 320 acre spacings are also reported in many instances. Well spacing is influenced by rock and fluid characteristics, reservoir heterogeneities, optimum injection pressure, time frame for recovery, and economics, among other factors. Generally, tight reservoirs require close well spacings.

Life of waterflood: The life of a waterflood project, reflecting the rate at which oil is recovered, depends upon the number of injector wells, water injection rate and well injectivity, distance between injectors and producers, and reservoir quality, among other factors. A drastic decline in water injectivity may be observed during the early period of water injection into a reservoir that is produced by solution gas drive. Once injected water fills up the pore space previously occupied by gas, the injectivity of water is dependent upon the mobility ratio.

Mobility ratio: The mobility ratio is one of the most critical parameters in the waterflood process. It is defined as the ratio of mobility of displacing fluid, i.e., water over the mobility of displaced fluid, i.e., oil in case of waterflood. The mobility of a fluid, oil or water, is defined as the permeability of the rock to the fluid in question over viscosity. The relative permeability to water should be obtained from the zone swept by water, while the relative permeability to oil should be based on an unswept region, which is located ahead of the displacement front.

Significance of mobility ratio: When the mobility ratio is less than one, water is less mobile than oil, which leads to higher oil recovery. On the contrary, when the mobility ratio is greater than one, water is more mobile than oil resulting in unfavorable conditions. Water breakthrough occurs prematurely at the producers resulting in poor recovery of oil.

Reservoir performance under waterflood: Performance of the reservoirs, whether primary, secondary, or tertiary, is greatly influenced by the heterogeneity of the formation. The heterogeneity of rock is attributed to changes in depositional environments and geologic events that occurred following deposition and formation of rock and oil. The heterogeneities that affect waterflood include:

- Significant variations in storativity and transmissibility of rock within the reservoir
- Presence of high permeability streaks and channels
- Stratification or layering of geologic formation with contrasting transmissibility
- Presence of natural fractures in rock
- Existence of sealing and nonsealing faults
- Compartmental reservoirs where noncommunicating sections are encountered

Generally, carbonate rocks exhibit a high degree of heterogeneity. Injected water is likely to flow in preferential channels leading to early water breakthrough and poor recovery of oil.

Waterflood recovery efficiency: The overall waterflood recovery efficiency is given by the product of pore-to-pore displacement efficiency and volumetric sweep efficiency as follows:

$$E_R = E_D \times E_A \times E_V \tag{16.3}$$

Displacement efficiency is influenced by rock and fluid properties, and pore volumes injected. It depends on fluid viscosity, reservoir dip, wettability of the rock, and interfacial tension. Volumetric sweep efficiency is given by the product of areal and vertical sweep efficiencies. The areal sweep efficiency is influenced by the flooding pattern type, mobility ratio, and throughput and reservoir heterogeneity. Vertical sweep efficiency is influenced by layer permeability variations and the mobility ratio.

Waterflood performance prediction methods: Modern prediction methods of waterflood performance are based on computer-based simulation of dynamic reservoir models, including streamline, finite difference, and finite element solutions. Classical methods for predicting waterflood performance include Dykstra−Parsons, Stiles, Prats−Matthews−Jewett−Baker, Buckley−Leverett, and Craig−Geffen−Morse. However, these methods have many restrictive assumptions. Most of the prediction methods are based on either laboratory investigations or simplified analytic solutions of waterflooding in an ideal or near-ideal situation. Dykstra−Parsons waterflood recovery correlations are widely used to estimate waterflood recovery efficiency. The method uses Dykstra−Parsons permeability variation factor, mobility ratio, and water cut.

Water flood surveillance: For a successful waterflood project, continuous monitoring of the waterflood operation followed by corrective action, if necessary, is imperative. In managing waterflood operation through reservoir surveillance, the following are considered:

- Waterflood data including reservoir pressure, well rates, water cuts, cumulative volumes and recovery
- Graphical diagnostic tools based on streamline simulation, bubble map and Hall plot
- Well conversion from injector to producing, well drilling scheduling, waterflood pattern balancing and changes in well spacing
- Reservoir characterization including identification of hydraulic units, thief zones, presence of fractures, watered-out layers, high permeability channels, and other heterogeneities
- Identification of viscous fingering and gravity underriding by injected water
- Monitoring of formation damage or skin around the wellbore, selection of target layers to optimize oil recovery
- Well testing to determine fluid flow characteristics under water flow
- Waterflood pattern − realignment or repositioning of injectors and producers for better waterflood efficiency
- Tracer studies to characterize the injected water flow paths
- Installation of observation wells to monitor the progress of waterflood
- 4D seismic surveys to monitor the dynamic oil saturation profile and bypassed oil

A general workflow for the design, implementation, and management of waterflood is presented in Figure 16.12.

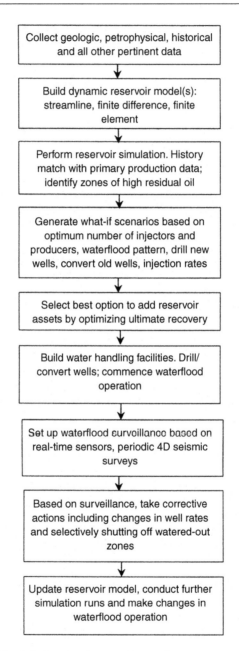

Figure 16.12 Waterflood design, implementation, and management workflow.

Questions and assignments

1. What is waterflooding? Why is it implemented in oil reservoirs?
2. Why is waterflooding referred to as secondary recovery?
3. How does the waterflood process work? What rock properties are important in analyzing waterflood?
4. What are the types of waterflood patterns used? How would you calculate the distance between wells in 80 acre spacing?
5. What are the deciding factors in selecting well pattern and spacing?
6. What is well conversion? How is a well conversion schedule determined?
7. Can horizontal wells be injectors? Why or why not?
8. Why might multistage fracturing be required in waterflood operations?
9. Define waterflood recovery efficiency in terms of volumetric sweep and displacement efficiency. What is the expected range of recovery efficiency by waterflood?
10. Describe the significance of mobility ratio. How can you estimate mobility ratio prior to waterflood?
11. What factors would be detrimental to areal and vertical sweep during waterflood?
12. What is displacement efficiency? Why does water not displace all the oil it comes in contact with during waterflood?
13. Why should injection pressure gradient not exceed the fracture gradient of the formation?
14. What reservoir heterogeneities would influence recovery efficiency and how?
15. What are the prediction methods for waterflood performance? What data are required in predicting the performance? Describe the advantages and disadvantages of empirical methods.
16. Describe Buckley−Leverett frontal displacement theory. List the assumptions.
17. Why is reservoir simulation used in waterflood design? What information is sought from simulation studies?
18. Why is waterflood surveillance necessary? What are the waterflood surveillance techniques?
19. What are the aspects to consider in monitoring the performance of waterflood?
20. What is Hall plot? How is the plot used in diagnosing potential issues in waterflood?
21. What is a spinner survey? How does it aid in waterflood management?
22. How is a bubble map used as a visual diagnostic tool?
23. What is pattern balancing? How can streamline simulation aid in balancing a waterflood pattern?
24. Why is water quality maintained during waterflood?
25. Why is the salinity of produced water monitored in waterflooding?
26. Your company is planning to waterflood a heterogeneous carbonate reservoir with limited production history. Describe the steps you would propose in designing a successful waterflood and reservoir management once the project is implemented.
27. Provide a field example where a transient well test is used to monitor waterflood.
28. Based on a literature review, describe waterflood projects in the following reservoirs and explain the contrasting strategies:
    a. Fractured reservoir with tight rock matrix
    b. Compartmental reservoir
    c. Stratified reservoir with interlayer communication

References

[1] History of petroleum engineering. American Petroleum Institute: Washington (DC); 1961.

[2] Thakur GC. Waterflood surveillance technique – a reservoir management approach. J. Petrol. Technol. 1991;October:1,180–8. Society of Petroleum Engineers.

[3] Craig FF Jr. The reservoir engineering aspects of water flooding. SPE monograph, vol. 3. Richardson (TX): Society of Petroleum Engineers; 1971.

[4] Muskat M. Physical principles of oil production. New York: McGraw-Hill Book Co., Inc.; 1950.

[5] Deppe JC. Injection rates – the effect of mobility ratio, areal sweep, and pattern. Soc. Petrol. Eng. J. 1961;June:81–91.

[6] Rose HC, Buchwalter JF, Woodhall RJ. The design aspects of waterflooding. SPE monograph 11. Richardson, (TX): Society of Petroleum Engineers; 1989.

[7] Dykstra H, Parsons RL. The prediction of oil recovery by waterflooding. Secondary recovery of oil in the United States. 2nd ed. Washington, DC: American Petroleum Institute; 1950. p. 160–74.

[8] Dyes AB, Caudle BH, Erickson RA. Oil production after breakthrough as influenced by mobility ratio. Trans. AIME 1954;201:81–6.

[9] Stiles WE. Use of permeability distribution in water flood calculations. Trans. AIME 1949;186:9–13.

[10] Buckley SE, Leverett MC. Mechanisms of fluid displacement in sands. Trans. AIME 1942;146:107–16.

[11] Craig FF Jr, Geffen TM, Morse RA. Oil recovery performance of pattern gas or water injection operations from model tests. Trans. AIME 1955;204:7–15.

[12] Prats M. Prediction of injection rate and production history for multifluid five-spot floods. Trans. AIME 1959;216:98–105.

[13] Talash AW. An overview of waterflood surveillance and monitoring. JPT 1988;December:1539–43.

[14] Muskat M, Wyckoff RD. A theoretical analysis of waterflooding networks. Trans. AIME 1934;107:62–77.

[15] Hall HN. How to analyze waterflood injection well performance. World Well 1963; October:128–30.

Enhanced oil recovery processes: thermal, chemical, and miscible floods

17

Introduction

Enhanced oil recovery (EOR) processes are implemented in most reservoirs worldwide to recover additional amounts of oil that are not otherwise recovered during secondary recovery by waterflood or gas injection. Worldwide statistics of petroleum reserves indicate that a vast portion of oil and gas, although discovered and quantified, is left underground following primary and secondary recovery due to the lack of available technology and unfavorable economics. Unrecovered oil in a conventional reservoir even having good reservoir quality often exceeds half of the amount of petroleum initially in place (PIIP). Furthermore, recovery from heavy oil reservoirs as well as from unconventional reservoirs, including tight shale formations, is quite low. According to an estimate by the Energy Information Administration, 300 billion barrels of oil remain untapped. Therefore, EOR techniques are employed as a tertiary recovery process to recover the unexploited resources to the extent possible in a technological and economic sense.

The EOR processes as practiced in the industry are quite large in number. This is due to the fact that the suitability of a specific EOR process is highly sensitive to reservoir and fluid characteristics. Economics also dictate the feasibility of a specific EOR process. Certain EOR processes are not viable unless crude oil reaches a price point.

This chapter presents an overview of the EOR processes and provides answers to the following:

- What is EOR? How does it differ from other types of oil recovery?
- What recoveries are expected during primary, secondary, and tertiary production of an oil reservoir?
- What are the main classifications of EOR methods?
- What reservoirs are suitable for EOR operations? What are the screening criteria to implement an EOR process?
- What are the mechanisms at work in the reservoir to produce more oil during an EOR process?
- What EOR methods are practiced worldwide? How much oil is produced by enhanced recovery processes?
- What is the cost of production by EOR processes? How does it compare against conventional and unconventional oil?
- What factors influence the success of an EOR method?
- What are thermal, miscible, immiscible, and chemical recovery methods?
- What are the relative advantages and limitations of each method?
- How is the EOR method designed, tested, and implemented in the field?

Reservoir Engineering. http://dx.doi.org/10.1016/B978-0-12-800219-3.00017-6

Table 17.1 **Primary, secondary, and tertiary recovery of oil**

Type of oil recovery	Recovery process	Cumulative recovery (%)	Notes
Primary	Natural reservoir energy	10–30	Lower recovery from certain unconventional reservoirs
Secondary	Water flood or gas injection	20–40	
Tertiary (EOR)	Thermal, chemical, gas injection, and others	30–60	Under favorable conditions, up to 80% of PIIP may be recovered

Note: The EOR processes are part of improved oil recovery (IOR), which includes both secondary and tertiary recovery projects.

Primary, secondary, and tertiary recoveries

Recoveries from various reservoirs are heavily dependent on rock and fluid properties, and reservoir characteristics. In general, based on worldwide statistics, the guidelines shown in Table 17.1 can be used.

Reservoir candidates for EOR

Unfavorable fluid properties and poor reservoir quality are the principal causes of limited success during primary and secondary recovery. The reservoirs that are prime candidates for EOR include, but are not limited to:

- Heavy oil reservoirs where oil viscosity is quite high and oil is not mobile
- Unconventional reservoirs including oil sands where hydrocarbons are ultraheavy
- Conventional oil reservoirs where high residual oil saturation exists due to the interfacial tension that exists between oil and water
- Heterogeneous reservoirs with significant variations in formation transmissibility and other issues where injection of gas or other fluids may result in further recovery

EOR processes (Figure 17.1) can be broadly classified as follows [1–3]:

- Thermal: The mechanism of recovery is primarily based upon reducing the viscosity by imparting heat to heavy oil and thus increasing its mobility in the reservoir. As thermal energy is introduced into the reservoir by steam or oil and combusted by air, other processes also occur in the reservoir that enhance the recovery of heavy oil. Thermal recovery methods are applied to both conventional and unconventional reservoirs.
  - Conventional reservoirs: Steam flooding, cyclic steam injection (huff-and-puff method), hot water drive, and *in situ* combustion.
  - Unconventional reservoirs: Steam assisted gravity drive, vapor extraction. The two processes are described in Chapter 21.
- Miscible: Light to medium gravity oil is recovered by nonthermal EOR methods, the most popular of the methods being miscible injection. Miscible displacement is a highly efficient method of oil recovery as it reduces interfacial tension between fluids and significantly enhances the microscopic displacement process. Miscibility is achieved when injected fluid mixes with *in situ* oil

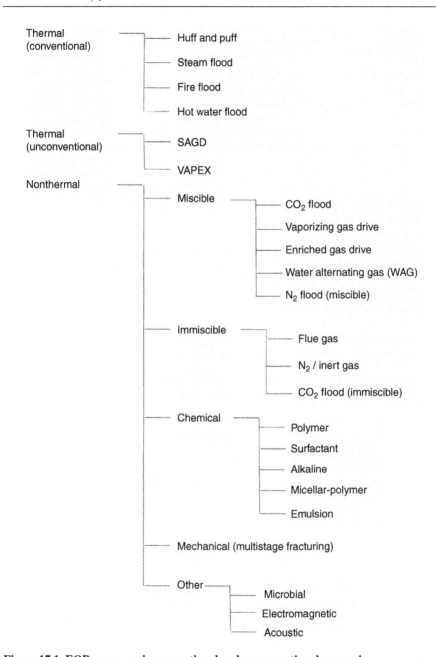

Figure 17.1 EOR processes in conventional and unconventional reservoirs.

completely; the individual fluids are indistinguishable within the miscible bank. The bank moves through the reservoir and recovers oil efficiently. EOR miscible processes primarily include carbon dioxide (CO_2) and hydrocarbon gas. Water alternating gas (WAG) injection is used where alternate slugs of hydrocarbon and water are injected into the reservoir. CO_2 flood has been a success story. It is widely implemented as a nonthermal method depending on adequate reservoir pressure and availability of CO_2. Miscibility between injected and *in situ* fluids occurs only when the reservoir pressure exceeds the minimum miscibility pressure (MMP). It is a function of fluid composition and reservoir temperature. Nitrogen is also used in miscible flood; however, it is not as effective as CO_2 due to its poor solubility and lower viscosity. Again, N_2 requires higher pressure to be miscible with oil.

- Immiscible: Inert and flue gas are used for immiscible displacement of oil. The process can be partially miscible depending on pressure and oil composition. When CO_2 is injected below the MMP, miscibility between injected fluid and oil is not attained; hence, immiscible displacement takes place as a recovery process. Recovery by immiscible displacement is less efficient than that by miscible EOR.
- Chemical: Polymer, surfactant, and alkaline flood are part of EOR processes that recover oil by adding the chemical substances to injected water. These are viewed as modifications of waterflood with an objective to recover additional volumes of oil. The mechanisms for recovering additional amounts of oil by chemical methods include, but are not limited to, increase in injected water viscosity, thus reducing its mobility, reduction in interfacial tension, emulsification of oil and water, solubilization of oil, and increase in volumetric sweep of the reservoir.
- Other: Microbial, acoustic, and electromagnetic. These methods are in an experimental stage with little documentation of any large-scale implementation or their economic feasibility. However, the microbial enhanced recovery process has been the subject of quite a number of studies.

Of note, the definition of EOR is treated in a much broader perspective in certain literature. In addition to the application of thermal energy, injection of fluids and addition of chemicals to water, mechanical and other methods for enhancing oil recovery are also included. For example, multistage fracturing of tight formations is a major innovation for recovering oil and gas that was not viable before, and the method is considered to be part of the EOR universe. A study of horizontal wells in the United States indicates that over 60% of the wells have been hydraulically fractured to enhance productivity.

Economics of EOR

The success of an EOR method depends on its suitability for a specific reservoir, recovery efficiency, availability of injected fluids, and costs. Projects pose technical challenges and risks are significant. EOR processes require substantial financial investments initially and are associated with high operating costs. The cost of production of one barrel of oil by tertiary recovery can be significantly higher than that by waterflooding or gas injection. Return on capital investments may be realized after a considerable period of time.

Table 17.2 is provided as a general guide for assessing the cost of oil production based on various EOR methods [4].

Table 17.2 **Cost of oil based on EOR methods; comparison with conventional and unconventional oil production cost**

Method of production	Cost ($/bbl)
EOR processes	
CO_2 injection	20–70
Thermal	40–80
Others	30–80
Conventional oil	10–30
Unconventional oil	50–90

A review of worldwide production attests to the fact that EOR projects are dependent on market conditions. As the price of oil increases, more efforts are seen to produce oil by various EOR methods.

Worldwide EOR perspective

According to one report that combined data from various sources, production from EOR projects is about 3 million barrels per day, contributing 3–4% of the total daily production worldwide [4]. By far, most tertiary oil (about 2 million barrels per day) is produced by thermal EOR methods worldwide. There are over 100 thermal recovery projects in various countries including Canada, the United States, Venezuela, Indonesia, China, and other countries. CO_2-EOR projects, contributing over 300,000 barrels of oil per day, are mainly concentrated in the Permian Basin of the United States and the Weyburn field in Canada. The number of CO_2 injection projects exceeds 100 as well. Besides EOR efforts, some of the projects are involved in CO_2 sequestration. Projects involving the injection of hydrocarbons are found in Venezuela, the United States, Canada, and Libya. The total recovery by hydrocarbon injection methods is similar to CO_2-EOR. However, the number of projects is far fewer, about 25 or less. Recovery by chemical methods is concentrated in China. The projects are just a few; however, oil production by chemical recovery is similar to that of CO_2-EOR and hydrocarbon injection.

A review of EOR projects in the United States indicates that steam flooding and CO_2 injection are currently the dominant processes. Gas injection, including CO_2 flooding, accounts for about 60% of all EOR-based production in the United States, while thermal methods account for about 40%. Tertiary recovery by chemical methods such as polymer injection or surfactant account for less than 1% of the total production by EOR methods due to the high cost of injection materials, process complexity, narrow range of applicability, and associated risks.

Thermal recovery: cyclic steam injection process

Enhanced oil production by thermal recovery methods, including steam injection and *in situ* combustion, is based upon the fact that when thermal energy is applied to heavy oil, its viscosity is reduced to a point where oil becomes mobile and can be produced

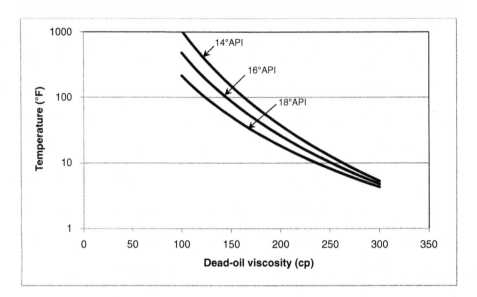

Figure 17.2 Significant reduction of heavy oil viscosity with increase in temperature.
Dead oil viscosity is plotted against temperature based on the samples of North Sea crude [5].
Once all the volatile components are liberated from oil due to decline in pressure, it is referred
to as dead oil.

at the wellbore with relative ease. The viscosity of oil is quite sensitive to temperature.
For example, oil viscosity can be reduced from hundreds of centipoise to 10 cp or less
when heated from 100°F to 300°F (Figure 17.2).

The method requires a single well to inject steam and subsequently produce in
a heavy oil reservoir. The cyclic steam injection process is very effective where the
reservoir has good permeability, sometimes in darcies, and oil is highly viscous oil.
The performance of this method drops as further cycles are carried out. Oil recovery
is generally small in this process, as it can impart thermal energy in a localized area
surrounding the well that acts as both injector and producer.

In the cyclic steam injection process, also referred to as the "huff-and-puff" method,
steam is injected into a single well at a high rate for a period of time, which can last
2–6 weeks. A soak period of 3–6 days ensues, when the steam is allowed to provide
thermal energy to oil. As a result, oil becomes less viscous and mobile. The well is allowed
to flow back and is pumped to produce oil. The oil rate increases initially, followed by a decline in rate as most of the oil with reduced viscosity is produced. Increased oil production
may last for several months to a year. When the rate of production declines, the entire process is repeated. This cyclic steam injection process, followed by production, is repeated
until the well becomes uneconomic to operate due to marginal production. In some cases,
however, the well can be converted from cyclic steam stimulation to direct steam flooding
where steam drives the less viscous oil to a producing well.

Cyclic steam injection or huff-and-puff is basically a well stimulation process
where oil properties are altered to facilitate flow. In the stimulation process, the steam

spreads through the oil around the wellbore, heating the oil. The soak period allows the oil to be heated even further. During the production cycle, the mobilized oil flows into the wellbore; the driving mechanism includes reservoir pressure and reduced oil gravity as the well is pumped.

Steam flooding

As the name suggests, the steam flooding process involves the continuous injection of steam to "flood" the reservoir in order to thermally stimulate heavy oil and alter its characteristics. As a result, oil becomes less viscous and mobile. Steam of 80% quality, i.e., steam at the surface containing about 80% steam and 20% water, is injected continuously into the reservoir to reduce the oil viscosity; consequently the oil is driven towards the producing wells due to the pressure differential between the reservoir and wellbore, among other factors. The injected steam forms a steam zone that advances slowly (Figure 17.3). The injected steam not only heats the viscous oil, but heat is transferred to the formation as well as to the adjacent cap and base rock. Steam flooding works better in relatively thick formation as the thermal energy can contact more oil rather than be dissipated in adjacent layers. Similarly, heat dissipation is more pronounced in shallow reservoirs. Due to heat loss, some of the steam condenses to yield a mixture of steam and hot water.

A hot oil bank is formed ahead of the steam and hot water zone, which moves towards the producing well. In many cases, the injected steam overrides the oil due to gravity, which can create certain issues. Eventually, steam breakthrough occurs at the producers following sustained injection. This results in increased steam injection rate and less recovery efficiency. The rate is reduced by recompleting the wells or by shutting off

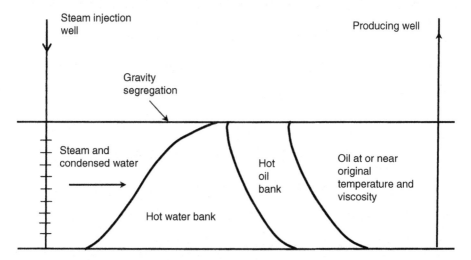

Figure 17.3 Simplified depiction of steam flooding process. Sharp edges of steam, water, and oil fronts are shown for illustration only. In reality, the fronts are diffused.

steam-producing zones in the formation. Steam flooding is an effective recovery process. The residual oil saturation in the steam zone can be as low as 10% or even less.

Besides increased mobility of less viscous oil under a pressure gradient, there are other drive mechanisms at work during steam flood. Some portion of heated oil is transported and recovered by steam distillation. The solvent extraction process is also involved in the recovery of oil.

The limitations of the steam flooding process include the following:

• Relatively thin oil zones and shallow reservoirs may not be suitable for steam flood due to the significant heat loss resulting in poor recovery efficiency.
• High oil saturation, good reservoir transmissibility, and relatively thick formations (20 ft. or more) are critical considerations in designing steam flood.
• Presence of bottom water and gas cap is detrimental to recovery efficiency during steam flood.
• Steam flood is not generally implemented in carbonate reservoirs. However, heavy oil reservoirs in unconsolidated formations are known to produce satisfactorily by steam injection.
• Steam flood does not have any additional advantage in the reservoirs having light to intermediate oil that can be water flooded to augment recovery.

Case Study: Steam Flooding in the Giant Duri Field, Indonesia

The Duri Steam Flood Project [6–8] is the one of the largest as well as highly successful thermal EOR projects in the world. Duri is the second largest field in the country. The OOIP is estimated to be in the billions of barrels. The field has an anticlinal structure with faults. Reservoir permeability is very good, ranging between 100 mD and 4 D. Average oil gravity is about 21°API and viscosity is about 330 cp at 100°F. Oil viscosity is reduced to only 8 cp when temperature is raised to 300°F. Initial reservoir pressure and temperature are 100 psi and 100°F, respectively. Initial oil saturation is 55%.

The field began its primary production in the late 1950s and reached its peak oil production rate in less than a decade. Poor mobility of viscous oil combined with low solution gas−oil ratio led to the rapid decline in rates. A weak aquifer provided marginal support and a mechanism of gravity drainage did not play any significant role in recovery.

The thermal EOR process, in the form of cyclic steam stimulation, started in 1967. Eight years later, a steam flood pilot project was started to test its suitability and effectiveness in the field. The pilot project involved 18 inverted five-spot patterns where each pattern was comprised of four producers and one steam injection well. Oil recovery from the patterns was about 30% of the OOIP. Following the favorable outcome of the pilot project, a field-scale operation was started in 1985 based principally on an inverted seven-spot pattern. After 14 years of steam flood, the recovery factor was 60−70% in one area of the reservoir. In the mid-2000s, Duri oil production exceeded 200,000 barrels of oil per day.

The Duri development project for thermal recovery was massive, involving 4,000 producing wells spread over an area of 15,000 acres of reservoir. The large area was divided in several sections, each requiring about 2 years to develop. Symmetric steam flood patterns of varying size and configuration were put in place. Other steam flood patterns involving inverted five- and nine-spots, approximately 15½ acres in size, were planned based on detailed reservoir simulation studies. In 2009, steam injection was deployed in 80% of the field. In that period, 185 production wells and 40 steam injection wells were drilled to enhance recovery. Duri field has produced more oil than what is produced from the giant steam flood projects in California, namely, Kern River and Belridge Fields located in the San Joaquin valley.

In situ combustion

In situ combustion, also referred to as fire flooding, is based on the combustion of a certain amount of heavy oil underground and injection of air, or oxygen-enriched air, to sustain the burning of oil. As much as 10% of OOIP is subjected to combustion in the process. A common technique is the forward combustion process, where oil is ignited at the bottom of the well and air is injected to propagate the combustion front further into the reservoir. A second technique of the *in situ* combustion process is a reverse combustion. In this process, a fire is started in a well that will eventually become a producing well, and air injection is then switched to adjacent wells. However, a literature review suggests that the process has not been economically viable.

In situ combustion is one of the oldest techniques of heavy oil recovery where the reservoir is ignited at the bottom of the injection well by a special heater, and air is injected to propagate the combustion front away from the well. The lighter hydrocarbon components of the heavy oil are mobile and propagate forward; these components also mix with the heavier oil ahead of the combustion zone to upgrade the oil. The heavier hydrocarbons are burned. The temperature rises to about 600°C generating a large volume of flue gas. Heat is generated within a combustion zone at a very high temperature, about 600°C. As a result of burning the crude oil, large volumes of flue gas are produced. Steam, hot water, combustion gas, and distilled solvent produced in the process further aid in driving oil toward the wellbore.

In certain reservoirs, the injection of air for combustion is followed by the injection of water, or air and water are injected simultaneously. The injected water improves the efficiency of the process by transferring heat from combustion front to oil. The process is referred to as wet combustion or combination of forward combustion and waterflooding.

The following drives the recovery of oil by *in situ* combustion:

- Thermal energy that lowers the oil viscosity and makes the oil mobile
- Increase in reservoir pressure as created by injected air that drives the less viscous oil to the wellbore
- Lighter hydrocarbon components, the product of steam distillation, and thermal cracking, which are carried forward and upgrade the heavy crude oil
- Burning coke, which is produced from the heavier component

The limitations of the process include the following:

- In general, the *in situ* combustion process could be complicated and suitability may depend on a case-by-case basis
- Combustion cannot be sustained if coke is not burned in sufficient quantities
- On the other hand, excessive coke deposits lead to a slow advancement of combustion front
- Adverse mobility ratio, as the hot gases produced as a result of combustion are much more mobile than *in situ* heavy oil
- The process is found to have poor vertical sweep as it is more effective in the upper part of the formation; hence, relatively thick formations may not have satisfactory recovery
- Environmental and operational issues related to the production of large amounts of flue gases, corrosion, oil–water emulsions, temperature-related failures of pipes, and increased sand production

Miscible methods

As noted earlier, miscibility is achieved when two fluids mix with each other completely and individual fluid phases are indistinguishable. In the miscible EOR process, a gas is injected into the reservoir, which is miscible with the oil under appropriate conditions of pressure and temperature. The injected fluid is either light to intermediate hydrocarbons or CO_2 in most cases. Since it acts as a solvent, the interfacial tension between the two miscible fluids is minimal. As a result, efficient microscopic displacement of oil takes place, enhancing recovery.

Light hydrocarbon components are injected into the reservoir to recover oil by the mechanism of miscible displacement. Hydrocarbon miscible flooding recovers crude oil by generating miscibility between light to intermediate components of oil and the injected fluid. Consequently, a miscible front is created ahead of the injected fluid that efficiently displaces *in situ* oil towards the producers. Other mechanisms of recovery are also at work, including the increase in oil volume (swelling), decrease in oil viscosity, and better sweep of the reservoir.

The three major types of miscible displacement by the injection of hydrocarbons are as follows:

Enriched (condensing) and vaporizing drive

- Enriched gas drive: In this method, a slug of 10–20% of pore volume (PV) of natural gas enriched with ethane through hexane (C_2–C_6) is injected in the reservoir. In certain cases, the slug is followed by lean gas and possibly water. Then the enriching hydrocarbon components are transferred from the injected gas to *in situ* oil. Hence, miscibility is attained between gas and oil. The miscible zone, also referred to as miscible bank or miscible flood front, displaces the reservoir oil toward the producing wells.
- Vaporizing gas drive: In the vaporizing gas drive method, lean gas such as methane is injected at high pressure into the reservoir. The injected gas leads to the vaporization of light to intermediate components from *in situ* oil resulting in miscibility. When light hydrocarbons are injected into the oil reservoir, the injected and *in situ* fluids dynamically exchange various components through multiple contacts with each other until miscibility is achieved. This

is referred to as multiple contact miscibility. Ultimately the miscible front drives oil to the producers.

- First contact miscibility: In contrast to multiple contact miscibility, first contact miscibility is referred to the process where injected gas becomes miscible with oil under appropriate conditions of pressure, temperature, and fluid composition at initial contact with each other.

Limitations of hydrocarbon miscible processes are as follows:

- Depth of the reservoir: Reservoir pressure increases with depth. Shallow reservoirs are not capable of generating the relatively high pressure required to attain miscible conditions within the reservoir. Liquefied petroleum gas (LPG) injection requires a minimum pressure of 1200 psi. High pressure miscible gas drives generally operate in a range of 3000−5000 psi depending on reservoir temperature and oil composition. Hence, the depth of the reservoir should support the minimum pressure to attain miscibility and be effective.
- Poor sweep: Viscous fingering of injected fluids may result during the flood due to the significant contrast in viscosity with *in situ* oil. As a result, horizontal and vertical sweep efficiencies could be poor leaving large portions of oil underground.
- Mobility ratio: Unfavorable mobility ratio can be prevalent lowering the ultimate oil recovery. Hence, the process may be more appropriate in a dipping formation where the advantage of gravity can be leveraged.
- Quantity of injected hydrocarbons: This process requires large quantities of expensive products including hydrocarbons, which could remain unrecovered.

Water alternating gas injection

WAG method: A slug of LPG of about 5% PV, such as propane, is injected followed by lean gas. Water is injected with the chase gas in a WAG mode to improve the mobility ratio between the solvent slug and the chase gas. Alternating gas and water injection improves sweep efficiency and reduces channeling of gas. WAG is widely practiced in CO_2 flooding described in the following.

Carbon dioxide flooding

CO_2 flooding plays a major role in the tertiary recovery of oil. It leads in significant numbers all the nonthermal EOR projects engineered and implemented successfully worldwide. The mechanisms of CO_2-EOR include:

- Attainment of miscibility between *in situ* oil and injected gas reducing interfacial tension between oil and injected gas
- Swelling of oil
- Reduction in oil viscosity
- Lowering of interfacial tension between the oil and the CO_2−oil phase in the near-miscible regions
- Efficient displacement in comparison to other miscible processes

The method is carried out by injecting large quantities of CO_2 into the reservoir under sufficiently high pressure to attain miscibility. The amount of CO_2 may exceed 15% of hydrocarbon pore volume. Although CO_2 is not miscible with the crude oil in the strict sense, it extracts the light to intermediate components from the oil under

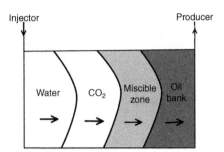

Figure 17.4 Schematic of CO₂-WAG process. The front end of the CO_2 slug forms a miscible bank under requisite pressure, which drives oil toward the producer. Water is injected following the CO_2 slug in order to improve mobility ratio and sweep efficiency. Sharp edges of various fronts are shown for illustration purposes only.

sufficient pressure. If the pressure is above MMP, miscibility develops to efficiently displace the crude oil from the reservoir (Figure 17.4).

Miscible displacement by CO_2 is analogous to the vaporizing gas drive described earlier. However, the advantage with CO_2 flooding is that a wider range of components, C_2-C_{30}, is extracted. Hence, CO_2 injection is effective at lower miscibility pressures as compared to vaporizing gas drive, and more reservoirs can be targeted for miscible EOR.

For CO_2 miscible flood, MMP ranges between 2200 psi and 3200 psi under typical conditions of reservoir pressure and temperature, and oil composition. MMP is lower for lighter crude oil (40°API or higher), and the minimum reservoir depth requirement is about 2500 ft. However, for heavier oil (21.9°API or less) at shallow depth (about 1800 ft.), miscibility between oil and CO_2 is not achieved. As a result, recovery is less due to immiscible displacement of oil. In that case, the reservoir depth should be greater than 4000 ft. to attain miscibility. MMP is usually determined in the laboratory by slim tube experiment. Analytic methods based on equations of state are also used. Generally, the requirement of higher MMP is associated with the heavier crude oil and greater reservoir temperature.

Besides miscibility, important factors that contribute to EOR include swelling of oil, reduction in oil viscosity, solvent extraction, and better sweep efficiency of the reservoir. Due to the solubility of CO_2 in oil at high pressure, swelling of oil and reduction in viscosity are observed even before complete miscibility is attained. At or above the MMP, both the oil and the CO_2 containing intermediate oil components flow together due the low interfacial tension between the phases.

In CO_2-WAG process, about 20–50% of the CO_2 slug is followed by a slug of water intended to improve mobility ratio between oil and injected fluids. Reservoir sweep efficiency is also enhanced in the process (Figure 17.4).

The limitations of CO_2-EOR include the following:

- Availability of CO_2
- Requirement of CO_2 in large quantities
- Poor mobility control due to the low viscosity of CO_2

- Premature breakthrough of injected gas
- Corrosion in wells
- Separation of CO_2 from oil in produced stream

Nitrogen and flue gas flooding

Miscible EOR processes based on the injection of nitrogen and flue gas recover oil by the vaporization of lighter hydrocarbon components of crude oil under sufficient pressure. The process can be either miscible or nonmiscible, depending on reservoir pressure and oil composition. The advantages of N_2 and flue gas flooding are:

- Utilization of inexpensive nonhydrocarbon gases
- Injection of large volumes of N_2 and flue gas due to low cost
- Availability of injection gas
- Attainment of miscibility under high pressure
- Can be used as a chase gas in hydrocarbon and CO_2 miscible drive

However, the limitations of N_2 and flue gas are as follows:

- N_2 has lower viscosity than CO_2 resulting in adverse mobility ratio
- Poor solubility in oil
- Much higher pressure requirements for attaining miscibility with oil
- Consequently, deeper reservoirs having light crude oil are candidates for N_2-EOR
- Viscous fingering bypassing large amounts of oil
- Poor horizontal and vertical sweep
- Dipping reservoirs are preferred to avoid the effects of gravity segregation
- Corrosion problems in wells when flue gas is used
- Separation of hydrocarbon gases in produced stream

Polymer flood and chemical methods

A polymer flood increases water viscosity and improves mobility ratio and sweep efficiency. Chemical flooding processes involve the addition of a chemical, such as surfactants, micellar−polymer, and caustic (alkaline) materials to water in order to achieve similar goals as well as lower interfacial tension. The chemical-EOR processes can be viewed as modifications of waterflood and require the same favorable conditions to succeed as in waterflood. However, processes are sensitive to additional parameters described in the following sections.

Polymer flooding

In this process, water-soluble polymers are added to the water before it is injected into the reservoir. Low concentrations (usually 250–2000 mg/L) of certain synthetic or biopolymers are used.

The advantages of polymer flooding are as follows:

- Increase in water viscosity
- Decrease in water mobility
- Better horizontal and vertical sweep
- Contact with large reservoir area

The displacement of oil is more efficient than conventional waterflood in the early stages. However, no significant reduction in residual oil saturation is observed. However, more oil can be produced in the early stages of polymer flood generating revenues.

The limitations of polymer flood include:

- Lesser injectivity of polymer flood compared to waterflood alone
- Loss of viscosity of certain polymers due to shear and microbial degradation
- Potential for wellbore plugging
- Adsorption of polymer by clays in the formation
- Sensitiveness to reservoir heterogeneities such as fractures and channels where cross-linked polymers or gel are preferable
- Requirement of large amounts of polymer to achieve desired results
- Cost of polymer material

Micellar–polymer flooding

The process of micellar–polymer EOR may include the injection of fluids and chemicals in the following sequence:

- A preflush of water having low salinity
- Micellar slug (surfactant in colloidal solution)
- Polymer as mobility buffer
- Water as a driving fluid

Mechanisms of recovery include the reduction of interfacial tension between oil and injected fluid, solubilization of oil, and emulsification of oil and water, which improves mobility ratio. Alteration of rock wettability may also take place. The size of the slug may vary from 5% to 15% of pore volume when the concentration of surfactant is high. However, for chemicals in lower concentration, the size of the slug is larger, as much as 50%. Oil may also be added to the slug.

Another method utilizes microemulsions for better oil displacement and recovery. Microemulsions are single-phase fluids that contain oil, water, and a suitable surfactant, and in some cases cosurfactant, to attain ultralow interfacial tension. Microemulsions can either be injected or developed *in situ*.

The limitations of EOR-surfactant flood are as follows:

- The process is chiefly applicable to light oil reservoirs
- High areal sweep is a requirement to make the process effective
- May not be effective in heterogeneous rock

- The chemicals may be adsorbed in certain types of rocks
- The complex process works only within a narrow set of operating conditions
- Degradation of chemicals under high temperature
- Potential reaction between polymer and surfactant
- Injection materials are expensive

Caustic or alkaline flooding

The caustic or alkaline EOR process works by creating surfactants *in situ*; the surfactants in turn lower the interfacial tension between oil and injected fluid. The alkaline materials that are injected include sodium hydroxide, sodium silicate, or sodium carbonate, which reacts with the organic acids in certain crude oils to produce the surfactants in the porous medium. The surfactants may also react favorably with reservoir rock to alter wettability.

Alkaline flooding has been applied to a wide range of reservoirs having light to intermediate oil gravity. Crude oil having sufficient quantities of organic acids is desirable. However, the process works better with light to intermediate oil where the mobility ratio is favorable.

Some of the limitations of the process are the following:

- Alkaline flooding is not suitable in carbonate reservoirs due to the high adsorption potential of anhydrite, gypsum, and clay.
- Consumption of injected chemicals can be high when reservoir temperature is greater.
- Scale formation in producing wells
- Cost of injected material
- Sensitivity to reservoir heterogeneities

EOR design considerations

The design of a successful EOR process for a reservoir depends on wide-ranging factors requiring multidisciplinary skills and experience. Some of the important design aspects are outlined in the following [9,10].

Thermal recovery:

- Suitability of the reservoir, including its characteristics such as thickness and heterogeneities
- Reservoir pressure favorable for EOR
- Oil properties including viscosity and composition
- Extent of heat loss to the adjacent formation, to the water present in the formation, and in the wellbore
- Controlled propagation of gas and steam fronts
- Need for new wells and conversion of existing producers
- Pilot flood project
- Reservoir simulation studies
- Reservoir surveillance including tracking of thermal front

Miscible processes:

- Rock and fluid properties
- Screening for specific process
- Phase behavior of injected and *in situ* fluids
- MMP and attainment of miscibility
- Reservoir depth to support required injection pressure and miscible conditions
- EOR performance under miscible and near miscible conditions
- Mobility ratio and sweep efficiency
- Slug size to optimize recovery
- Availability and cost of solvent
- Amount of solvent needed
- Injectivity of solvent and water
- Relative proportion of water and gas in WAG process
- Time lag between injection and response at producers
- Separation and recovery of solvents from produced stream
- Drilling of new wells
- Laboratory studies
- Pilot flood project
- Reservoir simulation studies
- Reservoir surveillance during miscible flood, including monitoring of well rates, water−oil ratio, and gas−oil ratio

Chemical processes:

- Concentration requirements of chemicals/polymer and costs
- Effective slug design; large slugs do not result in incremental recovery beyond a limiting value
- Mobility control of injected fluid
- Areal and vertical sweep by chemical bank
- Potential adsorption of chemicals/polymer with clay and other materials in formation
- Adverse effects of formation water salinity
- Potential degradation of chemicals in subsurface conditions including temperature
- Time lag between injection and response at producers
- Separation and recovery of solvents from produced stream
- Drilling of new wells
- Laboratory studies to demonstrate the effectiveness of chemical-EOR
- Pilot flood project to verify laboratory results
- Reservoir simulation studies
- Reservoir surveillance during miscible flood, including monitoring of well rates, water−oil ratio, and gas−oil ratio

Screening guideline for EOR processes

The success of an EOR application, usually requiring substantial capital investment, is rooted in proper screening of what exact method would be most suitable for the field in question. Careful review and analysis of both fluid properties and reservoir characteristics are needed to arrive at a decision. The selected EOR method is usually validated in laboratory studies and by pilot projects, if needed, before implementation

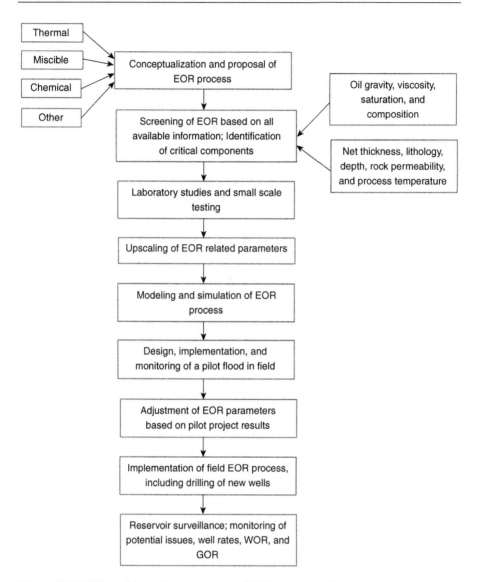

Figure 17.5 EOR workflow; from concept to field implementation.

on a field scale (Figure 17.5). Reservoir simulation is conducted to predict EOR performance based on the selected method. An economic analysis is made to determine whether the application of the EOR process would be feasible. The success or failure of the EOR method in similar reservoirs is also reviewed.

Taber, Martin, and Seright [1] made a comprehensive review of the implementation of EOR processes and their applicability in various types of reservoirs. Certain highlights of the study are shown in Table 17.3.

Table 17.3 EOR screening guideline

EOR method	Oil properties	Reservoir characteristics	Notes
Steam flood	Gravity: 8–25° API; viscosity: 100,000 cp or less	Net formation thickness: 20 ft. or more; oil saturation: 40% or higher; permeability: 200 mD or higher; depth: 5,000 ft. or less	
In situ combustion	Gravity: 10–27° API; viscosity: 5,000 cp or less	Net formation thickness: 10 ft. or more; oil saturation: 50% or higher; permeability: 50 mD or higher; depth: 11,500 ft. or less; temperature: 100°F or more	Presence of asphaltic components aid in coke deposition
CO_2 flood	Gravity: 22° API or more; viscosity: 10 cp or less	Oil saturation: 20% or higher; depth: 2,500 ft. or more depending on oil gravity; relatively thin formation unless dipping	Permeability not critical if injection rate can be maintained; effective in both sandstone and carbonate formations
Enriched and vaporizing gas drive (hydrocarbon miscible flood)	Gravity: 23° API or more; viscosity: 3 cp or less; high percentage of light hydrocarbons required	Oil saturation: 20% or higher; relatively thin formation unless dipping. Depth: 4,000 ft. or more	Not very effective in the presence of fractures and high permeability streaks
Micellar–polymer, alkaline, ASP	Gravity: 20° API or more; viscosity: 35 cp or less	Oil saturation: 35% or higher; permeability: 10 mD or more; depth: 9,000 ft. or less; temperature: 200°F or less; oil composition: light intermediate hydrocarbons are desirable. Organic acids are needed to create low interfacial tension	Sandstone formations work better. Adsorption issue with clays. Formation thickness is not critical
Polymer	Gravity: 15° API or more; viscosity: 10–100 cp preferred	Oil saturation: 50% or higher; permeability: 10 mD or more; depth: 9,000 ft. or less; temperature: 200°F or less;	Oil composition is not critical
Nitrogen, flue gas injection	Gravity: 35° API or more; viscosity: 0.4 cp or less	Oil saturation: 40% or higher; depth: 6,000 ft. or more	Not very effective in the presence of fractures and high permeability streaks. Reservoir temperature is not critical

Enhanced oil recovery workflow

Screening, design, testing, implementation, and monitoring of EOR processes require detailed technical and economic analyses. EOR screening guidelines were presented earlier. The process starts sooner rather than later in the life of a reservoir, although actual implementation of EOR may take place once waterflooding or gas injection has been utilized to recover oil at a comparatively lower cost. A generalized workflow is presented in the following.

Case Study: Enhanced Oil Recovery Projects in Oman

Oman is an example where dramatic turnaround in the country's oil production occurred due to the implementation of EOR projects [11]. Oman's average annual crude oil production peaked in 2000 at 970,000 bbl/d, but dropped to just 710,000 bbl/d in 2007 as the production from the reservoirs declined. However, the trend was reversed successfully. EOR techniques helped drive this turnaround along with some new discoveries. Annual crude oil production rose in each of the next 5 years, attaining a daily production of 919,000 barrels in 2012 (Figure 17.6).

EOR techniques and developments in those technologies are important to Oman's future production. After declining for several years in the early 2000s, EOR techniques have played a key role in reversing decline in oil production since 2007. Oman expects 16% of its oil production to come from EOR projects by 2016, up from just 3% in 2012. In late 2012, investments were also made in a solar-powered EOR process.

EOR techniques that are currently implemented include:

- Polymer injection: Polymer flood has been implemented at Oman's Marmul project for heavy oil recovery. The polymer-EOR process is found to be more effective than other EOR techniques such as steam injection. In 2012, the production from the Marmul project was approximately 75,000 bbl/d.
- Miscible gas injection: Miscible gas injection involves pumping gas, which creates miscibility with oil, resulting in enhanced recovery. Operators at Oman's Harweel oil field cluster use this technique in their operations. As a result, Harweel produced an additional 23,000 bbl/d in 2012, and production is expected to increase by another 30,000 bbl/d in the near term.
- Steam injection: Thermal EOR methods are implemented at the Mukhaizna, Marmul, Amal-East, Amal-West, and Qarn Alam fields, among others. Thermal EOR could increase production at both Amal-East and Amal-West to 23,000 bbl/d by 2018. Furthermore, steam injection at Qarn Alam is expected to increase production by 40,000 bbl/d by 2015 through steam assisted gravity drive in which the steam drains oil to lower producer wells. Steam assisted gravity drive is described in Chapter 21.

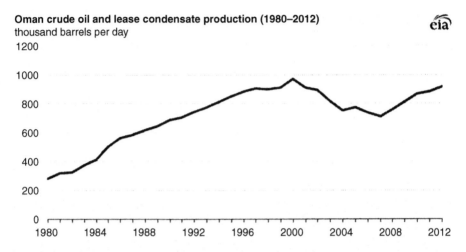

Oman crude oil and lease condensate production (1980–2012)
thousand barrels per day

Figure 17.6 **Production rate in Oman highlighting the initial increase in production until 1998, followed by steady decline, and finally a turnaround based largely on EOR projects since 2007.**
Source: US Energy Information Administration, International Energy Statistics.

Summing up

EOR, a part of IOR processes, is implemented to increase the ability of oil to flow to a well by injecting water, chemicals, or gases into the reservoir or by changing the physical properties of the oil. The ultimate objective is to produce additional amounts of oil left behind after primary and secondary production.

EOR processes include all methods that use external sources of energy or materials to recover oil that cannot be produced economically by primary and secondary recovery processes. Widely recognized EOR processes include the following:

- Thermal methods: Steam stimulation, steam flooding, and *in situ* combustion.
- Chemical methods: Surfactants, polymer, micellar−polymer, and caustic alkaline.
- Miscible methods: Hydrocarbon gas, CO_2, and nitrogen. In addition, flue gas and partial miscible/immiscible gas flood may be also considered.

EOR processes are designed to recover oil left after primary and secondary recoveries by improving oil displacement efficiency and volumetric sweep efficiency. Thermal methods are primarily used in heavy oil reservoirs to reduce oil viscosity and improve the mobility of oil toward the producing wells. Miscible methods, including the injection of CO_2 and light to intermediate hydrocarbons, create miscibility between oil and the injected fluid phase; as a result, interfacial tension is reduced between the fluids, and a miscible bank is formed, which drives oil efficiently to the producers. Chemical methods are used to decrease oil viscosity resulting in an increase in displacement efficiency. Polymer in the water is used to improve volumetric sweep efficiency by decreasing the water mobility.

EOR methods often require significant investment of capital and are generally associated with risks. No single method of EOR is effective for all reservoirs. The candidates for EOR are screened based on the reservoir fluid and rock property before a specific process is selected. Following the screening process, laboratory studies, reservoir simulation, and pilot flood are conducted before embarking on field-wide implementation.

EOR processes are summarized in Table 17.4.

Questions and assignments

1. What is EOR and how does it differ from primary and secondary recovery?
2. What are the principal categories of EOR methods? What criteria are important in deciding on the type of EOR process?
3. Name three most critical characteristics of a reservoir that determine the efficiency of steam flood.
4. What properties need to be screened for applications of thermal, miscible, and chemical processes? Provide a detailed explanation.
5. Distinguish between miscible and immiscible EOR processes. Which of the two methods is likely to recover more oil? Why?
6. Describe multiple contact miscibility. How does it differ from first contact miscibility?
7. Why is reservoir depth a critical factor in attaining miscibility?
8. Why is water injected alternately along with gas in the CO_2-WAG process?
9. Why are polymer, surfactant, and caustic material injected to recover more oil?
10. Can polymer flood be utilized in heavy oil reservoirs to recover oil? Why or why not?
11. How would you optimize slug size in chemical recovery?
12. What are the design considerations in the surfactant—polymer recovery process? Explain.
13. What EOR processes have gained dominance? Give reasons for each successful process.
14. Why would a pilot project be necessary in designing an EOR project?
15. What are the design considerations in chemical-EOR?
16. Why CO_2-EOR has been more successful compared to N_2 and inert gas injection?
17. Describe the role of reservoir simulation in planning an EOR process.
18. Based on a literature review, describe a large CO_2-EOR project, include reservoir depth, oil gravity, residual oil saturation, MMP, formation permeability, number of injectors and producers, well spacing and recovery factor, and any other important parameter.
19. Draw a phase diagram that depicts multiple contact miscibility between injected fluid and oil.
20. Your company is planning an EOR project in a reservoir with a bottom water drive having the following fluid properties and reservoir characteristics:
    a. Oil gravity: 21°API
    b. Reservoir depth: 2900 ft.
    c. Rock permeability: 18–25 mD
    d. Formation thickness: 23 ft.
    e. Residual oil saturation: 55%
    f. Reservoir heterogeneity: Presence of a few high permeability streaks
    g. Lithology: Dolomite
    h. Gas cap: None
    i. Well spacing: 80 acres
    j. Water cut in wells following waterflood: Moderate to high
    Select the most suitable EOR method based on the information above. Provide a detailed explanation in favor of your selection. Also explain why other methods are not likely to succeed. Make any other assumption necessary in presenting your case. Describe a step-by-step implementation plan of the EOR project you have selected.

Table 17.4 **Summary of EOR methods**

Category	EOR method	Mechanism of oil recovery	Design considerations	Limitations
Thermal	Cyclic steam injection (huff-and-puff)	Reduces oil viscosity and increases mobility	Effective heating of viscous oil. Minimal dissipation of heat to adjacent layers	In thin formations, dissipation of heat to adjacent layers making the process ineffective. Recovery from limited area around the well, which acts as both injector and producer
	Steam flood	Reduces oil viscosity and increases mobility. Vaporizes and extracts light hydrocarbons	Effective heating of oil. Formation of steam chamber. Minimal heat dissipation to adjacent formation	Ineffective in formations having a thickness <20 ft., poor reservoir permeability and low oil saturation. Presence of bottom water and gas cap are detrimental to recovery
	In situ combustion (fire flood)	Reduces oil viscosity and increases mobility. Vaporizes and extracts light hydrocarbons. Upgrades crude oil with liberated light hydrocarbons	Effective heating of oil. Control and sustainment of fire front. Minimal heat dissipation to adjacent formation	Can be complicated with little control over combustion front. Combustion cannot be sustained without sufficient burning of coke. Adverse mobility ratio
Miscible	CO_2 flood (miscible)	Attains miscibility and lowers interfacial tension between oil and CO_2. Vaporizes and extracts hydrocarbon components (C_2–C_{30}). Reduces oil viscosity and increases mobility. Causes swelling of oil. Enhances permeability in carbonate rocks. May increase injectivity	Attainment of miscibility; mobility control; better sweep	MMP sets the reservoir depth. Not applicable in shallow reservoirs. Unfavorable mobility ratio. Availability of CO_2 in large quantities. Potential corrosion issues. CO_2 separation from produced steam

	Enriched (condensing) and vaporizing gas drive	Light to intermediate hydrocarbons are injected under sufficient pressure to create miscibility with oil upon multiple contact. In enriched (condensing) gas drive, light to intermediate hydrocarbons from injected gas condense into the oil phase. In vaporizing gas drive, light to intermediate components from the oil phase vaporizes into the injected gas phase. Result of both drives is the formation of a miscible bank, which drives oil to the wells. Improved mobility ratio and better sweep	Attainment of miscibility; mobility control; better sweep	Depth of the reservoir to achieve MMP; unfavorable mobility ratio; poor sweep efficiency; cost of injected hydrocarbons
	Wateralternating gas injection (WAG)	Alternating slugs of gas and water injected to improve mobility ratio, displacement efficiency, and areal sweep during miscible displacement	Optimum slug size	Same as miscible flooding
Chemical	Polymer injection	Increases water viscosity and reduces mobility. Improves water–oil mobility ratio. Increases areal and vertical sweep	Optimum slug size; interaction with *in situ* fluid; potential reaction and degradation; mobility control; suitability of formation; separation of injected material in produced steam	Lower injectivity. Degradation of polymer material. Loss of valuable injectant in fractures and channels

(Continued)

Table 17.4 Summary of EOR methods (cont.)

Category	EOR method	Mechanism of oil recovery	Design considerations	Limitations
	Surfactant flood	Lowers interfacial tension between oil and water. Emulsifies oil and water. Solubilizes oil. Improves water−oil mobility ratio	Optimum slug size; interaction with *in situ* fluid; potential degradation; mobility control; suitability of formation; separation of injected material in produced steam	Heterogeneous reservoir leads to potential loss of surfactants
	Alkaline flood	Lowers the interfacial tension between oil and injected fluid containing caustic or alkaline material	Optimum slug size; interaction with *in situ* fluid; potential degradation; mobility control; suitability of formation; separation of injected material in produced stream	Alkaline flooding is not suitable in carbonate reservoirs due to the high adsorption potential of anhydrite, gypsum and clay. Consumption of injected chemicals can be high when reservoir temperature is greater. Scale formation in producing wells

References

[1] Taber JJ, Martin FD, Seright RS. EOR screening criteria revisited – part 2: applications and impact of oil prices. SPE 35385; 1996.

[2] Thermal recovery processes. SPE Reprint Series No. 7. Society of Petroleum Engineers: Richardson (TX); 1985.

[3] Thermal recovery techniques. SPE Reprint Series No. 10. Society of Petroleum Engineers: Richardson (TX); 1972.

[4] Kokal S, Al-Kaabi A. Enhanced oil recovery: challenges and opportunities. EXPEC Advanced Research Center, Saudi Aramco; 2011.

[5] Bennison T. Prediction of heavy oil viscosity. Presented at the IBC Heavy Oil Field Development Conference; 1998London. December 2–4.

[6] Gael BT, Gross SJ, McNaboe GJ. Development planning and reservoir management in the Duri steam flood. SPE Paper #29668. Presented at the SPE Western Regional Meeting; 1995Bakersfield, CA. March 8–10.

[7] Fuaadi IM, Pearce JC, Gael BT. Evaluation of steam-injection design for the Duri steamflood project. SPE Paper #22995. Asia-Pacific Conference; 1991Perth, Australia. November 4–7.

[8] Sigit R, Satriana D, Peifer JP, Linawati A. Seismically guided bypassed oil identification in a mature steamflood area, Duri field, Sumatra, Indonesia. Asia Pacific Improved Oil Recovery Conference; 1999Kuala Lumpur, Malaysia. October 25–26.

[9] Tunio SQ, Tunio AH, Ghirano NA, El Adawy ZM. Comparison of different oil recovery techniques for better oil productivity. Int. J. Appl. Sci. Technol. September 2011;1(5): 143–53.

[10] www.Petrowiki.org.

[11] Enhanced oil recovery techniques helped Oman reverse recent production declines. Energy Information Administration. Available from: http://www.eia.gov/todayinenergy/detail.cfm?id(13631 [accessed 10.12.13].

Horizontal well technology and performance

18

Introduction

The advancement of horizontal well technology in the latter part of the twentieth century is a game changer for the petroleum industry. Traditionally, wells were drilled as either vertical or deviated, which could contact only a limited portion of the reservoir drainage area. A relatively large number of wells, and resources, were required to produce effectively under a variety of circumstances, including tight formations, lenticular sands, fault blocks, and thin layers, to name a few. Again, in many heavy or ultraheavy oil reservoirs, traditional wells were found to be ineffective. In extreme situations such as ultratight shale formations, sustained commercial production of oil and gas based on the traditional drilling practices was simply not viable. Consequently, shale oil and gas were deemed as unconventional resources. Since a horizontal well replaces the requirement of drilling several vertical wells, the footprint on land surface is also minimized. In offshore platforms, only a limited number of slots are available to drill wells; hence, horizontal drilling is the obvious choice to reach the distant parts and multiple layers of a large reservoir and maximize the area of contact. During horizontal drilling, wealth of data can be obtained about the lateral variation in reservoir characteristics over thousands of feet based on measurement while drilling (MWD) or logging while drilling (LWD), which is impossible to collect in case of a vertical well.

Production from many reservoirs, both onshore and offshore, is economically feasible today only due to the advent of horizontal well technology. In many petroleum regions in the world, the overwhelming majority of wells drilled today are horizontal as the technology has matured in recent decades. It is a common industry practice to recomplete a vertical well as horizontal to target areas where oil in commercial quantities is left behind or as water production increases substantially. A horizontal well may have one or more lateral sections that can effectively contact, and produce from, thousands of feet of oil- and gas-bearing formation. Although the cost of drilling horizontal wells is more than that of vertical wells, well productivity is at least several hundred percent higher in most instances. Some horizontal wells would produce even more in complex geologic settings.

This chapter describes the application of horizontal drilling in petroleum reservoirs from the viewpoint of reservoir engineering, including how the horizontal wells recover efficiently where the reservoir quality or fluid properties are less than favorable. The following questions are answered:

- How did horizontal well technology evolve?
- What are the types of horizontal wells and are how the wells classified?
- What are the advantages of horizontal wells as opposed to vertical or deviated wells?

Reservoir Engineering. http://dx.doi.org/10.1016/B978-0-12-800219-3.00018-8

- What is extended reach drilling (ERD)? Why are these wells drilled?
- What reservoirs and geologic settings are best suited for horizontal drilling?
- What factors influence the design and trajectory of a horizontal well in a pay zone?
- How is horizontal well productivity calculated? How does it compare with the productivity of vertical wells?

History of horizontal drilling

Horizontal drilling dates back to the early part of the twentieth century. In 1929, the first known horizontal oil well was drilled in Texas. Fifteen years later, another well was drilled in the Franklin Heavy Oil Field, Pennsylvania, at a depth of 500 ft. However, horizontal drilling became commercially successful in the 1980s as improved downhole drilling motor and telemetry were introduced. Following the successes in drilling horizontal wells by the French company Elf Aquitaine in Europe and BP in Alaska, the first generation of horizontal wells rapidly grew in number. Horizontal drilling in tight chalk formation in Texas proved to be quite a success. The wells were effective in contacting and producing from a fracture network where the rock matrix had very low matrix permeability. Horizontal wells were also drilled in Bakken shale formation in North Dakota having low to ultralow permeability. In the 1990s, a huge technological stride was made in many areas of horizontal drilling, including the depth of wells, length of the horizontal sections, radius of curvature, and design of multilaterals accessing difficult to reach parts of the reservoir. Extended reach wells drilled in recent years are several miles long. Horizontal wells are currently drilled to produce from complex geologic structures, heterogeneous formations, coal beds, and shale formations. Horizontal wells are also drilled in older fields, where less than optimum recovery was encountered prior to the dawn of horizontal drilling.

By 2009, over 50,000 horizontal wells had been drilled in the United States alone. The present generation of horizontal wells are drilled deeper and extend further than ever before. Coupled with other technologies like multistage fracturing in shale formations, the wells produce effectively from the reservoirs that were not considered to be economically producible only a decade ago.

Types of horizontal wells

Horizontal wells can be broadly classified according to the radius of curvature as the well trajectory transitions from vertical wellbore to horizontal section. Four classes of horizontal wells based on radius of curvature are drilled as follows:

- Long radius
- Medium radius
- Short radius
- Ultrashort radius

Long radius horizontal wells have a build rate of $1-6°/100$ ft. The radius of curvature is typically about 1500 ft. The wells are drilled based on conventional tools and

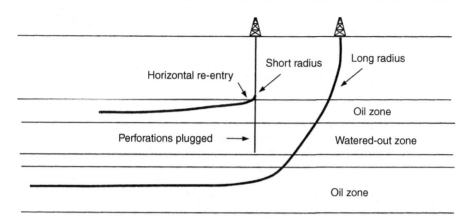

Figure 18.1 Horizontal re-entry well with short radius curvature. The old vertical well is plugged to avoid excessive water production. A long radius horizontal well is shown for comparison.

methods. As the technology matured, horizontal wells with medium, short and ultrashort radius of curvature were drilled to be more efficient. Medium-radius horizontal wells have build rates of 6−35°/100 ft., radii of curvature ranging between 1000 ft. and 160 ft., and lateral sections of several thousand feet. Short-radius horizontal wells have build rates of 1.5−3°/ft., which equates to radii of curvature of 40−20 ft. The length of the lateral section is less than the wells of medium radius of curvature. Ultrashort radius of curvature wells have even higher build rates and are usually associated with re-entry wells. Short-radius and ultrashort-radius wells are drilled with specialized drilling tools and techniques. The short-radius type is most commonly employed as a re-entry from any existing well (Figure 18.1).

Another classification of horizontal wells is based on the number of lateral branches. Horizontal wells can be single lateral or multilateral depending on the well location, target zones, and reservoir geometry (Figure 18.2). Two or more horizontal branches are drilled from one wellbore in the case of a multilateral well.

As mentioned earlier, certain horizontal wells are re-entry wells drilled from existing vertical wells, as opposed to the originally drilled horizontal wells. From an operational point of view, horizontal wells are drilled both as producers and injectors wherever a larger contact area between the well and the formation is a better solution.

Extended reach wells

Horizontal wells created by ERD have very long horizontal section, in tens of thousands of feet, to reach distant productive zones and multiple reservoirs. In most instances, the reservoirs are located offshore and are shallow. Drilling extended reach wells is quite challenging but may be the only solution where field locations are remote. Moreover, reservoir characteristics may be unfavorable requiring production by horizontal wells.

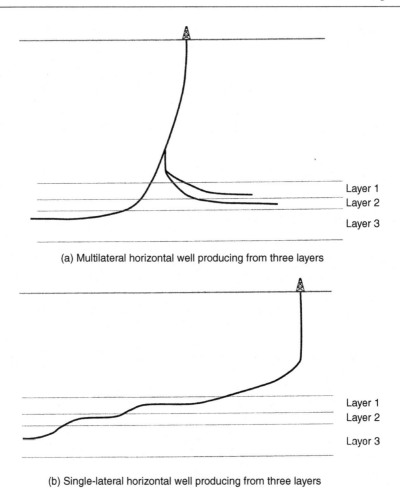

(a) Multilateral horizontal well producing from three layers

(b) Single-lateral horizontal well producing from three layers

Figure 18.2 (a) Multilateral and (b) single-lateral horizontal wells. Both configurations are capable of producing from multilayered reservoirs.

In 2008, a horizontal well was drilled in Al-Shaheen field, located in Qatar offshore. The total length of the well is 40,320 ft. MD while the horizontal section is 35,770 ft. long. The reservoirs are spread over a large area. Good permeability was observed only in localized areas. In the rest of the areas, thin and stacked carbonate formations of low permeability were encountered. Oil viscosity is relatively high. Lateral discontinuities also exist. Hence, oil and rock properties and reservoir characteristics were not deemed favorable for economic production before innovative methods in horizontal drilling were implemented [1]. Three years later, a horizontal well was drilled in Sakhalin Island [2], Russia, being 41,667 ft. (7.7 miles) long with a horizontal section of 38,514 ft. As of 2011, 7 of the 10 longest horizontal wells in the world were drilled in this location.

Advantages of horizontal wells

The advantages of horizontal wells over vertical wells are quite significant. These include, but are not limited to, the following:

- Enhanced oil and gas production rate: A horizontal well generally creates a much larger drainage area that enhances oil and gas production significantly in most geologic settings. Higher productivity of the horizontal wells leads to increased ultimate recovery in a relatively short period. The horizontal wells reach further into the reservoir, and add reserves.
- Reduction in water coning and sand production: The occurrence of water and gas coning is diminished because of the flow geometry, thereby reducing the remedial work required during the life of the well. In the case of horizontal wells, relatively reduced pressure drop around the wellbore associated with lower fluid velocities is observed. A general reduction in sand production is expected due to a combination of the above.
- Minimization of non-Darcy effects: In high permeability gas reservoirs, drilling of horizontal wells can minimize the adverse effects of non-Darcy flow and relatively high pressure drop encountered near the vertically drilled wellbore.
- Less operational issues: A properly designed horizontal well may effectively alleviate various production-related issues encountered with the traditional vertical well such as low rates or production, early water breakthrough and rate decline, cost intensive workover, low ultimate recovery, and premature well abandonment. Hence, horizontal wells may hold the key to better manage the reservoir and bring a higher return on investments.
- Avoidance of watered-out zones: It is a common practice to recomplete the existing vertical or deviated wells as single- or multilateral horizontal wells where the laterals are targeted to produce from zones that are not watered out by previous waterflood operations.
- Better thermal recovery: Horizontal wells can provide thermal energy to reduce the viscosity of heavy or ultraheavy oil over a large area in a reservoir, thus making it more mobile and producible. Horizontal wells are drilled in most heavy oil reservoirs and thermal methods are designed for economic and sustainable production.
- Reduction in the effects of rock heterogeneity: Localized effects of rock heterogeneity, such as the presence of a barrier, degradation of reservoir quality and facies change, may be diminished as a horizontal well produces from a large drainage area.
- Viable strategy in complex reservoirs: In many complex geological settings, drilling of horizontal wells is the only option to produce the reservoir economically. These wells may target the sweet spots or lenticular reservoirs, where pockets of good reservoir quality exist in terms of porosity, permeability, and fluid saturation.
- Reduced footprint on land: As stated earlier, horizontal wells have a significantly reduced footprint on land when compared to the large number of vertical wells required for oil and gas operation. This is likely to have a positive impact on any environmental issues, among others.
- Offshore field development: Since only a limited number of drilling slots are available from offshore platforms, horizontal wells are the obvious choice to contact a large portion of productive formation.
- Use of low density drilling mud: Since the producing zone is cased during horizontal drilling, low density drilling mud can be used that may reduce formation damage.

Field application of horizontal wells

The application of horizontal wells in the petroleum industry is numerous, from tight shale gas reservoirs to fractured heavy oil reservoirs to conventionally developed fields nearing the end of their productive life based on vertical or deviated wells. Broadly speaking, horizontal wells are drilled wherever the vertical and deviated wells have been found to be inefficient, uneconomic, near abandonment, or not deemed to be a viable option at all due to the unfavorable rock and fluid characteristics.

A literature survey indicates that horizontal well technology has been applied to the following reservoirs:

- Tight formations where permeability is low to ultralow
- Thin beds where the area of contact by vertical or deviated wells is very limited
- Fractured reservoirs where matrix permeability is low but the well laterals intersect high conductivity fractures
- Compartmental reservoirs having noncommunicating sections
- Formation comprising sealing fault blocks
- Heavy oil reservoirs where large areas in the reservoir require the application of thermal energy
- Offshore fields where limited slots are available to drill wells

Tight reservoirs having low to ultralow permeability, among others, are the prime candidates for horizontal drilling, as the horizontal wells are capable of contacting a much larger area and produce oil and gas in significant quantities. Benefits of horizontal well technology are quite astounding in certain types of reservoirs. It is found to be most effective in fractured reservoirs where rock matrix permeability is very low. Productivity can be enhanced by as much as 1200%. Next are the heavy oil reservoirs, where productivity can improve by 700% with horizontal wells.

Thin reservoirs are the obvious choice for horizontal drilling, as a vertical or deviated well can only contact a very small portion of the reservoir. In such cases, a single horizontal well may have one or more laterals thousands of feet long, eliminating the requirement of drilling a large number of vertical wells. Thin beds with low permeability require the drilling of horizontal wells to produce economically. The longest horizontal well is drilled in an offshore field in Qatar comprising a thin, tight, and laterally discontinuous carbonate formation.

Horizontal drilling technology is applied extensively in unconventional gas arenas including tight shale gas reservoirs and coalbed methane reservoirs. Gas is produced from the horizontal wells that are intersected by natural and induced fractures.

In certain heavy oil reservoirs, a steam assisted gravity drainage process based on a pair of horizontal injectors and producers is implemented to produce the highly viscous oil in commercial quantities.

In offshore reservoir development, horizontal wells are drilled to cover large areas of the reservoir from a relatively small number of slots in the offshore drilling platform (Figure 18.3).

Horizontal wells are highly efficient in producing from a fracture network when the fractures intersect the horizontal trajectory of the well.

Compartmental reservoirs or reservoirs with noncommunicating fault blocks are efficiently produced by drilling horizontal wells. In reservoirs having nonconnected areas, a lateral wellbore can penetrate multiple sections and produce effectively (Figure 18.4). Compartmental reservoirs can be identified by well testing, production behavior, flood front tracking, and variations in oil composition, among others.

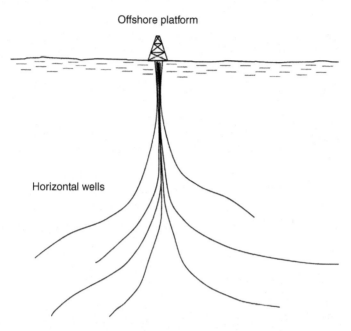

Figure 18.3 A large number of horizontal wells drilled from an offshore platform reaching distant parts of the reservoir.

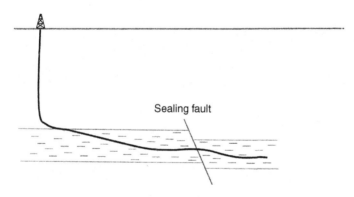

Figure 18.4 Trajectory of a single-lateral horizontal well producing from two noncommunicating fault blocks.

In reservoirs with water and gas coning problems, horizontal wells have been used successfully to minimize coning problems and enhance oil production.

In watered-out formations following prolonged water injection, existing vertical wells are sidetracked to drill horizontal wells that contact areas of high remaining oil saturation and enhance oil production. The strategy is depicted in Figure 18.1.

In all cases, however, oil saturation must be high enough to justify horizontal drilling. Furthermore, reservoir pressure must be capable of driving oil to the wellbore effectively. As in any project, the decision to drill horizontal wells is based on cash flow and rate of return, taking into account the cost of drilling and estimated recovery. The comparison [2] between a vertical well and a horizontal well is based on the cost of producing oil and gas ($/bbl or $/MCF) rather than the cost of drilling alone ($/ft.).

Horizontal well placement guidelines

The following serve as general guidelines in placing horizontal wells in conventional and unconventional reservoirs:

- In conventional oil reservoirs under waterflood, horizontal wells are placed in zones where residual oil saturation is at a maximum, while watered-out zones are avoided.
- In placing the horizontal laterals, vertical permeability, fluid viscosity, formation thickness, rock heterogeneities, and other factors must be taken into account to avoid water coning.
- In reservoirs where permeability is tight, the length and direction of horizontal well laterals are designed to contact the reservoir as much as possible.
- In naturally fractured reservoirs, horizontal wells are drilled transverse to the principal direction of fractures to contact as many as fractures possible that serve as highly conductive microchannels for oil and gas.
- Similarly, in formations where directional permeability exists, horizontal well trajectory is transverse to the principal direction of permeability to facilitate maximum fluid volume into the borehole.
- In heterogeneous reservoirs having faults and compartments, horizontal well trajectory is designed to produce oil from multiple sections of reservoirs that have either limited or no connection.
- Unconventional shale gas reservoirs are continuous in nature. Horizontal wells are targeted to contact localized "sweet spots" that are rich in organic content with desirable thermal maturity; additionally, natural fractures present in these spots serve as conduits for gas, and geomechanical rock properties should be conducive to effective multistage fracturing.

Horizontal well performance

The performance of a horizontal well can be evaluated by estimating the enhancement in well productivity as compared to that of a vertical well. A literature review suggests that a number of analytic equations have been proposed by various authors to calculate the productivity index of horizontal wells that take into account various reservoir and fluid properties, as well as the nature of fluid flow. The following equation proposed by Joshi [3] can be used to estimate the productivity index based on the assumption of steady-state flow of a single-phase fluid, uniform rock properties, and known drainage area, among others:

$$J_h = \frac{7.081 \times 10^{-3} \, h \, k_h}{\mu_o \, B_o \left[\ln(R) + (Bh/l) \ln\left(Bh/r_w (B+1) \right) \right]} \tag{18.1}$$

where r_{eh} = radius of drainage of the horizontal well in ft.

$$R = \frac{a + \left[a^2 - (0.5L)^2\right]^{0.5}}{0.5L} \tag{18.2}$$

$$a = 0.5L\left[0.5 + \left\{0.25 + \left(\frac{2r_{eh}}{L}\right)^4\right\}^{0.5}\right]^{0.5} \tag{18.3}$$

$$B = \left(\frac{k_h}{k_v}\right)0.5 \tag{18.4}$$

Example 18.1

Calculate the productivity index and production rate of a horizontal well with the data shown in Table 18.1.

Solution

$$B = \left(\frac{25}{3}\right)^{1/2} = 3.162$$

$$r_{eh} = (160 \times 43,560 / 3.14)^{1/2}$$
$$= 1,490\,\text{ft.}$$

$$a = 0.5L\left[0.5 + \left\{0.25 + (2r_{eh} / L)^4\right\}^{0.5}\right]^{0.5}$$
$$= 0.5(5000)\left[0.5 + \left(0.25 + (2 \times 1,490 / 5000)^4\right)^{0.5}\right]^{0.5}$$
$$= 2637.8$$

$$R = 1.3917$$

Table 18.1 Estimation of horizontal well productivity

Parameter	Value
Length of lateral (ft.)	5000
Bottom-hole flowing pressure (psi)	1250
Horizontal permeability (mD)	25
Vertical permeability (mD)	2.5
Estimated drainage area (acres)	160
Average pressure in drainage area (psi)	2100
Radius of wellbore (ft.)	0.42
Average formation thickness (ft.)	30
Oil viscosity (cp)	1.1
Oil formation volume factor (FVF) (rb/STB)	1.89

$$J_h = 6.287\,\text{stb/day/psi}$$

$$q_o = 6.287(2100 - 1250)$$
$$= 5344\,\text{stb/day}$$

In comparison, the productivity of a vertical well is much lower.

$$J = 7.081 \times 10^{-3}(25)(30) / \left[(1.1)(1.89)\{\ln(1490/0.42) - 3/4\}\right]$$
$$= 0.31\,\text{stb/day/psi}$$

In cases where the horizontal well productivity is lower than expected, there could be many possible issues. These include formation damage, excessive sand production, and ineffective sections of the lateral due to water encroachment or poor reservoir quality.

Horizontal well performance issues

When horizontal wells perform less than expected, the root cause may be associated with one or more of the following:

- Unknown reservoir heterogeneities
- Subpar reservoir quality than what was expected
- Low vertical permeability that does not facilitate fluid flow to the laterals
- Well placement near the oil–water contact
- Skin damage
- Borehole stability
- Poor completion
- Pressure drop along the horizontal section hindering fluid flow
- Invalid assumptions about the reservoir

Case Study: Heavy Oil Reservoir, California

In Kern River field located in Bakersfield, California, horizontal well technology has been successfully applied since 2007 to enhance the ultimate recovery of heavy oil [4]. Top of the reservoir is found at 50–1000 ft. below the surface. At least nine oil zones were identified in the reservoir. The viscosity of oil is 4000 cp at the initial reservoir temperature and the density is 13°API. Reservoir porosity is 29–33% and permeability is high, ranging from 1 D to 8 D.

The field is more than 100 years old. Due to the highly viscous nature of oil, a large number of vertical wells were drilled in the last century. In fact, over 20,000 wells were drilled. As part of the enhanced recovery operation, steam flooding was employed. In the mid-1980s, field production increased to 140,000 barrels of oil per day. However, production began to decline at an average annual rate of 6%. Drilling of horizontal wells began in 2007 and over 400 wells were drilled. The horizontal wells were found to be the largest producers in the field. The number of horizontal wells was 4% of total wells, but accounted for 24% of field production. As a result, the annual decline in field production was reduced ranging between 1% and 2%.

The strategy involved the drilling of the horizontal wells in the heated portions of the reservoir where oil recovery was poor. In order to pinpoint the areas of relatively low recovery, decline curve analysis predicting the ultimate recovery was compared against volumetric analysis. Estimates of hydrocarbon volume were based on a full-field 3D reservoir model with necessary data obtained from lithology and open-hole logs, among others. Most recent saturation information was obtained by C/O logs. There are about 700 observation wells in the field collecting saturation and temperature data.

Case Study: Bakken Horizontal Drilling, North Dakota

The Bakken formation was discovered as a significant resource of petroleum in North Dakota several decades ago. It is located in Williston basin, and extends from Montana to North Dakota to parts of Canada. Although oil was originally discovered in the early 1950s, Bakken has risen to worldwide prominence only in this century following the successful implementation of horizontal well technology and multistage fracturing techniques to produce economically from the very low permeability formation. In 2005, daily oil production in North Dakota was less than 100,000 bbl/day when the first hydraulic fracturing was demonstrated to be a success. In 2013, production has exceeded 700,000 bbl/day as Bakken development witnessed a spectacular rise in horizontal drilling and multistage fracturing (Figure 18.5). There are over 600 horizontal wells drilled, and the number of stages in fracturing is as high as 40. As a result, North Dakota is the second largest producer of oil in the United States next to Texas.

Bakken shale dates back to late Devonian and early Mississippian age in the geologic time scale [4]. Three distinct layers are identified in the formation, the upper and lower layers of shale, which are source rocks, with an intervening middle layer of dolomite, including Three Forks. Additionally a sandstone layer is also encountered, referred to as Sanish. Bakken shale oil is referred to as "tight oil" due to very low permeability, in the order of 10^{-2} mD. Porosity is also low, averaging about 5%.

The USGS has estimated the recoverable reserves to be 4 billion barrels. In fact, the Bakken is estimated to have 400 billion barrels of oil equivalent in place [5].

Summing up

Horizontal wells have ushered in a new era in the oil and gas industry in recent decades. Many reservoirs that were considered economically not viable or technically challenging have been developed successfully based on horizontal drilling technology.

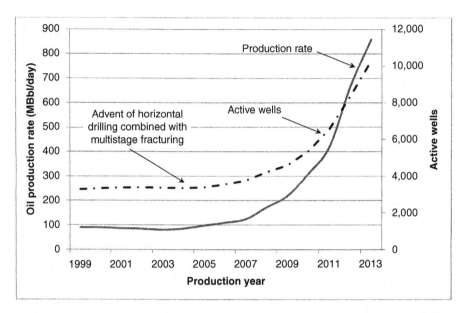

Figure 18.5 North Dakota production statistics. Most of the production is based on Bakken shale development following the advent of horizontal drilling combined with multistage fracturing. Courtesy: Department of Mineral Resources, North Dakota.

Horizontal drilling technology adds to reserves as the wells can produce from geologically complex settings where vertical wells are not effective. The outstanding advantages of horizontal wells include, but are not limited to, the following:

- Horizontal wells are suitable in producing reservoirs in complex reservoirs with compartments, sealing faults, and other lateral discontinuities where vertical or deviated wells cannot reach all the noncommunicating sections of the reservoir from a single location.
- Thin beds and "shoestring" reservoirs are prime candidates for horizontal drilling as the vertical wells have very limited contact and often do not have adequate productivity for economically viable operation.
- Horizontal wells reduce the effects of water coning and sand production in reservoirs as the pressure drop in the vicinity of a wellbore is lower than that of vertical wells.
- A popular application of horizontal drilling is the recompilation of vertical wells to produce from the zones that are not watered out during waterflood operation.
- Horizontal wells are solution to various well problems, including high water cut and frequent workovers.
- Similarly, in low permeability formations, horizontal wells can contact a large area for sustained production. Horizontal drilling is widely applied in both conventional and unconventional reservoirs including tight gas sands, coalbed methane, and shale gas reservoirs having ultralow permeability in micro- or nanodarcies.
- In most offshore fields, horizontal drilling is the only option as only a limited number of drilling slots is available to drill wells and develop the reservoir.
- Horizontal wells work effectively in producing heavy and ultraheavy oil reservoirs as thermal energy can be applied over a large area and makes the oil less viscous to flow to the surface.

Table 18.2 Classification of horizontal wells

Well type	Radius of curvature (ft.)	Build rate (deg)	Length (ft.)
Long	1500	1–6/100 ft.	1500+
Medium	160–1000	6–35/100 ft.	1000–4000
Short	20–40	1.5–3/ft.	100–800
Ultrashort	1–2	–	100–200

• LWD or MWD can collect a wealth of information about the rock properties in a lateral direction. In contrast, open-hole logs in a vertical borehole can only collect petrophysical information from a very limited area.
• Horizontal wells required to develop a reservoir are fewer in comparison to the number of vertical wells. Hence, horizontal wells have smaller footprint on land that may reduce or eliminate environmental and other issues.

Horizontal wells can be classified according to the radius of curvature as the well trajectory transitions from vertical hole to horizontal section. The categories are shown in Table 18.2.

The cost of horizontal wells is generally three to five times greater than vertical wells; however, an increase in well productivity by 700% is observed in most instances. In most cases, horizontal wells turn out to be the preferred option based on economic analysis when production cost per barrel of oil is considered.

Horizontal laterals are drilled and placed in a manner to avoid watered-out zones and produce from portions of the reservoirs with relatively high oil saturation. Horizontal wells are also drilled in a manner to intersect principal directions of fractures or the direction of preferential permeability transversely to facilitate maximum flow of fluid into the wellbore. In unconventional shale gas reservoirs, horizontal wells are targeted for "sweet spots" having good organic content and correct thermal maturity. Furthermore, natural fractures present in sweet spots should facilitate flow of gas, and rock properties should be conducive to effective multistage fracturing.

There are various analytic models available in the industry to predict the productivity of horizontal wells. An equation is presented in the chapter with example calculations for horizontal well productivity. The equation indicates that:

• Productivity of a well increases as longer horizontal sections are drilled. A relatively high value of vertical permeability would lead to better production provided all other parameters remain the same.
• Well productivity will be diminished when oil has high viscosity. However, horizontal wells are a better option in producing heavy oil reservoirs by thermal methods.

Besides, recent experience has shown that multistage fracturing can increase horizontal well productivity significantly. Multistage fracturing is a common practice in unconventional shale gas reservoirs in order to facilitate production where permeability is in micro- or nanodarcies.

The chapter includes a case study involving the application of horizontal well technology in recovering heavy oil from a field that is more than 100 years old. A large number of vertical wells have been drilled to recover oil followed by steam flooding.

However, the field started showing a decline in production. Since 2007, horizontal wells were drilled in areas where the recovery was poor but the formation was heated due to thermal recovery efforts. The horizontal drilling project was a success as it accounted for 25% of total production based on only 4% of the wells.

Questions and assignments

1. What is a horizontal well and how does it differ from a vertical well? What factors are considered for drilling a horizontal well?
2. List the advantages of horizontal wells. Cite from the literature three types of heterogeneous reservoirs that can benefit significantly from horizontal well technology.
3. How horizontal drilling aids in characterizing a formation that was not possible prior to its introduction in the industry? What rock properties are measured during horizontal drilling?
4. What are the benefits of multilateral horizontal wells? When would you consider drilling a multilateral well?
5. How do re-entry wells work? Give reasons for recompleting a vertical well as a horizontal well.
6. How is a horizontal well placed in a waterflooded reservoir? Explain.
7. Why is horizontal drilling technology implemented in unconventional shale reservoirs?
8. How can horizontal wells be effective in naturally fractured formations?
9. How does vertical permeability affect horizontal well performance?
10. Your company is considering the drilling of several multilateral oil wells in a fractured dolomite reservoir having low permeability and high water saturation. Perform a detailed analysis of cost and benefit. Make all necessary assumptions.

References

[1] Thomasen J, Al-Emadi IA, Noman R, Ogelund NP, Damgaard AP. Realizing the potential of marginal reservoirs: the Al Shaheen field offshore Qatar. IPTC #10854. Presented at the International Petroleum Technology Conference; 2005 Doha, Qatar. November 21–23.
[2] At the end of the earth: the longest, deepest oil wells in the world. Popular Science; 2011. Available from: http://www.popsci.com/technology/article/2011-06/end-earth-longest-deepest-oil-wells-world.
[3] Joshi SD. Horizontal well technology. Oklahoma: Pennwell; 1991.
[4] Bakken shale geology. Available from: www.Bakkenshale.com [accessed 12.08.15].
[5] Bakken history. Available from: http://www.undeerc.org/bakken/pdfs/BakkenTimeline2.pdf [accessed 14.08.15].

Oil and gas recovery methods in low permeability and unconventional reservoirs

19

Introduction

Petroleum reservoirs are discovered in a wide range of geologic settings across the continents. As a consequence, the reservoir engineering team, in collaboration with others, adopts various approaches and technologies in producing the reservoirs effectively and economically. In the previous chapter, one such technology, i.e., application of horizontal drilling, is described; horizontal wells have led to outstanding results in producing reservoirs of varying complexity at various phases of the reservoir life cycle. This chapter discusses another approach that gained popularity in the oil and gas industry early on, namely, infill drilling. The concept is quite simple; when new "infill" wells are drilled in between the original wells, additional quantities of oil and gas can be recovered from the formation as more oil and gas are exposed to flow conduits in wellbores. Lastly, various recovery methods are discussed in the context of a low permeability gas reservoir to demonstrate the tools and techniques available to the reservoir engineer in enhancing reservoir performance and adding value to the assets.

This chapter briefly discusses the various aspects of infill drilling, including applications in both conventional and unconventional reservoirs, as well as various strategies to develop low permeability reservoirs. Case studies are also presented at the end of the chapter. The following queries are addressed:

- What are some of the common strategies to recover oil and gas effectively?
- What is infill drilling?
- What are the benefits of infill drilling?
- What reservoirs are best suited for infill drilling?
- How did infill drilling evolve in the industry?
- What would be a comprehensive strategy to develop the low to ultralow permeability reservoirs?

Strategies in oil and gas recovery

The common strategies in producing low permeability reservoirs include, but are not limited to, the following:

- Infill drilling, resulting in relatively close well spacing.
- Horizontal drilling, single lateral or multilateral.
- Well stimulation by hydraulic fracturing and acidization.

Reservoir Engineering. http://dx.doi.org/10.1016/B978-0-12-800219-3.00019-X

- Reservoir management based on intensive monitoring of well performance, including the deployment of permanent downhole gauges. This leads to the evaluation of formation damage, well productivity, and reservoir performance as a whole on a regular basis.

Benefits of infill drilling

Infill wells are the wells drilled following the initial phase of development and production of an oil or gas reservoir. Infill wells are expected to contact the portion of the formation where oil saturation remains high following primary and secondary recovery. Infill wells can be either vertical or horizontal.

The benefits of infill drilling are multifarious and are listed as follows:

- Increase in oil and gas production where formation permeability is very low
- Enhancement of connectivity between injectors and producers
- Increase in horizontal and vertical sweep efficiency as more wells are available for displacing oil by injected fluid
- Increased reservoir contact where reservoir heterogeneities such as widely varying permeability and noncommunicating sections exist
- Overall increase in reservoir assets, reduction in well abandonment rates, and better reservoir economics

Target reservoirs for infill drilling

There are a number of reasons why large spaced wells are not able to attain optimum recovery; some of these are described in the following:

- Tight reservoirs having low to ultralow permeability where the drainage area per well is quite small and individual well production declines prematurely. A large number of tight gas fields are produced by implementing an infill drilling strategy successfully.
- Highly heterogeneous reservoirs where waterflooding and other enhanced recovery operations cannot contact large parts of the formation due to channeling of injected fluids.
- Heavy oil reservoirs where oil mobility is not sufficient to produce in commercial quantities from widely spaced wells.
- Matured reservoirs where pockets of high oil saturation remain even after implementing enhanced recovery operation.
- Compartmental reservoirs where wells with large spacings are inadequate to reach isolated portions of the reservoir.

Evolution of infill drilling methodology

Traditional practices involved the infill drilling of vertical wells as part of a regular pattern. For example, many fields in the United States were initially developed with 160-acre well spacing. As production dwindled, infill wells were drilled in the next phase reducing the well spacing to 40 acres. Certain low permeability reservoirs were drilled with 20-acre well spacing to augment production. The wells were drilled at regular intervals with little attention to the unique geology or high fluid saturation that may exist in a particular section of the field. As reservoir characterization became intensive leading to the simulation of robust reservoir models, certain areas of the reservoir and geologic intervals identified were not swept or contacted. Infill drilling was targeted for the specific areas rather than adopting a regular pattern drilling throughout the field (Figure 19.1). Again, following the introduction and game changing success of horizontal well technology in the industry, many infill wells were drilled horizontally to increase oil and gas production further.

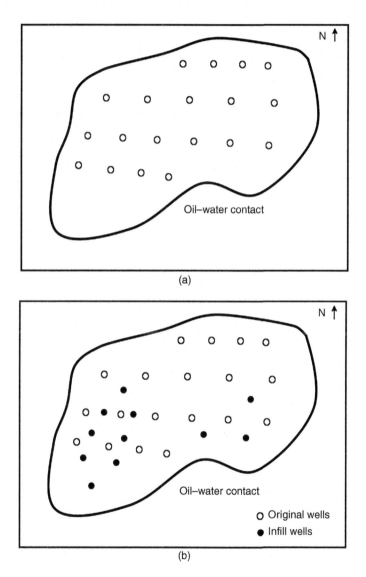

Figure 19.1 Depiction of infill drilling scenario in an oil reservoir. (a) The reservoir produced through a number of wells in the initial phase, which was followed by water injection. As production declined, an integrated study was conducted to identify the areas of high residual oil saturation. (b) The areas are predominantly located in the western part of the reservoir. Infill wells were drilled to recover the oil left underground.

Tight gas and unconventional gas

Significant accumulations of natural gas are found worldwide in sandstone, carbonate, and shale formations with characteristically low to ultralow permeability. In fact, such accumulations exceed in volume in comparison to what is found in conventional gas reservoirs having higher permeability. Typically, rock permeability ranges from a fraction of a millidarcy in tight sandstones down to nanodarcies in shale. Extraction of gas is feasible due to the low inherent viscosity of natural gas coupled with the high initial pressure of the reservoir. "Tight gas" is the term commonly used to refer to very low permeability reservoirs that are known to produce mainly dry natural gas [1]. In the 1970s, gas reservoirs having a permeability of 0.1 mD or less were defined as tight gas reservoirs by the relevant authorities in the United States. However, the definition was politically aligned as it has been used to determine which operators would receive tax credits for producing gas from tight reservoirs.

Shale gas and coalbed methane, referred to as unconventional gas, belong to a subset of tight gas reservoirs having extremely low matrix permeability. It is also noted that low permeability oil reservoirs having a permeability of 5 mD or less are often difficult to produce economically by conventional methods due to the high viscosity and decreased mobility of oil.

As Darcy's law suggests, a well in a tight gas reservoir will produce relatively less gas over a longer period of time than what is expected from a well completed in high permeability formation given all other conditions remain the same. Sufficient exposure of the low permeability formation to the wells is required to produce effectively. As part of the strategy to produce effectively, many more wells with closer spacings must be drilled to attain good recovery. Horizontal wells may be particularly effective in producing from tight gas reservoirs. Furthermore, permeability enhancement of rocks, in the form of well stimulation and fracturing, is required for long-term production on a commercial basis.

A vertical well drilled and completed in tight gas reservoirs must be successfully stimulated to produce at a commercial scale. Usually a large hydraulic fracture treatment is required to produce gas economically. In some naturally fractured tight gas reservoirs, horizontal wells can be drilled to produce gas economically, but these wells also need to be stimulated.

Development of low permeability reservoirs: tools, techniques, and criteria for selection

The development of low permeability reservoirs is based on multiple sources of data and relevant analyses. These include, but are not limited to, a detailed reservoir description, utilization of fracture propagation models, reservoir simulation, and economic analysis. One of the challenges in reservoir description is the estimation of the drainage area due to the very low permeability of the formation. The shape and size of the drainage region are influenced by the depositional environment and the length and orientation of the hydraulic fracture. The trajectory of a horizontal well may also determine the shape and extent of the drainage area.

The criteria for selecting economically feasible low permeability gas reservoirs are the following [2]:

- Reservoir pressure and rock permeability: Typically, formations having relatively high pressure and permeability in the range of 0.005–0.01 mD are candidates for development. Abnormally low pressured reservoirs, partially depleted formations, and reservoirs located at a shallow depth may require a higher rock permeability to produce economically.
- Potential formation damage: One other point to consider is the fluid retention or trapping phenomenon in low permeability formations, which may significantly affect reservoir performance. A low permeability reservoir can be subjected to considerable damage due to fluid retention or trapping during completion. For a reservoir having permeability in microdarcies, the initial water saturation can be quite low resulting in undersaturated rock. However, during completion, fluids can be trapped in the rock leading to severe formation damage. The combined effects of pore geometry, wettability of rock, depth of fluid invasion, drawdown pressure, capillary pressure, and relative permeability contribute to fluid retention. However, not all the low permeability reservoirs exhibit fluid retention during completion.

Case Study: Evaluation of Horizontal Infill Drilling in Complex Carbonate Reservoirs

Many carbonate reservoirs of Mississippian age in central Kansas are compartmental in nature due to the presence of vertical shale barriers. In addition, a number of factors such as the existence of thin pay zones, high water cut, low recovery, and lack of integrated reservoir characterization pose significant challenges for efficiently managing the matured fields. Conventional vertical wells are limited by design in contacting multiple compartments and producing oil in optimum quantities. A detailed study was conducted to evaluate the potential for drilling targeted horizontal infill wells to attain significant production potential from the reservoirs by linking the compartments [3]. Production from the reservoirs is also supported by the presence of strong water drive. However, once a well is drilled, the shale barriers could be unstable, causing damage to the well.

The study included the following:

- Screening of various reservoirs that are good candidates for horizontal drilling based on publicly available data
- Integrated reservoir characterization and construction of a reservoir model
- Reservoir simulation and validation of a reservoir model based on history matching
- Identification of zones of high residual oil saturation based on reservoir simulation
- Determination of production potential of targeted infill wells
- Three-dimensional seismic attribute analysis to delineate the reservoir compartments and estimate the rock properties in the compartments such as porosity and pay zone thickness

Fourteen fields were selected by initial screening based on cumulative primary production and pressure support as evident from drill stem test data. Further screening was performed to rank the suitability of the fields for horizontal infill drilling. The following parameters were considered in the process:

- Extent and thickness of the reservoir
- Average porosity of the formation
- Depletion in reservoir pressure
- Estimated remaining reserves per acre-ft.
- Average well spacing

In the subsequent phase of study, three fields were selected for reservoir characterization and simulation, and one of the fields was finally chosen to drill a pilot well. Based on log and core data, the reservoir was found to be more complex than what was previously modeled. Shut-in well tests performed at two nearby wells also confirmed the high degree of complexity of the reservoir in the form of compartmentalization. A 3D seismic survey was then conducted to characterize the compartments better. The reservoir model was then updated with the 3D seismic data. Results of the subsequent simulation with various trajectories from the pilot hole indicated less than expected production potential for the well. This was due to the reduced drainage volume as a result of compartmentalization and lack of evidence of pressure support. The horizontal infill well was not drilled in view of the associated risks involved and possibility of drilling in other locations.

The integrated study, based on a large amount of well and reservoir data, concluded that strong pressure support and well spacing in excess of 40 acres are critical in successfully producing from infill horizontal wells for the Mississippian fields.

Case Study: Evaluation of Horizontal Infill Drilling in Unconventional Shale Reservoirs

Bakken shale in North Dakota has witnessed intense drilling activities to produce unconventional oil and gas in recent years. Many horizontal wells having 5,000–10,000 ft. laterals are drilled on 640-acre and 1,280-acre spacings, respectively. Typical recoveries from the wells are in single digits, ranging between 3% and 7%. A reservoir simulation study based on a black oil model was conducted for three layers, namely, upper, middle, and lower Bakken to explore the potential of infill drilling with an objective to recover additional hydrocarbon [4]. The study incorporated fracture modeling, as multistage fracturing is routinely done to make the horizontal wells productive in ultratight shale formation. Fracture modeling, based on available rock mechanic data and planned trajectory of wells, was utilized to determine the dimensions and conductivity of the fractures. The number of fracturing stages can be quite high, in the range of 30–40, which are expected to create a large number of fractures in vertical and horizontal directions throughout the length of the lateral.

A three-phase black oil simulation model was developed to predict production performance. Wells considered in the study produced mainly oil with some associated gas. Rock properties used in the simulation model included relatively low porosity (6%) and ultralow permeability (0.002−0.04 mD), which are typical of unconventional shale reservoirs. As for mechanical properties of rock, Young's modulus of 500,000 and 150,000 psi for the Upper and Lower Bakken, respectively, were used. The closure stress gradient for Middle Bakken was 0.65 psi/ft. Total net pay thickness for the three layers was 42 ft. Oil properties included specific gravity of 42°API, bubble point pressure of 2398 psi, and gas−oil ratio of 700 ft.^{3}/bbl.

Three scenarios were studied where the average effective permeability of shale varied between 0.002 mD and 0.04 mD. Fracture stages were varied from 4 to 12 in the study. The study also considered three treatment sizes for sand and ceramic proppant.

The results of the study indicated that infill drilling potential in unconventional shale reservoir, along with fracture design and proppant selection, depends heavily on formation permeability. The infill drilling potential for the wells with 640-acre spacing is significant, particularly where rock permeability is favorable. There appears to be an optimum number of stages for fracturing, beyond which a diminishing return from production is anticipated.

Case Study: The Supergiant Sulige Gas Field, China

The following is a case study of implementation of low cost intensive drilling strategy in a giant low permeability field with significant heterogeneities. Sulige Gas Field [5], located in the central part of the Ordos Basin, was discovered in 2000 and began production a few years later. Total in-place proved gas reserves for the field are estimated at nearly 60 Tcf, making it the largest gas field in China. Production is reported to be about 1.3 Bcf/day. The produced gas is low in rich hydrocarbon components. The field is characterized by low reservoir pressure, low permeability, complex gas water distributions, and significant reservoir heterogeneity. The interplay between sedimentary facies and diagenesis has resulted in reservoir heterogeneities.

Geologically, the reservoir is bounded by large stratigraphic closure. Traps in the field are predominantly stratigraphic variations associated with facies change. Due to the resolution limit of seismic data, a two-row 800-m (28,250-ft.) spacing exploratory well pattern was used to delineate the gas reservoirs.

Reservoirs are dominated by coarse-grained sandstones of the Lower Permian Shihezi and Shanxi formations, especially those of the Xian Shihezi 8 and Shanxi 1 units. The sandstone formations, with an average burial depth of 10,500–11,480 ft., were deposited in a braided-river environment. Coarse-grained bar and basal channel sandstone facies provide effective reservoirs with porosity ranging from 5% to

12% and permeability in the range of 0.02−2.0 mD. Porosity and permeability are found to vary noticeably with depth. Due to deep burial depths, primary porosity was reduced significantly; however, secondary dissolution pores developed, accounting for 80% of the total pore spaces. Coarse-grained sandstones with rich quartzite rock fragments experienced less compaction due to abundant rigid grains, allowing the flow of fluid. As a result, secondary dissolution pores developed in these sandstones. Data from 16 densely drilled wells indicate that net pays are typically less than 8 m (26 ft.) in thickness with lateral continuity of several hundred meters.

The significant heterogeneity in the field suggests the necessity of a dense production well pattern. As of 2011, there were 4222 gas wells in operation, 3439 of which being opened per day on average. With daily gas production exceeding 353 MMcf in 2007, 706 MMcf in 2008, and 1059 MMcf in 2009, Sulige had been able to produce over 1300 MMcf per day, becoming China's largest uncompartmentalized gas field, and a sample of low cost development of tight gas reservoirs in the country.

Case Study: Simulation of Horizontal Well Performance in a Tight Reservoir

A simulation study highlighting the effectiveness of horizontal drilling in a tight oil reservoir is presented in the following. The study clearly indicates that horizontal wells recover substantially more oil over a longer productive life. The reservoir model is assumed to have an average horizontal permeability of 0.25 mD. Moreover, the vertical permeability is assumed to be half of the horizontal permeability. The porosity of formation is 15%. The initial oil saturation was 75%, and the rest was formation water. The entire field is divided into two sectors having the same rock and fluid properties. Nine vertical wells are placed in the first sector. Three horizontal wells are drilled in the second sector. The total cost of drilling the three horizontal wells is not much higher than drilling the nine vertical wells. The two sectors are noncommunicating as the transmissibility of grid blocks between the two sectors is set to zero.

The performance of a horizontal well is compared against a vertical well in Figure 19.2. The combined primary recovery from the nine vertical wells was about 8.8% in 12 years when the wells reached their economic limit of 13 bbl/day. In contrast, the three horizontal wells recovered about 18.5% of original oil in place in 29 years. The initial reservoir pressure was 4862 psi, which declined to 1800 psi in the sector having the three horizontal wells at the end of the simulation. The continued drop in average reservoir pressure resulting in the evolution of the gas phase and the rise in gas−oil ratio are shown in Figure 19.3.

The 3D simulation study is based on a black oil model [6]. The number of grids used is 17 × 14 × 4. Grid dimensions vary between 230 ft. and 520 ft. in a lateral direction. The height of each grid block is 19 ft. Local grid refinement was used to simulate the performance of horizontal wells.

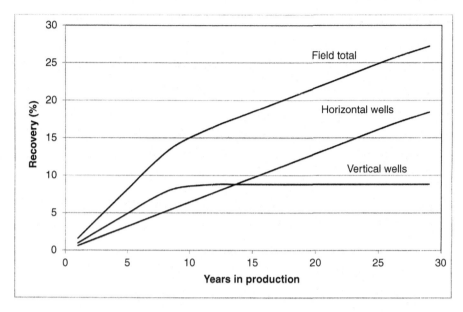

Figure 19.2 Comparison of oil recovery by vertical and horizontal wells.

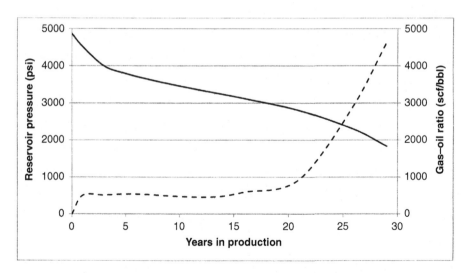

Figure 19.3 Typical decline in reservoir pressure and rise in gas–oil ratio during primary production.

Summing up

Reservoir engineers adopt various strategies in the development of low to ultralow permeability reservoirs, including the unconventional reservoirs. Some of the methods are as follows:

- Infill drilling, resulting in relatively close well spacing
- Horizontal drilling, single lateral or multilateral
- Well stimulation by hydraulic fracturing and acidization
- Reservoir management based on intensive monitoring of well performance, including the deployment of permanent downhole gauges. This leads to the evaluation of formation damage, well productivity, and reservoir performance as a whole on a regular basis.

Various aspects of horizontal well technology have been discussed in an earlier chapter. This chapter discusses infill wells that are drilled as a widely practiced strategy in the oil and gas industry. Infill wells contact the portion that was not contacted before and produce additional quantities of oil and gas. Some of the benefits of infill drilling are:

- Better reservoir economics and addition of assets as more wells are drilled in appropriate locations
- Long lasting production in commercial quantities from low to ultralow permeability formations
- Increased rock exposure to wellbores where severe reservoir heterogeneities exist
- Increase in horizontal and vertical sweep efficiency during waterflood
- Enhancement in connectivity between injectors and producers
- Longer reservoir life

The following types of reservoir are best suited for the implementation of infill drilling technology:

- Tight reservoirs having low to ultralow permeability. Infill drilling strategy is adopted to produce tight gas economically in many basins worldwide.
- Highly heterogeneous reservoirs where large portions of the reservoir are left untapped due to the presence of geologic discontinuity, facies change, compartments, sealing faults, and others.
- Heavy and ultraheavy oil reservoirs where economic oil production cannot be accomplished with limited number of wells.
- Matured reservoirs or reservoirs nearing abandonment where substantial quantities of oil are left behind.
- Compartmental reservoirs where infill wells are needed to produce from untapped portions.

Tight gas is the term commonly used to refer to low permeability reservoirs that produce mainly dry natural gas. Formation permeabilities range from a fraction of a millidarcy to microdarcies. Unconventional shale gas and coalbed methane are a subset of low permeability where rock permeability is in nanodarcies.

Wells drilled in tight gas formations cannot sustain economic production over a long period of time. Hence, infill wells with closer well spacing must be drilled to recover gas and attain economic viability. A vertical well drilled and completed in a tight gas reservoir must be successfully stimulated to produce a sufficient volume of gas economically. Horizontal wells drilled in low permeability reservoirs also require stimulation and fracturing.

In the 1970s, tight gas reservoir was defined by the US government as one in which the expected value of permeability to gas flow would be less than 0.1 mD. However, the definition was politically aligned as it has been used to determine which wells would receive tax credits for producing gas from tight reservoirs.

A comprehensive strategy for development of low permeability reservoirs is based on multiple sources of tools and techniques. Closer well spacing (infill drilling), horizontal drilling, and well stimulation by hydraulic fracturing and acidization are among the options available to develop low permeability reservoirs. The essential studies include, but are not limited to, a detailed reservoir description, utilization of fracture propagation models, reservoir simulation, and economic analysis. Reservoir management based on intensive monitoring of well performance is also required.

Criteria for selection of low permeability reservoirs include the range of permeability coupled with reservoir pressure. When the reservoir pressure is relatively higher, the reservoirs with permeabilities in microdarcies (0.005–0.01 µD) can be selected for development. For reservoirs having low pressure due to shallower depth, partial depletion and abnormal geologic setting, better rock permeability would be required for selection.

Potential damage of low permeability is also a concern in the selection of low permeability reservoirs for intensive drilling and completion. Certain formations are found to be undersaturated with connate water and fluid trapping occurs during completion resulting in loss of productivity. Various petrophysical and other properties are responsible for the phenomenon.

This chapter presents three field cases that give a firm grasp of how infill drilling and other technologies are implemented in the field.

- Evaluation of infill drilling of a conventional carbonate reservoir with heterogeneities
- Evaluation of horizontal infill drilling based on a simulation model in an unconventional shale gas reservoir
- Successful development and production of a supergiant gas field of low permeability based on dense drilling strategy

Questions and assignments

1. What is infill drilling and when are the infill wells drilled?
2. Why would infill horizontal wells be drilled in a reservoir?
3. What reservoirs are best suited for infill drilling? Name at least four types of reservoirs based on a literature review.
4. Are infill wells drilled randomly or in a definite pattern?
5. Are infill wells expected to produce more oil and gas than originally drilled wells? Why or why not?
6. Are all infill wells producers? Can an infill well be drilled to inject fluid for enhanced oil recovery?

7. How do low permeability reservoirs perform differently to a moderate to high permeability reservoir? Explain with the help of relevant equations of fluid flow in porous media. Do tight gas reservoirs produce any condensates?

8. How are the low permeability sandstone and limestone reservoirs differentiated from shale gas reservoirs and coalbed methane in terms of rock properties?

9. Do well logs aid in drilling infill wells? Explain how.

10. How does reservoir pressure affect the performance of low permeability reservoirs? Explain with examples.

11. Based on a literature review, compare the reservoir life cycle between a high permeability reservoir (50 mD or greater) and a low permeability reservoir (0.5 mD or lower). Include a comparison of the recovery efficiency of the two reservoirs.

12. You are an engineer with a company that is planning to drill infill wells in a tight but fractured carbonate gas reservoir following limited success with the initial wells. Describe all the tools and techniques you might use to identify target areas and pay zones.

References

[1] Holdtich SA. Tight gas sands. J Petrol Technol 2006;58(6).

[2] Bennion DB, Thomas FB, Imer D, Ma T. Low permeability gas reservoirs and formation damage – tricks and traps. Hycal Energy Research Laboratories Ltd. SPE-59753-MS Publisher: Society of Petroleum Engineers. Source: SPE/CERI Gas Technology Symposium, 3–5 April, Calgary, Alberta, Canada.

[3] Bhattacharya S. Field demonstration of horizontal infill drilling using cost-effective integrated reservoir modeling – Mississippian carbonates, Central Kansas. Open File Report. Kansas Geological Survey; 2005.

[4] Eleyzer PE, Cipolla CL, Weijers L, Hesketh RE, Grigg MW. A fracture modeling and multi-well simulation study evaluates down-spacing potential for horizontal wells in North Dakota. World Oil, vol. 231, No. 5.

[5] He D, Jia A, Xu T. Reservoir characterization of the giant Sulige Gas Field, Ordos Basin, China. Poster presentation at AAPG Annual Convention, Houston, Texas, April 9–12, 2006.

[6] IMEX data file. Computer Modelling Group.

Rejuvenation of reservoirs with declining performance

Introduction

Worldwide statistics suggest that the likely recovery from most conventional oil reservoirs is in the range of 25−50%. In unconventional oil and gas reservoirs, recovery is less than 10% in numerous cases. Furthermore, major petroleum basins have already been discovered in various regions of the world with little hope of finding giant new fields in the future. Obviously, there is scope for the reservoir team to improve upon recovery from reservoirs where well rates are declining but large amounts of oil remain untapped in the reservoir. With technological advancements in the oil and gas industry, matured fields near abandonment or reservoirs beset with declining performance issues may hold great potential for rejuvenation and add to assets. In general, reservoir areas and zones of high residual oil saturation are identified and innovative strategies are planned, tested, and deployed to produce oil. In certain cases such as compartmental reservoirs, specific areas may not be contacted at all by recovery efforts conducted earlier.

This chapter discusses some of the major rejuvenation efforts undertaken by the reservoir engineering team to augment reservoir performance including case studies. Attempts are made to respond to the following queries:

- What are the symptoms of matured or problematic reservoirs?
- What are the major approaches to rejuvenate reservoirs with declining performance?
- What specific efforts are undertaken to improve well productivity and redevelop the field?

Decline in reservoir performance

A reservoir may exhibit one or more signs of decline in performance following a few years or even a short few months of production. If the reservoir is managed effectively by intensive data collection and implementation of innovative techniques, many performance-related issues are either delayed or avoided altogether. A reservoir in decline may exhibit one or more of the following:

- Substantial decrease in oil or gas production rates
- Decline in well rate approaching the economic limit
- Reservoir pressure approaching abandonment pressure
- Rapid increase in water−oil or gas−oil ratios
- Premature breakthrough of fluids such as water, steam, or gas
- Cycling of injected fluid is inefficient, incremental oil recovery is minimal
- A sizeable portion of oil is left behind even with continued enhanced recovery efforts
- Oil or gas field operating cost is higher than expected due to well and reservoir issues

Reservoir Engineering. http://dx.doi.org/10.1016/B978-0-12-800219-3.00020-6

Major strategies in redeveloping matured oil fields

In recent decades, field rejuvenation efforts focused on the application of horizontal drilling technology, among a number of other approaches. Vertical wells producing from a reservoir are likely to decline in production due to a number of issues, including low permeability, limited exposure to the oil zone, geologic complexities, and excessive water production, among others. Once horizontal drilling technology gained popularity, it became a common practice to locate the areas in the reservoir where oil is left underground and drill new horizontal wells. Besides, many existing vertical wells were side-tracked to produce from zones and locations that were not tapped earlier.

Another field rejuvenation strategy focuses on infill drilling of wells in older fields. Wells drilled in earlier stages of reservoir life are found to be inadequate to produce oil or gas effectively from areas in between due to low permeability of the formation or poor waterflood sweep efficiency. Hence, infill wells are drilled in progressively reduced spacing to produce from untapped portions of the reservoir. Target areas for infill drilling in the reservoir are also identified by seismic and other methods where waterflooding and enhanced oil recovery methods have not been quite effective.

Rejuvenation efforts offer certain opportunities and advantages. Matured and problematic oil fields may usher great opportunity to experiment with new tools and techniques in engineering and management. Furthermore, necessary approvals from authorities to operate the field are already in place.

Revitalization efforts

As worldwide demand for oil and gas increases, not only are the new frontiers of petroleum deposits vigorously explored but also older fields are re-evaluated for added assets. New technologies are introduced on a regular basis that open the door for new possibilities. Primarily, efforts are directed toward recovering the remaining oil that was bypassed during secondary or tertiary recovery operations. In reservoirs with geologic complexity, such as compartmental or faulted reservoirs, efforts are made to identify areas where oil was not contacted by older wells. A literature review suggests that a wide variety of tools and techniques are used to revitalize oil or gas fields with declining performance or a reservoir nearing abandonment, some of which are listed as follows [1]:

- Detailed reservoir characterization based on production history and time-lapse seismic studies, among others. Reservoir characterization is discussed in Chapter 6.
- Review of existing reservoir data in the light of new information obtained about the reservoir or a potential application of new technology.
- Development of robust earth and dynamic simulation models. Scenario-based modeling examines infill drilling and enhanced oil recovery (EOR) strategies.
- Identification of reservoir complexities that hinder wells from effectively producing the reservoir. Examples are compartmental reservoirs and presence of fault blocks.
- Identification of high permeability channels that lead to high remaining oil saturation in various sections of the reservoir following waterflood.
- Estimation of remaining oil and gas volume by developing isoHCPV maps, among others. Development of an isoHCPV map is described in Chapter 12.
- Review of lessons learned from earlier efforts in rejuvenating similar reservoirs.

- Economic analysis of investment in redeveloping the field.
- Drilling of stepout wells to produce from adjacent areas within the reservoir that were not explored earlier.
- Recompletion of wells in zones where significant quantities of oil are trapped.
- Shutting off watered-out zones following water injection by using intelligent well technology.
- Completion of producers in multiple zones where single completion is not economical.
- Well stimulation and installation of gas lifts to boost productivity.
- Optimization and better management of waterflood operation. Realignment of injectors and producers during waterflood operation to increase areal sweep efficiency.
- Implementation of a reservoir surveillance program in order to evaluate the effectiveness of rejuvenation efforts on a regular basis.

A typical workflow related to the revitalization of a matured reservoir is presented in Figure 20.1.

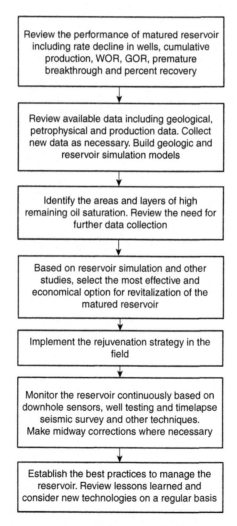

Figure 20.1 Workflow to rejuvenate reservoir with declining performance.

Case Study: Weyburn Field, Canada [2]

The Weyburn field with over 60 years of production history is an example of the confluence of many technologies that are available in the petroleum industry to revitalize a matured reservoir. Rejuvenation efforts included waterflooding, infill drilling, horizontal well injection, miscible CO_2-EOR, and time-lapse seismic study. In addition, the field served as a major test case for carbon dioxide sequestration. Revitalization of the field occurred in several stages over decades, starting from waterflooding, then to infill drilling, and finally to carbon dioxide injection, depending on the available technology, its effectiveness for the field, and economic feasibility at the time of implementation.

Located in the Williston basin in southeastern Saskatchewan, Canada, the field is spread over an area over 53,000 acres. The carbonate reservoir has two major layers, limestone at the bottom overlain by dolostone. Vugs are observed in the formation. Porosity ranges from 10% to 26%, while average permeability in various zones may vary from less than 3 mD to 50 mD. However, as expected in a typical carbonate reservoir exhibiting vugginess and other heterogeneities, the extreme values of permeability were reported to be 0.1 and 500 mD. The main producing zone is at a depth of 4750 ft.

Production began in 1954 and reached a peak of 46,000 bpd in about 10 years followed by a steady decline. In an effort to bolster production, waterflooding was initiated in the field. As the production from the field declined to less than 10,000 bpd in the 1980s, vertical infill wells were drilled. As horizontal well technology became popular, horizontal infill wells were drilled over the next decade. As a result of drilling the infill wells, production rose to about 24,000 bpd. However, production declined below 20,000 bpd in a few years. In 2000, injection of carbon dioxide into the reservoir was initiated as part of EOR and for the storage or sequestration of carbon dioxide. The injection rate of carbon dioxide was 95 MMscf/day. Water was also injected into the reservoir. Injectors include both horizontal and vertical wells. Carbon dioxide was recycled during the EOR operation and finally sequestered underground at the end of the project. In the mid-2000s, field production rose as high as 30,000 bpd. Time-lapse seismic monitoring was employed to track the movement of injected carbon dioxide and identify new well locations (Figure 20.2).

Case Study: Bahrain Field [3]

Bahrain field reached its peak rate of about 80,000 BOPD in the 1970s, following four decades of its discovery and subsequent production. The field experienced a decline rate of about 7% annually. The decline was reduced to 1.3% by adopting a number of revitalization measures. The field consists of 17 oil and 3 gas reservoirs in diverse geologic settings requiring a wide range of strategies that were implemented to arrest declining production to a large extent. Moreover, the carbonate reservoir has significant heterogeneities including intersecting faults posing challenges in successful implementation of redevelopment measures. Some of the rejuvenation efforts are listed as follows:

- Detailed reservoir characterization: This was accomplished by integrated study based on geologic, reservoir, and production data. Maps were generated to identify areas of good reservoir quality. Robust reservoir models were developed for attaining better accuracy in simulation and performance prediction.
- Implementation of horizontal well technology: Horizontal wells were drilled in two reservoirs to enhance oil production significantly.
- Drilling of infill wells: Infill wells drilled in certain areas, including tight zones near a geologic fault, proved successful in enhancing reservoir performance.
- Dual completion of wells: Certain wells were completed in two zones rather than one in order to tap oil from shallow reservoirs previously considered to be uneconomic.
- Recompletion of wells: Certain other wells were recompleted in different target zones in order to avoid high gas−oil ratio.
- Well management techniques: Deployment of gas lift, pumps, and gas production control devices.
- Implementation of improved oil recovery (IOR) projects: In suitable reservoirs, IOR projects were implemented for secondary and tertiary recovery.
- Asset management: Annual assessment of reserves was carried out as oil migrated across faults due to gas injection.

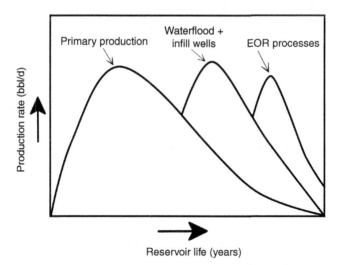

Figure 20.2 An example of incremental oil production at various stages of the reservoir life cycle.

Summing up

Worldwide experience in the recovery of petroleum indicates that substantial amounts of oil are left underground due to the lack of appropriate technology and favorable economics. In most cases, ultimate recovery of oil from conventional reservoirs is 25−50% depending on reservoir quality and cost. In unconventional reservoirs, the recovery is even less. Hence, there is scope for reservoir engineering professionals to enhance recovery based on detailed reservoir study and implementation of innovative techniques. Rejuvenation of matured oil fields requires the identification and mapping of high oil saturation that remains in porous rock following primary, secondary, or tertiary recovery. Appropriate technology, including horizontal drilling and recompletion of existing wells, is then implemented in targeted areas and zones to recover the oil left behind.

Reservoirs with declining performance may exhibit one or more of the following: declining oil or gas rate, declining reservoir pressure, increasing water cut, premature breakthrough of injected fluid, inefficient cycling of injected fluid, high remaining oil saturation, and increased cost due to well and reservoir issues.

Workflow for rejuvenation of matured reservoirs includes: (i) review of reservoir performance including declining well rates, reservoir pressure, increasing water and gas cuts, and any other relevant issues; (ii) review of existing geologic, geophysical, and other related data; (iii) update of static and dynamic reservoir models; (iv) history matching of simulation models; (v) identification of areas and zones of high remaining oil saturation; (vi) build scenarios for revitalization of the reservoir including drilling of new infill and stepout wells, recompletion of older wells, design of enhanced recovery, intensive reservoir surveillance, and others.

Some of the strategies used by the reservoir team in rejuvenating a matured reservoir are as follows:

- Detailed reservoir characterization based on production history and time-lapse seismic studies, among others. Reservoir characterization is discussed in Chapter 6.
- Review of existing reservoir data in the light of new information obtained about the reservoir or a potential application of new technology.
- Development of robust earth and dynamic simulation models. Scenario-based modeling examines infill drilling and EOR strategies.
- Identification of reservoir complexities that hinder wells from effectively producing the reservoir. Examples are compartmental reservoir and presence of fault blocks.
- Identification of high permeability channels that lead to high remaining oil saturation in various sections of the reservoir following waterflood.
- Estimation of remaining oil and gas volume by developing isoHCPV maps, among others. Development of isoHCPV maps is described in Chapter 12.
- Review of lessons learned from earlier efforts in rejuvenating similar reservoirs.
- Economic analysis of investment in redeveloping the field.
- Drilling of stepout wells to produce from adjacent areas within the reservoir that were not explored earlier.
- Recompletion of wells in zones where significant quantities of oil are trapped.
- Shutting off watered-out zones following water injection by using intelligent well technology.
- Completion of producers in multiple zones where single completion is not economical.

- Well stimulation and installation of gas lifts to boost productivity.
- Optimization and better management of waterflood operation. Realignment of injectors and producers during waterflood operation to increase areal sweep efficiency.
- Implementation of a reservoir surveillance program in order to evaluate the effectiveness of rejuvenation efforts on a regular basis.

Questions and assignments

1. What is matured field rejuvenation and why is it important?
2. What criteria must a reservoir meet to be a candidate for revitalization?
3. What are the major steps in reservoir rejuvenation efforts? Would the same strategy be applicable in all types of reservoirs?
4. What are the methods in estimating residual oil saturation?
5. List the major causes of performance decline in the following cases:
   a. A fractured reservoir with low matrix permeability
   b. A sandstone reservoir under solution gas drive
   c. A reservoir with high vertical permeability and long transition zone
   d. A heterogeneous carbonate reservoir under waterflood for several years
6. Do gas condensate reservoirs require revitalization?
7. Would you consider rejuvenating a stratified sandstone reservoir where 50% of original oil in place is already recovered? If yes, develop a detailed plan.
8. Why would you recommend the drilling of horizontal wells in a matured reservoir? Explain with examples from the literature.

References

[1] Satter A, Iqbal GM, Buchwalter JA. Practical enhanced reservoir engineering: assisted with simulation software. Tulsa, OK: Pennwell; 2008.
[2] Verdon JP. Microseismic modelling and geochemical monitoring of CO_2 storage in subsurface reservoirs. New York: Springer; 2012.
[3] Murty CRK, Al-Haddad A. Integrated development approach for a mature oil field. SPE #78531. Tenth Abu Dhabi International Exhibition and Conference: Abu Dhabi, UAE; October 13–16, 2002.

Unconventional oil reservoirs 21

Introduction

Unconventional oil reservoirs cannot be produced economically by traditional methods. The reservoirs require innovative technologies to develop, produce, and manage. Unconventional oil and gas have drawn worldwide attention due to the ever increasing demand for energy and dwindling resources of conventional oil. Based upon geologic and other evidences, many analysts believe that most of the conventional giant oil fields have already been discovered worldwide and conventional oil production is expected to reach a peak within the foreseeable future. On the other side of the spectrum, unconventional oil is more difficult to extract due to the relatively high cost per barrel of production and unique environmental issues. Above all, the various extraction technologies employed to produce unconventional oil are evolving. However, the development and production of unconventional reservoirs are becoming the center of attention as innovative technologies are introduced in the oil and gas industry and as long as the price of oil supports commercial production. For example, Canada became a world leader in heavy oil production by unconventional methods in recent times. Canadian oil reserves are currently estimated to be second only to Saudi Arabia. Unconventional shale gas, as discussed in the next chapter, is becoming a major source of energy supply along with tight oil. This is occurring due to the advancements in horizontal well drilling combined with multistage fracturing (Figure 21.1).

Unconventional oil resources can be viewed in two major categories. In the first category, oil having extremely high viscosity is hardly mobile unless certain thermal and nonthermal techniques are applied for extraction at an economic scale. The other category of unconventional oil resources has unfavorable reservoir characteristics in the form of ultralow rock permeability, which hinders the implementation of conventional methods to extract oil. Besides the above, oil shale is a significant unconventional resource. Oil shale refers to shale or other types of rock rich in kerogen, which is the precursor of oil and gas as described in Chapter 2. Oil shale requires thermal processes to extract the organic-rich kerogen for conversion into various types of fuel.

Unconventional oil resources are shown in Table 21.1.

This chapter presents an overview of current unconventional oil production technologies and answers the following questions:

- What is unconventional oil? How does it differ from conventional oil?
- What are the characteristic rock and fluid properties in unconventional oil reservoirs?
- What are the principal methods of extraction of unconventional oil?
- What are the factors that influence the recovery efficiency from unconventional oil reservoirs?
- In what countries are unconventional, large-scale oil production processes developed and implemented?

Reservoir Engineering. http://dx.doi.org/10.1016/B978-0-12-800219-3.00021-8

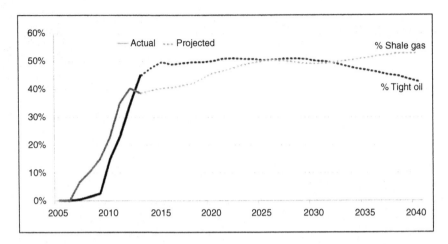

Figure 21.1 Projected supply of tight oil and shale gas in the United States.
Source: Congressional Research Service, 2014. http://fas.org/sgp/crs/misc/R43148.pdf.

Table 21.1 **Unconventional oil resources**

Unconventional oil resource	Attributes	Major extraction technologies/processes
Heavy oil	Oil specific gravity ranges between 10°API and 20°API.	Thermal, nonthermal, horizontal drilling
Extra heavy oil	Oil specific gravity is 10°API or heavier. Typically reservoirs are at a shallower depth with high permeability.	Thermal, nonthermal
Oil sands, tar sands, bitumen	Oil specific gravity is 10°API or heavier. Typically reservoirs are at a shallower depth with high permeability.	Thermal, nonthermal
Tight oil	Reservoir permeability in fractions of mD. Oil characteristics are similar to conventional oil.	Horizontal drilling, multistage fracturing
Shale oil	Reservoir permeability in fractions of mD. Oil characteristics are similar to conventional oil.	Horizontal drilling, multistage fracturing
Oil shale	Kerogen-enriched rock	Thermal. Oil extracted by retorting and distillation

Unconventional reservoir characteristics

There are other major differences between unconventional and conventional reservoirs, some of which are discussed in Chapter 2. For example, unconventional shale reservoirs are pervasive in nature, extending over hundreds of miles in many cases, while conventional reservoirs have distinct boundaries and limited areal extent. Due to the continuous nature of shale gas reservoirs, the probability of drilling a productive well is high, although the productive life of the well could be relatively short requiring increased drilling. In contrast, exploration of conventional oil may turn up a sizeable number of dry holes, while the productive wells may have commercial production over a long period of time. In another example, oil sands reservoirs can be located at very shallow depths and be mined to extract the highly viscous crude. Furthermore, unconventional oil extracted from these reservoirs needs upgrading before the usual processing in the refinery. The differences in reservoir properties as well as in oil characteristics influence how the unconventional oil reservoirs are to be explored, planned, developed, and managed.

The types of unconventional heavy oil, including oil sands and shale oil, and the methods of extraction are described in the following sections.

Oil sands and extra heavy oil

It is interesting to note that a very significant amount of world's oil reserves are in the form of oil sands, also referred to as tar sands [1]. The constituents of tar sands are bitumen, clay, sand, and water. A thin film of water envelops the sand particles. The outer envelope that surrounds sand and water is bitumen. It is estimated that over 1.75 trillion barrels of oil are deposited in the form of tar sands; the major areas of accumulation being in Canada and Venezuela.

The deposits of oil sands in Alberta, Canada, cover an area larger than England. Venezuelan oil sands, sometimes referred to as extra heavy oil, may be as high as over 235 billion barrels. In the United States, tar sands are found in Utah; the estimated resources are about 19 million barrels. Tar sands are also found in the Middle East.

Due to the high viscosity and presence of various constituents in oil sands, it requires appropriate processing for oil extraction and upgrading to synthetic crude. The viscous liquid requires mixing with conventional oil before transporting through pipelines.

The viscosity of oil sands is extremely high and runs into thousands of centipoise or more. Hence, oil sands cannot be pumped to the surface utilizing conventional oil well technology. Techniques to produce oil sands include [2–8]:

- Open pit mining
- Steam injection
- Injection of solvents
- Fire flooding

Open pit mining applies to oil sands found at shallow depths. Following mining, bitumen is extracted from oil sands by adding hot water and agitating the slurry. Bitumen, which rises to the top, is then separated by skimming from sand and other materials.

Oil sands buried deep below the surface are usually recovered by applying thermal energy in the form of steam injection, which results in sufficient reduction in viscosity. About 2 tons of oil sands are required to produce one barrel of oil. The recovery of bitumen from tar sands is about 75%.

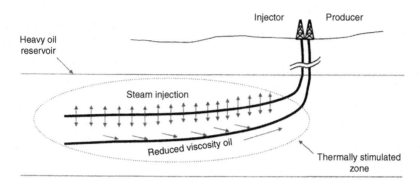

Figure 21.2 Steam assisted gravity drive (SAGD) process.

Steam assisted gravity drainage (SAGD) has successfully been applied to produce the highly viscous oil and bitumen in Canada since the 1990s. Two single-lateral horizontal wells are drilled in the target location (Figure 21.2). The lateral of one horizontal well is drilled directly above the other; the typical spacing between the two laterals in a vertical direction is 4–6 m (13–20 ft.). Steam is injected through the upper horizontal well, which reduces the viscosity of heavy oil and bitumen. Consequently, oil becomes mobile due to reduced viscosity and flows downward by the forces of gravity. The heated oil is then collected by the lower lateral for production through the other horizontal well. Both wells operate near the reservoir pressure. Condensation of injected steam takes place as thermal energy is transferred to oil and the formation. Hence, a large volume of water is also produced along with heavy oil. Heavy oil and condensed water are recovered at the surface by the use of progressive cavity pumps, which are designed to handle viscous fluids.

Substantial oil recovery can be attained by the SAGD process, up to 70% or more, where formation characteristics and operating conditions are favorable. Recovery efficiency is not affected significantly due to the presence of shale streaks or other heterogeneities in the formation. Cracks in the formation are created due to the application of thermal energy through which steam may rise and circulate.

The mechanism of production of highly viscous oil involves the formation of a virtual steam chamber around the lateral section of the injector. The chamber becomes enlarged, in both vertical and horizontal directions, with sustained steam injection. Steam flows to the periphery of the chamber to heat the viscous fluid. Due to thermal stimulation, light hydrocarbons and others gases such as CO_2 and H_2S are released, which rise in the steam chamber and fill the void created by the produced oil. Such gases can also act as a thermal insulator above the steam chamber limiting the loss of thermal energy. The rise of steam due to gravity also ensures that it is not produced through the lower horizontal well. However, in a heterogeneous formation, a steam chamber may not grow evenly; hence, some portion of steam is allowed to enter the producer to maintain adequate heat throughout the well, which ensures that bitumen remains less viscous and mobile. Colder parts of the formation are also heated in the process. In another method referred to as partial SAGD, steam is deliberately circulated through the producer after a lengthy shutdown of the producer or during startup. One other advantage is that the expanded circulation of steam ensures that the steam chamber is sustained and does not collapse in the case where steam condenses and enters the injector. The above may lead to the collapse of the steam chamber.

Other thermal recovery methods of heavy oil and bitumen are described in the following section.

Cyclic steam stimulation process

The cyclic steam stimulation (CSS) process, also referred to as the huff-and-puff method, has been used for many decades to recover heavy oil and bitumen. The method utilizes the same well for both injection and production in sequence. At the first stage of the huff-and-puff cycle, steam is injected into the formation to reduce the viscosity of heavy oil or bitumen. In the next stage, oil having relatively less viscosity is produced at the surface by the same well. Production continues until the reservoir cools to the point where the flow of oil ceases. Once the cycle is complete, the next cycle commences by injecting steam into the formation followed by the production of oil. There are several variations of the CSS process.

High pressure cyclic steam stimulation process

In this process, steam is pumped into the formation to reduce the viscosity of bitumen, then a mixture of bitumen and steam, referred to as bitumen emulsion, is pumped to the surface. Water condensed from steam aids in diluting the bitumen and separating sand from it. High-pressure steam creates cracks in the formation, which facilitates bitumen production. The high-pressure cyclic steam stimulation (HPCSS) process differs from SAGD, where steam injection takes place at a relatively low pressure and bitumen is produced by a gravity drainage mechanism through a horizontal producer drilled beneath the steam injector. Another important difference is both horizontal and vertical wells are utilized in the HPCSS process. Vertical well spacings range between 2 acres and 8 acres. Horizontal wells are placed 60–80 m apart. The process is suitable where water at the bottom or gas at the top of the formation is not significant. In relatively thin formations, heat losses may be significant, reducing recovery efficiency. Roughly one-third of oil sands in Alberta are produced by the HPCSS process.

Vapor extraction process

In the vapor extraction (VAPEX) process, steam injection is replaced by vaporized solvents such as propane mixed with noncondensable gas. As in SAGD, two horizontal wells, injector and producer, are utilized. Viscosity of bitumen is reduced as it comes in contact with solvent. It is finally recovered by gravity drainage through the producer. The solvents are injected to reduce the viscosity of heavy oil and bitumen to produce them with the expectation of better recovery. The process is suitable where injection of steam may not be very efficient. Examples include thin formations where heat loss is relatively high. Furthermore, in low permeability formations, reservoir heat capacity is relatively high, which may also reduce the effectiveness of steam injection.

Cold heavy oil production with sand (CHOPS)

As the name suggests, CHOPS is a nonthermal method for producing oil sands where sand is allowed to produce along with oil. As a result, "wormholes" or cavities, i.e., channels of high permeability, are created in the reservoir that lead to a dramatic increase in the flow of extra heavy oil. The productivity of the wells is enhanced by the extraction process. It is interesting to note that attempts to screen the sand at the well, as commonly practiced in conventional oil wells, may completely shut-in production from oil sands deposits. Shallow unconsolidated sandstone reservoirs located at a depth of 2700 ft. or less having a porosity of around 30% and permeability often in darcies are prime candidates for the CHOPS process. In typical cases, the viscosity of oil ranges between 500 cp and 15,000 cp. Production is typically low in wells where the process is applied, averaging about 150 barrels a day.

In Canada, CHOPS technology is quite successfully applied and thousands of wells operate by this method. The technology is based on relatively low capital investment. Hence, other countries including Venezuela, Russia, and China have implemented the CHOPS method in unconventional heavy oil reservoirs where the formation is unconsolidated, permeability is high, and condition is suitable for implementing the technique. However, one disadvantage of the method is relatively poor recovery, 5–15% in most cases, stressing the need for further study and development.

It must be borne in mind that the critical reservoir properties including porosity, permeability, and compressibility change constantly as the network of wormholes is formed in the vicinity of the wellbore and then propagated deep into the reservoir. These affect the dynamics of flow in the reservoir and at the well. Another important contributing factor is the foamy nature of the oil leading to greater mobility. The gas bubbles that are produced from oil do not coalesce into a free continuous phase but remain dispersed in the oil phase. The swelling of oil results in lower viscosity and addition of driving energy for enhanced production. Since there is no free gas present in the reservoir, production of oil is not hindered by the flow of gas along with oil. However, production of sand increases with oil production; the latter eventually reaches a peak production level. Hence, the ultimate recovery of oil is limited.

The production of oil and sand by the CHOPS process generally exhibits the following trend (Figure 21.3).

The initial oil rate is dependent on viscosity, presence of bubbles, and pumping rate, typically ranging between 60 bbl/day and 190 bbl/day. At the initial stage, the production of sand is quite high, as much as 40% of the total volume of liquid and solids. The peak of sand production is a function of oil viscosity. In highly viscous oil, the peak production is higher. In weeks or months, the production of sand drops to single digits and goes through a plateau period before declining further. The oil rate increases over several months while sand production declines. The peak oil rate could be as high as 60% more than the initial rate. The typical oil rate varies between 130 bbl/day and 250 bbl/day. Following a period that may last from a year to several years, the oil rate declines to a level that is no longer economic. Well workover is carried out to enhance productivity. Following a successful workover operation, the same cycle of sand and oil production is repeated; however, the oil rate does not reach the levels seen in the earlier cycle.

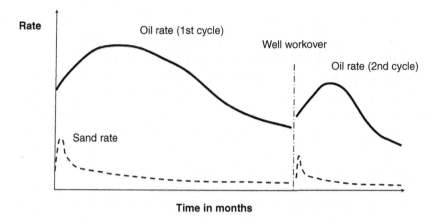

Figure 21.3 Oil and sand production characteristics in the CHOPS process.

Over short term, the production rate can vary significantly; however, in the long term, a steady decline rate can be identified from the production history. Often, the influx of water leads to a decline in the oil rate.

Common issues related to the CHOPS process include:

- Marginal reservoirs with shaley formations where the production of sand cannot be initiated and a successful recovery of heavy oil cannot be achieved.
- Substantial and early water influx that is encountered during production hampers oil productivity. Reservoirs having strong bottom water drive are likely to suffer in this regard.
- Collapse of overlying shale in ultraheavy oil formations where a high pumping rate is employed for recovery.

Case Study: Oil Sands Industry in Canada

Canadian oil sands reserves, primarily located in Alberta, are estimated to be about 175 billion barrels, making the oil reserves of Canada the third largest in the world following Saudi Arabia and Venezuela [9–11]. The estimate of recoverable reserves is about 10% of total oil sands in place. The major accumulations of oil sands in northeastern and northwestern Alberta are as follows:

- Athabasca deposits
- Cold Lake deposits
- Peace River deposits

As noted earlier, the largest deposits of oil sands, also referred to as tar sands and bitumen, are found in Canada and Venezuela, and the rest is found in various other countries including the United States. In Canada, the development of tar sands and heavy oil began in the 1960s. Notable is the Cold Lake project based on cyclic steam injection and stimulation in the early 1970s. Shallow deposits of oil sands are mined for extraction. In 1978, the Syncrude mine was started, which is known as the largest mine of oil sands in the world. Another mine in the Athabasca oil sands is the Albian sands. In the heavy oil belt, CHOPS technology was used to produce oil in the late 1980s and 1990s.

Since the refineries are only designed to process a conventional grade of petroleum, oil sands (bitumen) that are produced from these reservoirs require upgrading before processing in the refineries. The objectives of upgrading include the reduction in oil viscosity and sulfur content, and maximization of distillable content leading to synthetic crude oil. Hence, a regional upgrader was built in the Lloydminister. Daily production from the upgrader facility is in the range of 130,000 barrels of synthetic crude from the feed that averages about 15°API.

Technologies that have led to the commercial production of oil sands in Canada are as follows:

- Horizontal well drilling for shallow deposits of oil sands below 3000 ft., where both nonthermal and thermal methods are utilized.
- The SAGD process, based on two horizontal wells. One well is used to impart thermal energy to heavy oil to reduce its viscosity and the less viscous oil is then produced through the other.
- CHOPS technology, where the sand from the unconsolidated formation is allowed to be coproduced with viscous oil.

Since the 1990s, SAGD and CHOPS technology added hundreds of thousands of barrels of oil on a daily basis to production in Alberta and Saskatchewan.

Besides the technologies mentioned above, pressure pulse technology is also used, which involves pulses of pressure applied during well workovers to augment production of heavy oil.

The development of Canadian as well as Venezuelan heavy oil deposits has accelerated significantly in the late 1990s. The relatively high price of oil supported the requirement of huge capital investments in the development of oil sands and heavy oil reservoirs. It is also recognized that the world has a limited supply of conventional oil having light to intermediate gravity, and "peak oil" may be reached within a matter of a few decades.

With the development of new *in situ* production techniques such as SAGD, with oil prices increasing in recent times, there were several dozen companies planning nearly 100 oil sands projects in Canada. Capital investment is estimated to exceed $100 billion. The current cost of production of oil sands is estimated between $65 and $70 a barrel.

Case Study: Extra Heavy Oil Production in Venezuela

Venezuela is reported to have oil sands deposits similar in size to those of Canada, and is comparable to the world's reserves of conventional oil. Venezuelan oil sands are frequently referred to as extra heavy oil. The distinction between Venezuela's extra heavy oil and Canada's oil sands is made in terms of the level of degradation by bacterial action and weathering of the originally formed crude oil [12]. Venezuela's extra heavy oil deposits are relatively less degraded and may require fewer efforts in production and upgrading. Moreover, permeability of Venezuelan extra heavy oil reservoirs is better, ranging between 2 D and 15 D, compared to 0.5 D and 5 D in Canadian oil sands deposits.

Huge deposits of oil sands or tar sands are found in Venezuela's Orinoco oil belt. According to a study by the United States Geological Survey made public in 2009, recoverable reserves in the Orinoco belt were over half a trillion barrels, while the total proved and unproved reserves are estimated to be 0.9–1.4 trillion barrels. Based on the estimates, Venezuelan reserves exceed those of Saudi Arabia. However, the technology needed to produce the oil would be more complex than that used to produce conventional oil from giant Middle Eastern fields having favorable fluid and rock characteristics. Venezuela's oil sands production is reported to have increased several times since 2001.

The major developments in extra heavy oil technology are centered on the following Orinoco projects [13]:

- Cerro Negro Project
- Ameriven Project (Hamaca)
- Petrozuata Project
- Sincor Project

A number of multinational oil companies in cooperation with the Venezuelan State Oil Company are involved in the development. The above projects are based on long horizontal wells with multilaterals placed in the optimum zones for heavy and extra heavy oil recovery.

Case Study: Evaluation of Recovery Methods for a Heavy Oil Reservoir in Russia

This study highlights a methodical approach for efficient development of a heavy oil reservoir in Russia [14]. A large number of thermal and nonthermal methods are evaluated in the study. As a whole, the study demonstrates the thinking process for developing a new reservoir with unique attributes based on established as well as novel methods. Finally, the ideas are put to the test by reservoir simulation and field tests.

The Russkoe field, one of the largest in Russia, is located in the Arctic region. The reservoir differs from unconventional Canadian heavy oil reservoirs where the development methods are well established in certain important aspects. The differences are as follows:

- Oil is relatively less viscous (210 cp) than what is typically encountered in Canada
- Oil is richer in volatile components

- High clay content of reservoir rock with potential damage in permeability due to clay swelling
- Reservoir depth is twice as much, about 950 m, requiring greater thermal energy for effective steam chamber
- Presence of gas cap and bottom water above and below the pay interval
- Potential effects of viscous fingering during steam or hot water injection

The study explored the potential for primary and enhanced recovery methods as practiced for heavy oil reservoirs and combinations thereof, followed by the ranking of their suitability in the particular case. The methods are summarized in the following:

- SAGD: The method, requiring a pair of horizontal wells in close proximity to each other, grew in popularity for extracting unconventional ultraheavy bitumen deposits with the advent of horizontal technology. As noted earlier in the chapter, SAGD works by injecting steam in the upper horizontal wells while collecting and producing oil having reduced viscosity through the other well placed lower in the formation. The forces involved in the recovery process include gravity and capillarity. Oil recovery efficiency is reported to be high, between 50% and 70%, due to the absence of any viscous fingering effects. Hence, the SAGD is rated to be the most successful and efficient thermal recovery process for oil sands. However, the important difference with the field in the study is the depth. While bitumen deposits are typically located at about a depth of 400 m, the depth of the field in Russia is much greater. Consequently, reservoir pressure is higher, about 8.5 MPa. At the elevated pressure, steam quality is low, and it will tend to condensate near the wellbore hindering the propagation of a steam chamber. In order to enhance steam quality, higher pressure and temperature are required, which in turn would require higher flow rate of steam resulting in higher water−oil ratio. Further adverse effects may also be encountered due to higher flow rates, including pronounced viscous fingering, and the production of gas and water from a gas cap and a bottom water zone, respectively. Steam condensate has relatively low salinity, which may lead to significant clay swelling and reduction in formation permeability.
- Solvent enhanced SAGD: In this method, incondensable gas is injected along with steam in order to reduce the steam flow rate for ultraheavy oil recovery. Due to the high gas−oil ratio of the field in question as compared to the bitumen deposits of Canada, the process may not be as effective as in traditional cases. At higher temperatures in the field under study, the solubility of gas would decrease and dissolved methane is likely to come out of the solution. In conclusion, further studies were deemed necessary in the absence of detailed data.
- CSS: This method is widely practiced to recover heavy oil in a highly efficient manner in various parts of the world. Since a single well is utilized both for injection and production, the adverse effects of viscous fingering are nonexistent. On the contrary, the phenomenon has positive effects in increasing the contact area between steam and oil, and the rate of heat transfer. The presence of rock heterogeneities is not considered to be extremely adverse during recovery by the CSS process. Clay swelling, detrimental in other thermal processes, may offer some advantages during recovery. The phenomenon occurs at relatively high temperatures, which are attained only after multiple

cycles of steam injection resulting in sizeable recovery. Permeability of the formation is reduced due to the eventual swelling of clay, and steam may propagate to new areas in the oil zone, thus enhancing the ultimate recovery. Since the 1990s, multiple horizontal wells are utilized in a field in a synchronized manner to recover heavy oil by employing the CSS process, where recovery is reported to be much higher than the older generation vertical wells working singly. For the field under evaluation, a similar CSS operation was deemed to be a relatively safe and reliable method for recovering oil. However, single well implementation of the method in an unsynchronized manner was considered to create viscous fingering lowering overall recovery.

- Cyclic hot water stimulation (CHWS): The method is similar to CSS and thought to have similar advantages. However, a bottom-hole pump would likely be required to lift oil to the surface.
- Cyclic hot water flood (CHWF): The scheme involves two horizontal wells, one injector and one producer. The role of the two wells can be reversed after a certain period, thus converting the injector to producer and vice versa. Due to injection pressure, any viscous fingers formed will propagate horizontally rather than in an upward or downward direction. Adverse effects of clay swelling can be controlled, as the damage around a producer will be mitigated, as clay would migrate away from the well during the injection cycle. Hot water may be forced to take new pathways due to the migrated clays, thus improving areal coverage.
- Steam flood: This method has not been proved to be inefficient due to the adverse effects of viscous fingering leading to poor recovery. A significant quantity of thermal energy is lost in the vicinity of the steam injection well, while oil near the producer is not stimulated and recovered.
- Foamy oil recovery: This is a nonthermal method based on the dissolution of gas bubbles from oil that coalesce to form foam like fluid as a result of a downhole pump action. As the pressure is lowered, foamy oil is produced to the surface. The method can be suitable for the reservoir under evaluation; however, recovery could be limited by high initial water saturation and clay swelling.
- CHOPS: As mentioned earlier, the method works by producing unconsolidated rock along with heavy oil, which creates a high permeability channel or wormhole for the transport of liquid and solids to the well. The CHOPS process can be tried in the Russkoe field in portions of the reservoir where the process is capable of generating and maintaining wormholes. One significant advantage is that the CHOPS is a low cost operation along with high recovery, which makes it economically feasible in a large number of heavy oil fields in Canada and elsewhere.
- Capillary imbibition method: As a follow-up of recovery by the CHOPS process, the capillary imbibition method can be applied to recover additional oil. The wormhole created during the process can be injected with hot water, which will imbibe into the fractured rock by capillarity, displacing the *in situ* oil from the rock pores into the wormhole. Oil can eventually be produced from the wormhole by pumping. Oil recovery by capillary imbibition has been implemented in certain Norwegian fields.
- Carbon dioxide flood: Carbon dioxide, when miscible with *in situ* oil, provides a highly efficient recovery mechanism in many reservoirs. However, under the operating pressure and temperature of the Russkoe field, carbon dioxide can only remain in a liquid state, leading to limited miscibility and lower recovery efficiency.

Tight oil

Tight oil refers to oil trapped in reservoirs where rock permeability is low or ultralow leading to unfavorable reservoir quality. Permeability of rock matrix is often in the order of 10^{-1} mD or less. Some sandstone and carbonate reservoirs can have very low permeability for conventional development [15]. The industry approach to produce the unconventional oil includes horizontal drilling multistage fracturing leading to the stimulation of formation. The amazing rise in horizontal well drilling and multistage hydraulic fracturing to stimulate the ultratight formation resulting in a 700% rise in oil production in Bakken shale within a short period of time is presented in Chapter 18. The presence of natural fractures aids significantly in producing oil and gas from tight formations. The lithology of tight oil reservoirs includes shaley sandstone, siltstone, carbonates, and dolomites. Prominent tight oil reservoirs include Middle Bakken (carbonate), Three Forks (dolomite), Austin Chalk, Eagle Ford, and Niobrara.

Shale oil

As stated earlier, tight oil is difficult to produce on a commercial scale due to the very low transmissibility of reservoir rock. Similarly, shale oil refers to oil trapped in semi-pervious shale where permeability is in microdarcies or even less. Shale is the source rock where oil was generated in geologic ages and trapped with little or no migration to the reservoir rock. The presence of natural fractures in rock also aids in the recovery of shale oil.

Summing up

Unconventional oil reservoirs are developed and produced by utilizing nontraditional and innovative methodology. The underlying reason is that unconventional oil is not mobile under the circumstances encountered due to fluid or rock characteristics. Unconventional oil can be viewed in the following categories:

- Heavy and extra heavy oil and oil sands that cannot be produced easily due to extremely high viscosity
- Tight oil and shale oil that cannot be recovered by conventional methods due to low or ultralow permeability of the reservoirs
- Oil shale, which refers to kerogen-rich rock. Various types of fuel are extracted by retorting and distillation processes.

As the conventional oil reserves dwindle due to ever increasing demands for petroleum, unconventional oil deposits exceed that of the former. However, unconventional oil production is based on evolving technologies and has higher production cost per barrel. The production process is also associated with potential environmental issues.

Oil shale refers to kerogen-rich shale, which must be heated to extract the hydrocarbon-rich material. Kerogen is ultimately converted to various types of fuel.

Table 21.2 summarizes the unconventional heavy oil extraction processes that are currently being used economically.

Table 21.2 **Unconventional heavy oil extraction processes**

Thermal/nonthermal process	How it works	Notes
Steam assisted gravity drive (SAGD)	Two horizontal wells are drilled with a vertical separation of about 13–20 ft. Steam is injected through the upper horizontal well, which reduces the viscosity of heavy oil and bitumen. Less viscous oil is produced through the lower horizontal well.	Oil recovery can be as high as 70%
Cyclic steam stimulation (CSS)	Single horizontal or vertical well is used for both injection and production in sequence. Steam is injected initially into the formation to reduce the viscosity of heavy oil or bitumen. In the next stage, relatively less viscous oil is produced by the same well.	The process is also known as the huff-and-puff method. There are several variations of the process.
High pressure cyclic steam stimulation (HPCSS)	Steam is pumped into the formation to reduce the viscosity of bitumen, then a mixture of bitumen and steam, referred to as bitumen emulsion, is pumped to the surface. High pressure steam also creates cracks in the formation, which facilitates bitumen production.	Vertical well spacings range between 2 acres and 8 acres. Horizontal wells are placed 60–80 m apart. About one-third of oil sands in Alberta is produced by the HPCSS process.
Vapor extraction process (VAPEX)	Steam injection is replaced by vaporized solvents such as propane mixed with noncondensable gas. As in SAGD, two horizontal wells, injector and producer, are utilized.	The process is suitable where injection of steam may not be very efficient, such as thin and low permeability formation.

Questions and assignments

1. Define unconventional oil. Distinguish it from conventional oil. Explain the role played by reservoir characteristics and oil properties in defining unconventional oil resources.
2. Describe the major technologies in producing unconventional oil. Discuss their advantages and disadvantages.
3. Distinguish between shale oil and oil shale. Which of the two is thermally mature?

4. How does multistage fracturing facilitate tight oil production? Based on a literature review, describe with examples the typical length of horizontal wells and number of fracturing stages required in producing oil from tight reservoirs where permeability is in the order of 10^{-3}–10^{-1} mD.

5. How does SAGD work? What type of reservoirs are best for implementing SAGD? What might happen when the reservoir has a strong aquifer influence?

6. Which was the first commercial SAGD project? What is its current production level? What is upgrading? Why is upgrading of oil sands needed?

7. Describe the CSS and HPCSS processes. What additional benefits can be derived from the latter process?

8. Describe the VAPEX process with a field example.

9. Describe the principal design considerations in developing extra heavy oil reservoirs in Canada and Venezuela.

10. What is CHOPS? Explain the production characteristics of both heavy oil and sand during the process? Under what conditions could the production of sand be excessive?

11. Based on a literature review, describe the upcoming technologies and research projects in economically extracting unconventional oil worldwide.

References

[1] What are oil sands? Canadian Association of Petroleum Producers. Available from: http://www.capp.ca/canadian-oil-and-natural-gas/oil-sands/what-are-oil-sands [accessed 10.10.13].

[2] Jiang Q, Thornton B, Houston JR, Spence S. Review of thermal recovery technologies for the ClearWater and Lower Grand Rapids formations in the Cold Lake area. In: Alberta Canadian International Petroleum Conference. Osum Oil Sands Corp; 2009.

[3] CHOPS – cold heavy oil production with sand in Canadian heavy oil industry. Available from: http://www.energy.alberta.ca/OilSands/pdfs/RPT_Chops_chptr3.pdf [accessed 30.09.13].

[4] Cyclic steam stimulation. Thermal in situ oil sands. CNRL. Roger Butler, unlocking the oil sands. Schulich School of Engineering. University of Calgary: Calgary, Alberta; 2013.

[5] Drilling in oil sands. Cenovus Energy. Available from: http://www.cenovus.com/news/drilling-in-the-oil-sands.html [accessed 17.11.13].

[6] Open pit mining. Oil Sands Today. Available from: http://www.oilsandstoday.ca/whatareoilsands/Pages/RecoveringtheOil.aspx [accessed 20.08.13].

[7] In situ methods used in the oil sands. Regional Aquatics Monitoring Program. http://www.ramp-alberta.org/resources/development/history/insitu.aspx. Accessed 9/9/2013

[8] Butler RM, Mokrys IJ. A new process (VAPEX) for recovering heavy oils using hot water and hydrocarbon vapour. J Can Petrol Technol 1991;30(1).

[9] Oil sands. Alberta Energy. Available from: http://www.energy.alberta.ca/OurBusiness/oilsands.asp [accessed 06.02.14].

[10] Wiggins EJ. Alberta Oil Sands Technology and Research Authority. The Canadian Encyclopedia. Historical Foundation of Canada 2014.

[11] Facts about Alberta's oil sands and discovery. Oil Sands Discovery Center. Available from http://history.alberta.ca/oilsands/resources/docs/facts_sheets09.pdf [accessed 03.12.14].

[12] Dusseault MB. Comparison of Venezuelan and Canadian oil and tar sands. Canadian International Petroleum Conference 2001; June 12–14, 2001. Calgary, Alberta.

[13] Talwani M. The Orinoco Heavy Oil Belt in Venezuela; 2002. Available from: http://bakerinstitute.org/media/files/Research/8bb18b4e/the-orinoco-heavy-oil-belt-in-venezuela-or-heavy-oil-to-the-rescue.pdf.
[14] Babchin A 2015. Heavy oil recovery methods in application to Russkoe oil field. Available from: www.researchgate.net.
[15] Understanding tight oil. Canadian Society Unconventional Resources. Available from: http://www.csur.com/sites/default/files/Understanding_TightOil_FINAL.pdf [accessed 29.08.14].

Further reading

Deutsch CV, McLennan JA. Guide to SAGD (steam assisted gravity drainage) reservoir characterization using geostatistics. Centre for Computational Geostatistics; 2005.

Jiang Q, Thornton B, Russel-Houston J, Spence S. Review of thermal recovery technologies for the ClearWater and Lower Grand Rapids formations in the Cold Lake Area. In: Alberta Canadian International Petroleum Conference. Osum Oil Sands Corp.

Glassman D, Wucker M, Isaacman T, Champilou C, Zhou A. Adding water to the energy agenda (Report). A World Policy Paper; March 2011.

Speight JG. The chemistry and technology of petroleum. Boca Raton, Florida: CRC Press; 2007. 165–167.

Wiggins E.J. Alberta Oil Sands Technology and Research Authority. The Canadian Encyclopedia. Historical Foundation of Canada. Retrieved December 27, 2008.

Czarnecka M. Habir Chhina keeps Cenovus Energy Inc. running smoothly. Alberta Oil; 2013.

Yedlin D. Yedlin: showing cynics how oil business is running smoothly. Calgary Herald. Retrieved June 19, 2013.

Hall RM. Statement to the Committee on Science and Technology for the Produced Water Utilization Act of 2008. 110th Congress 2nd Session, Report 110–801.

Water use breakdown in Alberta 2005. Government of Alberta. Retrieved 2005.

Volume and quality of water used in oil and gas 1976–2010. Government of Alberta. Retrieved October 4, 2011.

Cyclic steam stimulation. Thermal *in situ* oil sands. CNRL. 2013.

Severson-Baker C. Cold Lake bitumen blowout first test for new energy regulator; 2013.

Alberta Energy Regulator orders enhanced monitoring and further steaming restrictions at Primrose and Wolf Lake projects due to bitumen emulsion releases. AER; July 18, 2013.

Unconventional gas reservoirs

22

Introduction

A new era has emerged in the petroleum industry in which substantial recovery of natural gas is expected from unconventional reservoirs based on innovative technology. In the broadest sense, unconventional gas is the natural gas that is more difficult to recover because the technology has not been fully developed, or is not economically feasible. A major category of unconventional resources stems from extremely tight reservoirs having permeability in microdarcies and nanodarcies. Typical recovery from unconventional resources of gas is quite low compared to that of conventional reservoirs. However, unconventional deposits of shale gas, a major source of unconventional gas, are continuous in nature; the locations as well as the extent of the shale formations are known.

Historically, conventional natural gas reservoirs with favorable porosity and permeability have been the most practical and easiest deposits to produce. However, with technological advancement and increasing market demand, unconventional natural gas constitutes an increasing larger percentage of the supply picture. As National Energy Technology Laboratory (NETL) observes [1]:

> *Actually, the term 'unconventional' has lost its original meaning. Currently (as of 2013), gas from shales, tight sands, and coalbeds accounts for 65% of U.S. natural gas production. By 2040 that share is expected to rise to 79%. The unconventional has become the conventional.*

This chapter describes the reservoir engineering and related aspects of major resources of unconventional gas and answers the following questions:

- What are the types of unconventional gas?
- What are their estimated reserves?
- What role does unconventional gas play in production and consumption?
- What are the mechanisms of storage and flow of gas in porous media?
- How are the unconventional gas reservoirs developed and produced?
- How much recovery is expected from unconventional gas reservoirs?
- How are horizontal wells modeled to optimize design of length and number of hydraulic fracturing stages?

Types and estimated resources of unconventional gas

Huge quantities of unconventional gas resources are known to exist across many regions of the world. The untapped resources can be exploited in the near future by innovative methods and at moderately higher cost. Currently, types of

Reservoir Engineering. http://dx.doi.org/10.1016/B978-0-12-800219-3.00022-X

Table 22.1 **Types of unconventional gas resources**

Unconventional gas resource	Description	Reservoir development methods
Shale gas	Dry and wet natural gas trapped in ultratight shale reservoir. Typical shale permeability is in nanodarcies (10^{-9} D). Lithology is predominantly organic-rich shale low in clay content.	Shale gas has been developed in significant quantities based on advances in horizontal drilling and multistage fracturing.
CBM	Gas deposits in micropores and seams of coalbed. Formation permeability is usually low, between 1 mD and 25 mD. Production mainly occurs through coal cleats or seams.	Reservoir development is based on vertical and horizontal well drilling combined with hydraulic fracturing.
Tight gas	Gas trapped in very low permeability formations, chiefly sandstones and some carbonates. Reservoir permeability typically ranges in microdarcies (10^{-6} D). Lithology is predominantly marine shale rich in clay content.	Reservoir development methods are similar to those of shale gas.
Arctic and subsea hydrates	Molecules of methane trapped in the ice lattices. Abundant in arctic and subsea environments.	Methane is released from ice under heat or reduced pressure. The resource is not yet commercially developed.
Deep reservoir gas/ geopressured gas	Gas accumulations in various basins at significant depths, below 15,000 ft.	Poses technological and economic challenges to drill and produce.

unconventional gas resources identified in various regions of the world are shown in Table 22.1.

The estimated major unconventional resources in various regions of the world are shown in Table 22.2.

In the following section, the two major unconventional resources, shale gas and coalbed methane, are discussed in detail.

Shale gas reservoirs

Shale gas refers to the vast unconventional resources of natural gas trapped in extremely tight shale formations where the matrix permeability ranges in hundreds of nanodarcies (10^{-7}–10^{-9} D). Even in the recent past, these formations were considered to be source rock of petroleum from which oil or gas cannot be produced economically due to ultralow permeability. Development of shale gas reservoirs has been spectacular since the dawn of the twenty-first century. Shale gas production has increased by 1400% within a short span of time. Studies indicate that only a small fraction of hydrocarbons (10–20%) from source rock migrate to, and accumulate in, conventional

Table 22.2 Unconventional gas resources in trillion cubic feet (TCF)

Region	Coalbed methane	Shale gas	Tight sand	Total resources
North America	3,017	3,840	1,371	8,228
Former Soviet Union	3,957	627	901	5,485
China and Central Asia	1,215	3,526	353	5,094
Pacific (OECD)	470	2,625	1254	4,349
Latin America	39	2,116	1,293	3,448
Middle East and North Africa	0	2,547	823	3,370
Sub-Saharan Africa	39	274	784	1,097
Western Europe	275	548	431	1,254
World	9,012	16,103	7,210	32,325

Note: All numbers are estimates and provided as a guide only.
Source: Energy Information Administration, 2009.

reservoirs. However, the major portion of oil and gas generated in the process is trapped in source rocks, chiefly shale, where traditional means of production are not adequate. Hence, shale gas is referred to as an unconventional resource. The success of shale gas production is based on the ever-expanding knowledge of shale as reservoir rock rather than source rock, innovations in horizontal drilling, multistage hydraulic fracturing, and microseismic monitoring of fracture network. Shale gas production has leaped several fold in just over a decade, accounting for a significant portion of total natural gas production in the United States. Significant increases in unconventional gas development are expected in the near future (Figure 22.1).

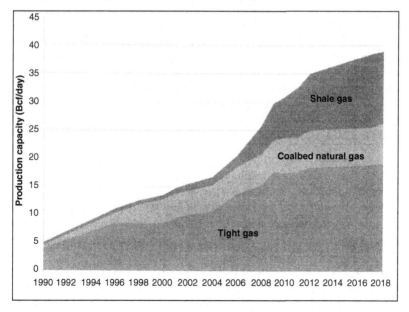

Figure 22.1 Shale gas, CBM, and tight gas production forecast in the United States.
Source: Energy Information Administration.

Table 22.3 **Global estimates of TRR of shale gas**

Country	TRR (TCF)	Percentage
China	1115	19.3
Argentina	802	13.9
Algeria	707	12.3
United States	665	11.5
Canada	573	9.9
Mexico	545	9.5
Australia	437	7.6
South Africa	390	6.8
Russia	285	4.9
Brazil	245	4.3
Total	5764	100

According to various estimates, technically recoverable resources (TRR) of shale gas are 665–750 TCF in the United States alone. The global estimates of TRR based on shale gas deposits are 7299 TCF. Countries with the top 10 resources are shown in Table 22.3 [2].

History of shale gas production

The first shale gas well was drilled in 1821 in the Appalachian Basin at a very shallow depth, about 70 ft., which provided lighting and other requirements of energy in Fredonia, New York, for several decades. Interest in shale gas in the United States continued to grow at a rapid pace with large numbers of wells drilled in a number of petroleum basins in the states of New York, Pennsylvania, Ohio, Kentucky, and Virginia. Older wells that produce from shale were vertical and subjected to traditional fracturing. In recent decades, shale gas development activities gained significant momentum in a relatively short period of time with the horizontal wells with well lateral section contacting thousands of feet of organic-rich shale formation, multistage fracturing at various points in the horizontal section of wells, and application of geophysical and other tools to monitor the effectiveness of the fractures. As shale gas development technology matures, the length of horizontal wells and the number of stages in fracturing and EUR increase (Figure 22.2).

Shale gas reservoirs and characteristics

Major shale gas reservoirs in the United States include, but are not limited to, Marcellus shale in the Appalachian Basin, Barnett shale in the Fort Worth Basin, Fayetteville shale in Arkansas, Woodford shale in Oklahoma, Eagle Ford in Texas, and Haynesville-Bossier shale in Gulf Coast, Texas and Louisiana (Figure 22.3).

One of the most important characteristics of shale is its ultralow permeability, often measured in nanodarcies (10^{-9} D). Hence, shale formations act as source, seal, and reservoir rock, where natural gas is generated and contained without any obvious trapping mechanism as in conventional reservoirs. A gas–water contact is also not

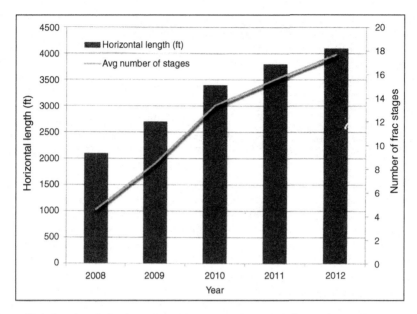

Figure 22.2 Increase in horizontal well length and hydraulic fracturing stages in various areas of Marcellus shale between 2008 and 2012.
Source: Energy Information Administration.

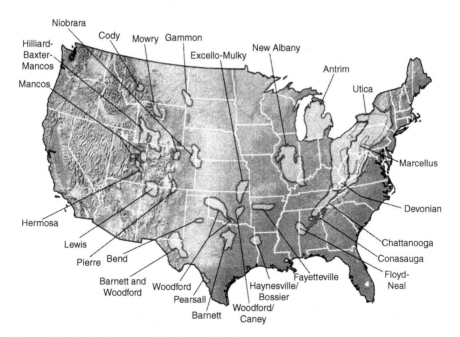

Figure 22.3 Shale gas map of the United States.
Source: Energy Information Administration.

encountered. Certain shale reservoirs are overpressured. Typical shale gas reservoirs extend over a very large area compared to the conventional reservoirs and are tens to hundreds of feet thick. For example, Marcellus shale covers an area of more than 100,000 square miles spanning several states at an average thickness of 50–200 ft. The important differences between unconventional shale reservoirs and conventional petroleum reservoirs are described in Chapter 2.

It is quite apparent that shale gas reservoirs are of high potential resources. The production cycle of the individual wells is short compared to that of wells in conventional reservoirs; however, the productive life of the entire field may continue for many decades as new wells are drilled. Shale gas reservoirs have relatively low recovery, often ranging from less than 10% to about 20%, as compared to conventional gas reservoirs where recovery may exceed 80%. However, shale gas recovery is expected to increase substantially in the foreseeable future as the technology matures.

Important characteristics of the unconventional shale gas reservoirs in the United States as well as economic data are presented in Table 22.4. It must be borne in mind that over 90% of the areas in each shale play remain unexplored and untested.

Geology of shale gas reservoirs

As noted in Chapter 2, shale is a sedimentary rock composed of finely grained silt and clay; the clay particles are compacted and hardened under pressure with lamination or interlayering. The fine sediments are deposited in relatively calm environments, including deep marine. Deposition of plant- and animal-based organic matter occurred simultaneously leading to the formation of dark organic-rich shale. It is relatively low in clay content and found to be a good producer of natural gas.

Furthermore, in sufficiently brittle shale formation, microfractures are created by the regional stresses. These natural microfractures can be clustered in certain areas in the geologic formation providing storage for natural gas; the fractures act as highly conductive channels for gas in combination with hydraulically created fractures, which facilitates the production of shale gas in commercial quantities.

The geologic age of shale gas formations is found to vary notably. For example, while the Devonian age shale rocks in the Appalachian Basin were deposited about 385 million years ago, the Haynesville-Bossier shale in the US Gulf Coast is much younger, dating back to the Jurassic period around 185 million years ago.

Geochemical properties

Total organic content (TOC), thermal maturity (R_o), and the type of kerogen are important properties of shale for the economic development of the shale gas reservoirs. Shale rocks having TOC greater than 2% (usually 4–10%) and thermal maturity over 1.1% are good candidates for reservoir development.

Petrophysical properties

In shale, pore sizes are in micrometers. Porosity is rather low, 2–10%, and matrix permeability is typically in hundreds of nanodarcies (10^{-9} D) or even lesser.

Table 22.4 Major shale gas reservoirs in the United States [3–5]

Basin	Barnett	Fayetteville	Haynesville-Bossier	Marcellus	Woodford	Eagle Ford
Location	Fort worth (TX)	Arkoma (AR)	Gulf Coast (TX, LA)	Appalachian (PA, WV, OH, NY, MD)	Anadarko-Arkoma (OK)	Western Gulf, Maverick (TX)
Est. area (square miles)	5,000	5,900	9,300	104,100	6,350	7,600
Geologic age	Mississippian	Mississippian	Upper Jurassic	Middle Devonian	Late Devonian	Upper Cretaceous
Depth (ft.)	6,500–8,500	1,000–7,000	10,000–13,500	4,000–8,500	6,000–11,000	4,000–12,000
Net thickness (ft.)	50–100	20–200	200–300	50–200	120–220	250
TOC (%)	4.5	4.0–9.8	0.5–4.0	3–12	1–14	4.5
%R_o	1.0–1.3		2.2	1.3	1.1–3	1.5
Porosity (%)	4–5	2–8	8–9	8–10	3–9	11
Matrix permeability (nD)	250	n/a	658	100–450	145–200	1,100
Pressure (psi)	4,000		8,500	4,000	3,000–5,000	5,200
Gas content (scf/ton)	300–350	60–220	100–330	60–100	200–300	n/a
Adsorbed gas (%)	20	n/a	n/a	n/a	n/a	n/a
Well spacing, acres	80–160	80–160	40–560	40–160	640	65–120
Number of wells	16,743	4,678	3,300	8,982	2,890	10,020
GIIP (TCF)	327	52	717	1,500	23	270
TRR (TCF)	44	13.2	251	356	21.7	50.2
Avg EUR (Bcf/well)	2.0	1.3	2.67	1.56	1.97–2.89	2.36
R.F. (estimated, %)	8–15	<20	<20	<20	<20	<20
Well cost (million $)	3.0	2.8	8.0	4–7	6.7	
Breakeven price ($/MCF)	3.74	3.65	6.1–6.95	4.02	5.5	6.24

Note: All figures are approximate and provided as a guide only.

Geomechanical properties

The presence of natural fractures plays a significant role in effectively producing shale gas reservoirs. Young's modulus, Poisson's ratio, and fracture stress are some of the geomechanical properties determined to evaluate the fractures in shale. Furthermore, data related to the length, height, orientation, and conductivity of hydraulically created fractures are needed.

Sweet spots

Conventional petroleum reservoirs have distinct geologic or hydrodynamic boundaries and are limited in extent. In contrast, shale formations with unconventional gas deposits are continuous in nature and may extend over a vast area. As noted earlier, Marcellus shale formation extends over hundreds of miles in several states. However, not all the areas of gas accumulation are capable of sustainable production. The economic success of shale gas production depends on the identification of "sweet spots," where wells are drilled based on state-of-the-art technology. Sweet spots are of keen interest to the industry in the exploration and production of shale gas. Sweet spots are expected to have favorable geological and geochemical characteristics as in the following:

- Relatively high TOC to yield significant quantities of gas, referred to as "black shale"
- Desirable thermal maturity for rock to produce gas as indicated by vitrinite reflectance (gas window)
- Better porosity and permeability for storage and mobility
- Presence of natural fractures in the formation
- Good hydraulic fracturing characteristics of rock as indicated by Young's modulus, Poisson's ratio, and fracture stress

Navarette [6] proposes the following guidelines to identify sweet spots:

- Reservoir thickness >200 ft.
- TOC >1.0%
- Porosity for storage of free gas >4%
- Permeability of shale >100 nD
- Brittleness index >25%

In addition, sweet spots must have sufficient reservoir pressure and appropriate thermal maturity of rock.

Gas accumulation and sorption

In shale, accumulation of gas under high reservoir pressure occurs as follows:

- Miniscule pores and microfractures in rock contain free gas
- Gas is stored in an adsorbed state onto the solid organic matter contained in rock
- Under high pressure, gas molecules are densely packed and exist in liquid-like films. A certain amount of gas is also absorbed in solution

The amount of adsorbed gas, chiefly consisting of methane, depends on a variety of factors, including pore size, type of organic material, mineral composition, and thermal maturity of the rock. Studies have indicated that about 15–80% of total gas

Table 22.5 **Storage and transport of gas in unconventional shale reservoirs**

State of gas	Location	Flow characteristic	Notes
Free	Porous network	Darcy flow	
Free	Hydraulically created fractures	Non-Darcy flow	Non-Darcy flow encountered in high velocities
Adsorbed	Surface of solid organic matter	Diffusion. The process is described by Fick's law	Desorbed gas eventually flows to the wellbore where Darcy or non-Darcy flow may occur

can remain in an adsorbed state. At high reservoir pressure, adsorbed gas may undergo condensation.

Mechanism of gas transport

As indicated earlier, natural gas is stored in a free state in pores and fractures of shale. However, certain portions of gas are trapped in an adsorbed state onto the solid organic matter contained in shale. As a well is drilled and fractured hydraulically, free gas stored in fractures flows toward the well due to the pressure differential that is created between the shale formations and wellbore. Once the reservoir pressure decreases sufficiently, desorption of methane from the surface of solid organic matter would occur resulting in further flow of gas. The relationship between reservoir pressure and the volume of adsorbed gas is characterized by the Langmuir isotherm, as described in Chapter 12.

In pores and fractures, transport of free gas is characterized by both Darcy flow and non-Darcy flow. Non-Darcy flow can be observed in hydraulically induced fractures due to the high velocity of gas. The transport of desorbed gas from the solid organic matter involves diffusion, which can be represented by Fick's law. Flow of desorbed gas can later be governed by Darcy's law. Once in highly conductive fractures, non-Darcy flow can occur due to high velocity. Table 22.5 summarizes the storage and transport of gas in shale.

Expected ultimate recovery and stimulated reservoir volume (SRV)

The methods of EUR from shale gas reservoirs include decline curve analysis, rate-transient analysis, reservoir simulation, and analogy. The decline in production may show two distinct trends. Production rate may decline rapidly in the initial phase as the formation is ultratight. In a few to several months, the production rate may fall significantly but the rate of decline can become rather small thereafter. Decline curve analysis and reservoir simulation methods are described in Chapters 13 and 15, respectively.

Furthermore, the estimation of gas initially in place (GIIP) requires consideration of the following:

- Free gas stored in micropores and fractures
- Gas stored in an adsorbed state onto the solid organic matter in shale

Recovery from shale gas reservoirs is primarily dependent on SRV, i.e., the portion of the reservoir that is stimulated by multistage fracturing, and connected to the naturally occurring fractures. The rest of the reservoir volume cannot be produced in the absence of conductive microchannels that the fractures create.

Shale gas reservoir development and management

The objectives of unconventional shale gas reservoir development include proper design and placement of horizontal wells followed by multistage hydraulic fracturing. The following questions must be answered in developing shale gas reservoirs:

- Where are the "sweet spots" that must be targeted for sustainable production?
- Is the formation naturally fractured?
- Do geotechnical properties of rock support effective hydraulic fracturing?
- What would be the length, orientation, and spacing of horizontal wells?
- What would be the SRV?
- What would be the design parameters for multistage fracturing in terms of liquid and proppant requirements?
- What is the EUR and well or reservoir life?
- What would be the cost of drilling horizontal wells and the operating costs?
- What are the potential environmental issues?
- What would be the payout period and discounted rate of return?

The process of unconventional reservoir development begins with collection of all available data related to the reservoir, including seismic and microseismic data (Figure 22.4). Some data are not available where assumptions are made based on experience and analogy. For example, performance of nearby wells and hydraulic fracturing data can be incorporated in the analysis. Next, horizontal well design, multistage fracturing design and reservoir simulation are performed. The process is completed by performing economic analysis, accounting for environmental aspects and determining the optimum approach.

Assessment of shale plays

The viability of shale gas reservoirs in a specific region is determined by a large number of factors, including the identification of sweet spots in otherwise noncommercial areas and implementation of various techniques, including decline curve analysis, to estimate the range of EUR that can be expected from wells drilled in the sweet spots. A methodology adopted by a major oil company that rapidly identifies sweet spots and economic potential of future wells is outlined in the following [7].

Identification of principal factors in shale gas field exploration and development as follows:

- Thermal maturity and TOC of source rock
- Net thickness of shale gas zone; large thicknesses point to large gas in place
- Lateral distribution of rock heterogeneity, including permeability; heterogeneity is tied to engineering decisions and risk factor
- Favorable fracture characteristics of rock; shale reservoirs require artificial stimulation to be productive

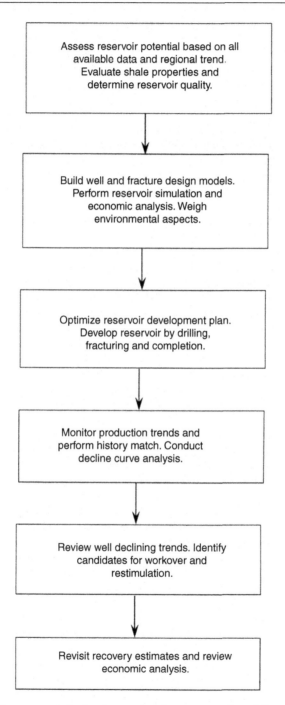

Figure 22.4 Shale gas reservoir development and management workflow.

Identification of reservoir property modifiers, which include:

- Existence of natural fractures and the characteristics of fractures
- Quantification of gas in free and adsorbed states
- Characterization of rock matrix porosity
- Reservoir depth
- Lithology of shale in terms of clay, silica, and carbonate contents
- Distribution of nonhydrocarbon gases, such as carbon dioxide

Identification of promising plays based on the following parameters that meet favorable criteria:

- Richness of organic matter in shale
- Net thickness of gas zone
- Thermal maturity of rock
- Depth
- Basin history
- Reservoir pressure
- Mineralogy

At the initial stage when a few wells are drilled, the following maps are generated:

- Measured depth in feet
- Net isopach in feet
- Thermal maturity of rock
- Cost of individual wells
- Land use, city versus rural

The following data, among others, may have limited availability and introduce a certain degree of uncertainty during the generation of maps:

- TOC
- Abnormal pressure gradients
- High gas saturation and extent of carbon dioxide present
- Lithological description, for example, presence of siliceous shale
- Presence of type II kerogen
- Economics of operation

Based on all available information, maps are drawn with the aid of a geographical information system that points to the location of sweet spots. EUR from potential wells is estimated based on Monte Carlo simulation of anticipated production decline. The data for decline curve analysis are obtained from wells producing from similar sweet spots in the area. The entire process utilizes software applications enabling rapid identification and evaluation of future shale gas reservoirs.

Multistage hydraulic fracturing

Hydraulic fracturing design is based on sophisticated 3D modeling that takes into account the depth, thickness, lithology, fracture stress, and other properties of formation, among others. Models allow an optimized design of fracturing operation where the height, length, and orientation of the fractures can be most effective as well as economic. For example, horizontal well EUR was plotted against well lateral orientation in a study concerning Barnett shale development [8]. A minimum of EUR could be identified for well laterals oriented at northeast 55° azimuth. This was the direction of principal horizontal stress and the fractures readily propagated along the above direction. Horizontal wells are drilled transverse to the direction of fracture orientation for maximum performance.

Hydraulic fracturing is accomplished by injecting the fracturing fluid under very high pressure and at a predetermined rate that would prop open fractures in the ultra-tight shale and facilitate flow of gas. The fracturing fluid is mostly water, about 98%. Sand is added to the fluid, which acts as proppant to keep the fractures open. Lubricants are added to facilitate the transport of various additives; hence, the fracturing fluid is referred to as slick water. A small amount of chemical additives is also added. The chemical agents reduce friction; hence, the injected water is referred to as slick water. Furthermore, certain other chemicals are added to prevent the growth of microorganisms and clogging of fractures, and minimize corrosion.

Typical constituents of hydraulic fracturing fluids are as follows [9]:

- Water: The major component injected into the formation to create, propagate, and enhance fractures. The amount of water needed for each stage in the fracturing operation is quite substantial and water requirements for multistage fracturing can easily exceed a few million gallons.
- Sand: Used as proppant to keep the new fractures open and conductive in order to maintain the flow of gas. Three to five million pounds of proppants are required in each hydraulic fracturing operation.
- Resins: Resins are used to hold the proppants (sand and other materials) in place and prevent any loss.
- Ceramics: When a stronger proppant is needed to obtain the desired properties of fractures, ceramics can be used. Ceramics are lighter than sand and may be transport with relative ease.
- Gels: Used to transport proppants.
- Acids: Usually dilute hydrochloric acid is used to clean up perforations by removing the cementing materials.
- Biocides: Prevent bacteria from growing and fouling the wellbore.
- Potassium chloride: Prevents swelling of clay.
- Peroxydisulfates: Used as breakers to reduce the viscosity of gel and release proppant into the formation.
- Corrosion inhibitor: Acts to prevent any corrosion of metallic casing and tubing as acid is used.

The length of the horizontal wells is usually long, 5,000–10,000 ft., where creating and maintaining high pressure by injecting fluid along the entire length is difficult. Hence, a hydraulic fracturing operation is carried out in multiple stages by isolating small portions of horizontal wellbore at a time in order to maintain the requisite pressure along the entire length of the well. The spacing between stages can range from a few to several hundred feet along the lateral. A typical horizontal well several thousand feet long may have as many as 30–40 fracturing stages. Each stage consists of substages where predetermined quantities of water, sand, and chemicals are used. Microseismic surveys can be conducted to determine the extent and orientation of the fractures created in the formation. A schematic of a horizontal well with multistage fracturing in a shale gas reservoir is presented in Figure 22.5.

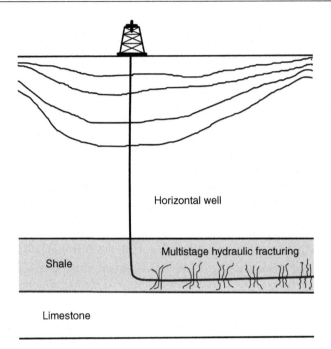

Figure 22.5 A horizontal well with six-stage fracturing.

Well performance

However, the later part of gas production may follow an exponential decline pattern, and the life of a well may last over 20 years. During the later period, well workover may be required to enhance productivity. EUR of shale gas is mostly based on decline curve analysis of production data of a well. Decline curve analysis is described in Chapter 13 (Figure 22.6).

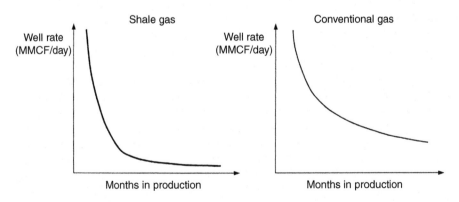

Figure 22.6 Decline curve of typical shale gas production as compared to decline in conventional gas production.

Significant enhancement in shale gas recovery is achieved by drilling horizontal wells. A study based on production data obtained from Barnett shale in the mid-2000s indicated that the EUR from horizontal wells is about 1.27–1.44 BCF (billion cubic feet); however, EUR from vertical wells ranged between 0.37 BCF and 0.49 BCF.

Production figures vary significantly from one well to another in a reservoir, and in various shale gas deposits. For example, in Marcellus shale, initial well rate can be 5 MCF (million cubic feet) a day or more, and the EUR from highly productive wells can be as be as high as 7 BCF. TRR of various shale gas reservoirs in the United States were presented earlier in Table 22.4. Typical recoveries from shale gas reservoirs are expected to rise in the future as the technology matures and new techniques are employed for extraction.

Challenges in shale gas development

There are several challenges in developing unconventional shale gas reservoirs outlined as follows:

- Shale is highly heterogeneous and laminated. Rock properties vary in both horizontal and vertical directions, and in macro- as well as microscale, which introduces significant uncertainties in drilling, completion, and production. Success of shale gas production largely depends on target drilling at sweet spots where reservoir characteristics are favorable. Performance of one well may vary significantly from the neighboring wells.
- Presence, direction, and conductivity of natural fractures add to reservoir complexities and affect well performance. Furthermore, the effectiveness of hydraulic fracturing largely depends on the geomechanical properties of rock.
- Shale gas reservoirs have extremely low permeability, often in the range of 10^{-9} D. In conventional reservoirs, the ultralow permeability rocks in the above-mentioned range are usually considered to be nonpermeable and viewed as a seal. Obviously, shale gas reservoirs require the implementation of groundbreaking technology. Considerable risk and uncertainty is involved in development.
- Shale gas development and production, including multistage fracturing, require detailed consideration to environmental issues including the potential contamination of groundwater through fractured pathways, induced seismic activity, migration of produced gas through casing leaks, and air and noise pollution.

Case Study: Marcellus Shale – Development of a Super Giant Unconventional Resource [10–28]

Introduction

The Marcellus shale is an unconventional gas resource in the United States that extends over an area 100,000 miles across several states, including Pennsylvania, New York, Ohio, West Virginia, and Maryland (Engelder and Lash, unpublished draft, 2007). Currently, natural gas production has leaped from about 0.53 BCF/day to over 17.5 BCF/day just in 5 years (Figure 22.7). As of 2012, over 6400 wells were drilled in Marcellus shale. More than 3600 wells were reported producing over 8 BCF of unconventional gas per day. In Pennsylvania, wells are clustered

Figure 22.7 Marcellus shale gas production.
Source: Energy Information Administration.

either in the northeast, producing dry gas, or in the southwest, producing gas and condensate.

Marcellus shale gas in place and recovery factor

Various estimates of the total amount of gas, including both discovered and undiscovered resources trapped in the Marcellus shale, have been reported in recent years. The resources are expected to be as much as 2700TCF. Assuming a recovery factor of 10–20%, the supply of natural gas from Marcellus alone can meet the demand for the entire United States for years to come. According to one estimate, the TRR of Marcellus shale gas is 490 TCF, second only to the conventional natural gas deposits in North Field and S. Pars located in the Middle East. Natural gas in the Marcellus shale is either trapped in pore spaces or adsorbed on minerals and organics present in the rock. Gas is also stored in naturally occurring fractures. The shale formation serves as the petroleum source, seal, and reservoir due to the very low permeability of rock.

The technologies that propelled the spectacular growth in the development and production of shale gas in the Marcellus formation include horizontal drilling, multistage fracturing, and microseismic surveys that identify the presence and characteristics of fractures. However, production of gas also takes place through vertical wells.

Marcellus shale geology and geochemistry

Located in Appalachian Basin, the Marcellus shale, a marine sedimentary rock, was formed about 390 million years ago during the Middle Devonian period. The deposition of fine-grained sediments along with organic matter occurred in a deep-water and anoxic environment where dissolved oxygen content is very low. The TOC in Marcellus shale ranges between 1% and 13%. Marcellus shale is located at a depth of 4000–8500 ft. from where the unconventional gas deposits are extracted. However, outcrop of Marcellus formation can be found in central New York state. The net formation thickness varies anywhere between 50 ft. and 200 ft. The thickest formation is found in northeastern Pennsylvania. As a result of intense subsurface temperature and pressure of burial, the organic content was transformed into natural gas over millions of years. The gas is accumulated in the micropores and fractures, and also in an adsorbed state in the organic content of the rock. As shale is fine grained, porosity is relatively low, less than 10%. Certain studies found that the porosity is as low as 0.5–5%. Permeability of shale is ultralow, in the order of nanodarcies. In one study, Marcellus shale permeability is reported in the range of 100–450 nD. It is interesting to note that, in conventional reservoirs, geologic formations having permeability in nanodarcies are considered to provide a seal. The pressure gradient of the Marcellus shale is about 0.4 psi/ft.

Marcellus shale gas economics

In contrast to conventional reservoirs where gas is trapped within distinct reservoir boundaries, the unconventional resource in Marcellus shale occurs continuously over a very large area. However, the limited connectivity between micropores leads to ultralow permeability, which is not adequate to produce gas by conventional means. Due to the unfavorable reservoir quality of shale and the depth of the Marcellus formation, development of the unconventional resource only became feasible following the introduction of horizontal drilling and multistage fracturing. Gas is mostly produced through hydraulically induced and natural fractures that are interconnected. Development of Marcellus shale, like any other unconventional resource, is also attributed to economic conditions, i.e., the rising price of natural gas. According to various estimates, horizontal wells may cost between $4 million and $7 million to drill and complete, and the breakeven price of Marcellus shale gas is about $4 per MCF.

Fracture characteristics

Two primary joint sets are observed in Marcellus shale. The joint sets provide a network of fractures through which gas can flow in otherwise tight rock. The first set, referred to as $J1$, is observed to trend in an east-northeast direction. $J1$ is also parallel to the direction of maximum horizontal stress. The other joint set, $J2$, trends in a northwest direction. For the efficient extraction of shale gas, taking advantage of the $J1$ joints is the preferred option as the fracture system runs parallel to the direction of principal horizontal stress. Again, the joints are closely spaced. Both the factors facilitate the production of gas from tight shale. Horizontal well laterals drilled in the direction transverse to the principal fracture direction are able to intersect large numbers of fractures, thereby they facilitate flow of gas from the tight formation.

Prospecting for Marcellus gas

Seismology is used in the prospecting for shale gas to determine the stratigraphy of the basin. Seismic waves reflected from the subsurface indicate stratigraphic boundaries and gas content may be indicated by correlating with the known occurrences in certain layers. Geophysical surveys include studies related to shale stratigraphy, porosity, and geomechanical properties of rock including fracture properties and orientation. Well logs, including gamma ray logs, pinpoint shale formation, including depth and thickness, as the rock exhibits higher radioactivity than sandstones and carbonates. Log information obtained from existing wells in the region may aid significantly in finding new deposits that can be developed economically. A significant contrast between conventional and unconventional gas deposits lies in the fact that the latter is continuous in nature; the prospect of finding natural gas in Marcellus shale and similar unconventional reservoirs is much higher than conventional accumulations; however, sustained production from shale gas reservoirs may not be attained due to poor reservoir quality.

Drilling for unconventional gas

Both vertical and horizontal wells are drilled in Marcellus shale. Although the vertical wells cost less, the wells are able to contact a very limited portion of the formation along the vertical direction and intersect a limited number of fractures serving as conduits for flow. Consequently, the productivity of the vertical wells is significantly less than that of the horizontal wells. Currently, the horizontal wells drilled in Marcellus shale formation are 5000–10,000 ft. in length; hydraulic fracturing is conducted in 30 stages or more in order to create fractures every few hundred feet along the length of the lateral. Since the primary joint set, referred to as $J1$, is observed to run in the east-northeast direction, the horizontal well laterals are aligned perpendicular to the direction of the joints, i.e., in the northeast direction. Furthermore, certain fractures in the formation are vertically oriented, which requires the horizontal wells to be drilled to intersect them to attain maximum productivity. The ultimate recovery from a horizontal well in Marcellus shale is expected to be as high as 4 BCF.

Horizontal well productivity is generally 300–500% higher than that of vertical wells. A significant development in recent years is that several horizontal wells are drilled from a single pad, which reduces the footprint of intense drilling on land and minimizes any risk to the environment (Figure 22.8).

Well completion

Wells are completed by setting the casing and when the pay zone is perforated. There are four types of casing that are set in a well as follows:

- Conductor: The uppermost casing set to contain surface soil.
- Surface casing: Set to isolate aquifers from drilling fluids.
- Intermediate casing: Set to isolate the flow of oil and from zones located at shallower depths.
- Production casing: Set at the pay zone; it is perforated by carefully designed charges to allow natural gas to flow from pay zone to the surface facilities.

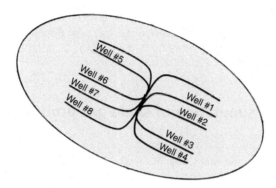

Figure 22.8 Pad drilling with a small footprint on the surface. It can replace several drilling sites from which individual wells are drilled requiring significant footprint.

The most brittle areas of shale formation are conducive to fracture creation and propagation; hence, these intervals are targeted for perforation. Usually, three perforations are made per foot over intervals that run about 2 ft.

Hydraulic fracturing of shale

The ultratight Marcellus shale is stimulated by hydraulic fracturing of wells to attain sustained gas production. In various sections of a horizontal lateral, hydraulic fracturing, referred to as fracking by many, is conducted in multiple stages, one section at a time. Hence, the entire operation is referred to as multistage fracturing. The horizontal sections are a few hundred feet in length. Water along with certain additives is injected under high pressure to create new fractures. Additionally, existing fractures can be enlarged and fracture density may increase. As stated earlier, Marcellus shale has a system of joints referred to as $J1$, which is aligned to the direction of principal stress. The horizontal wells are drilled in the transverse direction of $J1$, hence any hydraulically created fractures would be parallel to the direction of the joints maximizing the recovery potential. During fracturing operation, care is taken not to fracture the overlying or underlying formations, which could result in the loss of fracturing fluid and shale gas in adjacent layers. However, the limestone formation, known as Onondaga limestone, located beneath the Marcellus is resistant to such accidental fracturing.

Water disposal

The water disposal issues associated with shale gas reservoir development are noted in the following.

Water disposal: Multistage hydraulic fracturing requires the use of significant quantities of water, ranging between a few to several million gallons of water. According to an estimate, the total use for water is about 19 million gallons a day. Over 60% of the water used originates from surface water accumulations. A sizeable portion of water used in fracturing, between 30% and 70%, is returned to the surface requiring disposal in a safe and efficient manner. The water also contains the additives used in fracturing and dissolved solids from the formation. The wastewater can either be reused in fracturing operations or released in the environment following proper treatment to remove contaminants. The wastewater can also be injected into the ground where it can be contained by confining geologic strata in order to prevent any contamination of aquifers.

Modeling and simulation of shale gas production

Shale gas development requires substantial capital investment in horizontal drilling, multistage fracturing, and managing the wells. Moreover, high risk factors are associated with the commercial viability of the wells, which depends on the favorable reservoir quality and fracture characteristics, including length, numbers, and conductivity. Hence, modeling and simulation of shale gas production play a critical role in designing, drilling, and fracturing the wells. The objectives of simulation study include:

- Optimization of the horizontal well length
- Determination of optimum number of fracturing stages
- EUR under best and worst case scenarios
- Economic analysis based on the results of simulation

The following are considered in modeling shale gas [29]:

- Mechanism of storage: Gas is stored in shale in both a free and adsorbed state. Free gas is found in pores and fractures, while a sizeable portion of gas is trapped in an adsorbed state on the solid organic matter in shale. The amount of gas desorbed under declining pressure is estimated by the Langmuir isotherm that correlates the amount of gas adsorbed with pressure. The Langmuir isotherm is discussed in Chapter 12. A multicomponent adsorption model can be used for the different components present in shale gas. Rate of gas production is higher when desorbed gas is taken into account.
- Mechanism of flow of gas: Within the matrix of ultralow permeability shale, transport of gas is characterized by both diffusion and Darcy flow. The diffusion of gas is modeled by using Fick's law. Coefficient of diffusion and tortuosity are specified to model the diffusion process. Both Darcy flow and non-Darcy flow can occur in the fracture network of shale.
- Hydraulic fractures: Flow of fluid at high velocities such as encountered in hydraulic fractures cannot be represented by Darcy's law. The fractures are assumed to be about 2 mm wide having very high permeability. Hence, the Forchheimer equation is used in the model to represent the flow through hydraulic fractures accurately. The nonlinear equation is described in Chapter 3.
- Natural fractures: Naturally occurring fractures have much less conductivity than hydraulic fractures. Hence, these fractures are modeled by a relatively simple dual permeability model.
- Formation characteristics and model representation: Shale consists of both matrix and fractures. The fractures can either be natural or created by hydraulic fracturing. In modeling the flow of gas, a dual permeability system can be utilized to represent flow through rock matrix and fractures, the latter having significantly higher permeability. However, in shale and in other tight formations having ultralow permeability, pressure transient response is quite slow, and traditional dual permeability models may not be adequate. Flow through hydraulic fracturing is modeled explicitly by using a modified dual permeability model, which is logarithmically spaced and locally refined.

- Grid refinement: A logarithmic grid refinement scheme may be employed to capture the fine details of flow characteristics in the vicinity of the fractures. The grids become progressively coarser away from the fractures, which optimize the resource requirements of the simulation model. At least five layers of refinement are recommended for a model. Grid sensitivity analysis can be performed to determine the degree of refinement necessary.

Workflow to build a shale gas model is presented in Figure 22.9. Production history of a shale gas well is shown in Figure 22.10, which is matched with a simulation model for validation.

Coalbed methane

Introduction

Coalbed methane (CBM) refers to the natural gas, chiefly methane, which is stored in the cleats and micropores of coal [30,31]. Cleats are natural fractures that develop due to the prevailing stresses on the subsurface geologic formation. CBM is a significant resource of unconventional gas in the world. Natural gas, chiefly methane, is trapped in coal seams and micropores of coal and is extracted by drilling vertical and horizontal wells. Large deposits of coal are found in basins and produced commercially. CBM deposits are found in over 60 countries. According to one estimate, CBM reserves in the top 20 countries are 1800 TCF. Large CBM development projects have been undertaken in China, India, and Australia, among others.

In the United States alone, CBM accumulations have been discovered in more than a dozen basins (Figure 22.11). Total reserves are quite substantial. Proven reserves of CBM have increased the total reserves of natural gas in the United States substantially. The unconventional resource of CBM provides about 7% of the total natural gas consumption in the country.

History of coalbed methane

Production of CBM in the United States dates back to the early twentieth century. In the 1920s and early 1930s, wells were drilled to produce methane from coalbed deposits in Kansas, although the gas was thought to originate from another geologic formation. Some of the wells were 1000 ft. deep. In the 1950s, the first CBM well was fractured to enhance production. The CBM industry grew at a very rapid pace in the 1990s. Between 1992 and 2000, CBM production leaped from 1.5 BCF/day to 3.7 BCF/day. The number of producing wells grew significantly, from 5,500 to about 14,000.

Geology

Coal formations, with their origin dating back hundreds of millions of years in the geologic time scale, are rooted in the deposition of plant matters from swamp forests in shallow waters. The depositional environment was oxygen deficient. The continued deposition of sediments over a long period of time resulted in the formation of

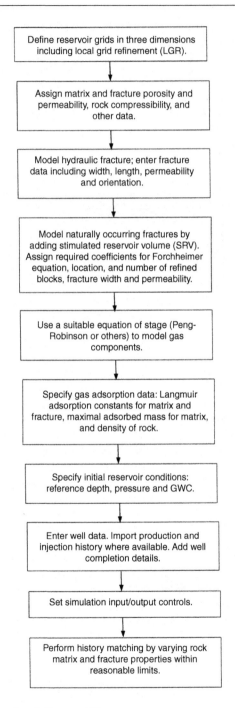

Figure 22.9 Shale gas simulation workflow.

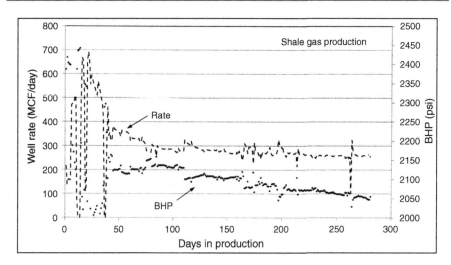

Figure 22.10 History matching of shale gas production, where simulated bottom-hole pressure is matched against field data.

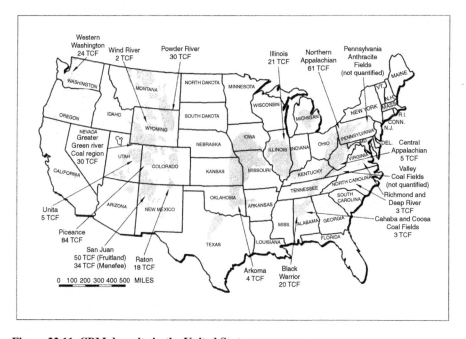

Figure 22.11 CBM deposits in the United States.
Source: http://www.halliburton.com/public/pe/contents/Books_and_Catalogs/web/CBM/ CBM_Book_Intro.pdf [accessed 22.04.14].

coal by the combined action of temperature, pressure, and various geochemical and other processes. Eventually, methane gas is formed in the coalbeds due to the intense thermal energy. Most coal deposits were formed during the Carboniferous Period, 300–360 million years ago in prehistoric times. The Carboniferous Period, referring to the formation of vast amounts of carbon-rich formations, is part of the Mississippian and Pennsylvanian age.

Distinctive features of CBM reservoirs

Coal formations act as source rock, where methane was originated in prehistoric times, as well as reservoir rock, from where the gas is produced. CBM is an unconventional resource as the reservoir characteristics differ significantly from those of conventional accumulations of natural gas, and nontraditional technologies are used for the extraction of CBM. The storage of methane in the coalbed, the mechanism of fluid flow, rock characteristics, and reservoir development are unique to CBM reservoirs. The major differences between CBM reservoirs and conventional gas reservoirs are as follows:

- As stated earlier, source rock and reservoir rock are the same in CBM reservoirs. In contrast, source rock and reservoir rock are not the same in conventional gas reservoirs. Following migration from source rock, conventional gas is accumulated in reservoir rock by one or more trapping mechanisms; however, the process takes place over a long period of time and distance.
- In conventional gas reservoirs, natural gas is stored in a free state in pores and fractures; however, CBM is trapped mostly in an adsorbed state. Free gas in cleats and fractures of coal only amounts to a few percent of total volume of gas in place. According to a study, coalbeds can contain as much as 98% of the total volume of gas in an adsorbed state. Hence, in estimating the total gas in place in a CBM reservoir, the amount of gas adsorbed in micropores must be known. This can be accomplished by direct measurements or be based on known correlations for the basin.
- In conventional reservoirs, the volume of gas in place can be calculated by applying the real gas law requiring the knowledge of hydrocarbon pore volume and prevailing reservoir pressure. In CBM reservoirs, the quantity of natural gas stored in coalbeds cannot be determined by using traditional methods of volumetric analysis. The adsorption capacity of coal is needed to estimate the gas in place. CBM reservoirs can store several times more gas in an adsorbed state than a conventional gas reservoir.
- Unlike conventional sandstone and limestone reservoirs where free gas is stored in relatively large pores, CBM is stored in the micropores of coal. Cleats or fractures in coal formations facilitate the flow of gas as the wells are drilled.
- The pore sizes in CBM reservoirs are smaller by orders of magnitude in comparison to conventional sandstone and carbonate reservoirs. Unconventional CBM reservoirs are usually characterized by relatively low porosity and permeability.
- Transport of CBM in micropores of rock is characterized by diffusion governed by Fick's law; however, flow of CBM though fractures and seams obeys Darcy's law. In contrast, the transport of fluids in conventional reservoirs is characterized by Darcy and non-Darcy flow.
- The unique characteristic of CBM production is the initial production of water contained in the cleats of coal formation. Water contained in cleats and fractures flows at a high rate during the initial production period.

- Water−gas ratio (WGR) decreases with time as a sizeable amount of water is produced; an increase in gas flow is observed with time. In contrast, WGR may increase rather than decrease with time in conventional reservoirs under strong water influx from adjacent aquifers.
- CBM production from a well eventually attains a peak rate following a certain period of time. In conventional reservoirs, however, gas rate is usually at a peak during the initial stage.
- The permeability of CBM reservoirs is dependent on *in situ* stress. However, in conventional reservoirs, the dependence of permeability on *in situ* stress is negligible.
- The presence of nearby wells is beneficial to CBM production, while in conventional reservoirs, other wells located in the vicinity of a well may interfere with its performance. CBM reservoirs need multiple wells to be drilled for development.
- Most CBM reservoirs are generally of low permeability requiring hydraulic fracturing to produce at a commercial scale. However, in conventional reservoirs, gas production rate can be quite high from unfractured formations where reservoir quality is excellent.

Coalbed cleats and porosity

Coalbed formations are typically dual porosity systems, which are comprised of pores and cleats. Generally two types of cleats or seams are observed, namely, face cleats and butt cleats. The cleats form due to shrinkage of the formation and regional stresses. The face cleats are oriented at about 90° to butt cleats. The cleats have apertures ranging from a fraction of millimeter to a few millimeters. The butt cleats are shorter than face cleats and terminate at face cleats. Besides, secondary and tertiary cleats are found in coalbed formations (Figure 22.12).

The cleats and other natural fractures hold less than 10% of the total volume of gas. Studies have shown that the porosity of coalbed based on macropores is in low single digits. Formation water is contained in both macropores and cleats. The porosity of coalbed also indicates the total volume of water that must be handled during the production of gas. Under subsurface pressure, a portion of gas is also dissolved in water.

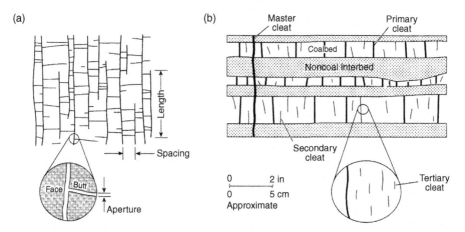

Figure 22.12 Cleats in coal, which provide a natural pathway to the production of gas.
Horizontal drilling and hydraulic fracturing contribute significantly to the development of CBM reservoirs [32].

Unlike conventional reservoirs where free gas is stored in the pores of rock, coalbeds can contain very large amounts of methane in micropores in an adsorbed state. The micropores are of molecular dimension, ranging from less than 5 A° to 50 A°. As stated earlier, coalbeds can contain a major portion of natural gas in an adsorbed state. Hence, the actual stored volume of methane in coalbeds can be several times more than the coalbed reservoir porosity would indicate.

Adsorption characteristics of CBM

The quantity of natural gas trapped in micropores of rock in an adsorbed state is estimated by the Langmuir isotherm, which basically states that the adsorption capacity of a solid surface increases as the prevailing pressure increases, and diminishes as the pressure is reduced. The relationship between pressure and adsorption capacity of coal suggests that the gas will be desorbed (and produced eventually) as the reservoir pressure declines. The rate of change is nonlinear. Desorption occurs at a slow pace initially, followed by an accelerated rate as the pressure declines further. A typical Langmuir isotherm for shale gas and CBM is presented in Chapter 12 describing how the quantity of adsorbed gas is determined.

The Langmuir equation can be expressed as:

$$V_a = \frac{V_L P}{P + P_L}$$

where V_a = volume of adsorbed gas, ft.^{3}/ton; V_L = Langmuir volume; the volume of gas adsorbed at infinite pressure; P = pressure, psi; P_L = Langmuir pressure; the pressure corresponding to half of Langmuir volume.

The above is presented in Chapter 12.

The adsorption characteristic is also dependent on subsurface temperature. It is observed that adsorption of gas in shale reservoirs is reduced at higher temperatures.

Estimation of gas in place

CBM is stored in micropores, macropores, cleats, and fractures of coal as follows:

- Micropores having a major portion of methane in an adsorbed state
- Cleats and natural fractures having a fraction of water and gas volumes
- Macropores containing free and dissolved gas in water

The natural gas content of CBM can be determined by canister test on site. The formation is cored, and the cores are brought to the surface and transferred to a sealed canister where subsurface reservoir temperature is maintained. Care is taken to minimize the loss of any gas from the cores. Gas content of the cores is determined by summing the volumes of desorbed gas, residual gas, and unaccounted or lost gas. The methods of measurement of the three components of total gas content are described in the following.

Desorbed gas: The quantity of desorbed gas inside the canister is measured. The rate at which gas desorption occurs is also noted.

- Residual gas: When desorption of gas is complete under atmospheric pressure, the cores are crushed and the volume of gas remaining in pores is measured. The measurement leads to the quantity of residual gas.
- Lost gas: The quantity of gas lost during transportation of the core is estimated from a plot by extrapolating a plot of desorption against the square root of time, where the time denotes the interval between core extraction and start of the test.

Coalbed permeability

CBM reservoir development is largely dependent on the permeability of rock. Relatively high reservoir permeability favors commercial extraction of CBM. A number of factors affect coalbed permeability as follows:

- Effect of *in situ* stresses: Higher stresses in subsurface formation reduces the permeability of the coalbed.
- Depth of reservoir: Coalbeds located at shallower intervals have much better permeability. A study of coalbed permeability in three basins has indicated a clear reduction in permeability with depth. While rock permeability can be in tens or hundreds of millidarcies in formations located at a depth of 1000 ft. or less, it may be reduced to less than 0.1 mD at depths below 5000 ft. This occurs due to the increase in overburden pressure.
- Characteristics of fracture network: The abundance and connectivity of fractures enhance overall permeability.
- Orientation of cleats in the coalbed: The orientation of determines permeability anisotropy and principal axis of permeability aligned to the direction of cleats.
- Production of water: As water is produced from CBM reservoirs, a reduction in matrix permeability occurs due to the reduction in pressure in cleats, increase in *in situ* stress, and consequent decrease in permeability.
- Matrix volume reduction due to desorption of gas: A reduction in matrix volume leads to the enhancement of permeability in cleats.
- Klinkenberg effect: Apparent permeability enhancement due to slippage as desorption of gas occurs rapidly near the abandonment pressure. A good reservoir management practice involves the prolonging of CBM production near abandonment as permeability is enhanced and a sizeable amount of gas is trapped in an adsorbed state.

During the production of gas, coalbed permeability can change significantly. The net effect of decline in reservoir pressure on permeability includes the following:

- In the initial period, production of water held in cleats and fractures takes place in significant amounts and reservoir pressure declines. As a result of decline in pressure, there is an increase in *in situ* stress, which acts to close the cleats. The net effect is the reduction of the effective permeability to gas. Rock permeability can also be reduced by the swelling or rock.
- However, at subsequent periods, increase in permeability is observed due to the shrinkage of coal matrix and Klinkenberg effect on the flow of gas. As gas is produced, coal matrix volume decreases, and the permeability in cleats is increased. The Klinkenberg effect originates from the phenomenon of gas slippage with the adjacent layer at low pressure. The effect can be significant when the CBM reservoir is near abandonment and reservoir pressure is low (Figure 22.13).

Coal formations typically indicate permeability anisotropy. Permeability along face cleats is substantially higher than butt cleats that are formed orthogonal to face cleats. Studies have indicated that the formation permeability along the principal direction

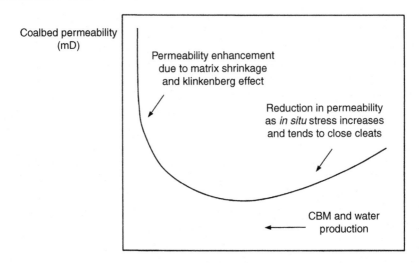

Figure 22.13 Permeability changes in coalbed with pressure decline due to changes in *in situ* stress, closure of cleats at the earlier stages, followed by the shrinkage of coal matrix and Klinkenberg effect at later stages.

can be higher by a factor of 1000% or more in comparison to permeability in the transverse direction.

Field studies suggest that most coalbed reservoirs have permeability values ranging between 1 mD and 100 mD when reservoir depth is below 4000 ft. In some basins, coalbed permeability is in the order of several hundred millidarcies at shallow depths. Production is marginal where the permeability is less than 1 mD, and unsustainable below 0.1 mD. Hence, hydraulic fracturing is generally a requirement to produce natural gas from coalbeds.

Measurement of permeability

Laboratory measurements can be inaccurate as coalbed permeability is stress dependent and small core samples may not be representative of the high permeable cleats present in the system. As a consequence, the values of core permeability are likely to be underestimated. Well tests such as drill stem tests, pressure buildup tests, and multiwell interference tests provide better options to measure *in situ* permeability of coalbeds. Initially, the cleats are completely filled with water and only water is produced. At this point, well test results are based on single-phase flow leading to the determination of absolute permeability. However, as gas is desorbed with pressure decline, two-phase flow of gas and water is observed. The relative permeability of coalbed to gas is low at first, followed by a rapid increase as water production diminishes.

Last but not least, reservoir simulation studies are routinely performed to match the production history in order to determine the effective permeability of the unconventional reservoir as it is observed to change with production.

Mechanism of flow

As indicated earlier, the mechanisms of transport of natural gas in coal formations differ significantly than what is observed in conventional reservoirs. The mechanisms include desorption, diffusion, and Darcy flow. The mechanisms are outlined in the following:

- With the reduction in reservoir pressure, desorption of natural gas from the surface micropores takes place.
- Following desorption, gas moves by diffusion through the micropores. The phenomenon of gas diffusion is represented by Fick's law.
- Once the desorbed gas enters the network of fractures and cleats, Darcy flow occurs, which results in transport of gas to the wellbore.

Reservoir characterization

Although resource intensive, well interference tests can be very useful in determining interwell permeability, storativity, reservoir heterogeneities such as permeability anisotropy, optimum well locations, and well spacings.

Coalbed methane production characteristics

The contrast between a conventional gas reservoir and an unconventional reservoir is highlighted by the production characteristics of CBM. At the initial stages, water production is nonexistent in most conventional reservoirs. However, in the case of CBM, the bulk of water trapped in cleats and fractures of the coalbed must be produced in order to attain significant gas production. The phenomena of dewatering of the formation and desorption of gas from the surface of micropores may continue for several months before the maximum rate of gas production can be attained. Finally, continuous decline in CBM production rate is observed, the characteristic of which depends on rock properties, water saturation, and gas content of the coalbed. Studies have shown that many CBM reservoirs undergo exponential decline. The life of a typical well may range from short few years to 20 years or more (Figure 22.14).

Darcy's law for the flow of gas in porous media is presented in Chapter 3.

Coalbed methane reservoir economics

In the United States, about 7% of the natural gas is CBM produced from various basins. In recent years, horizontal well technology has been applied extensively to enhance the production of this unconventional gas. In each basin, the viability of commercial production depends on the following:

- Permeability of coalbed: 1 mD or greater
- Gas content: 150 scf/ton of coal or greater
- Thickness of seams
- Depth of coalbed; deeper coalbeds may store a relatively high volume of gas under higher pressure
- Rank of coal, bituminous coal with high volatility is a good source of CBM

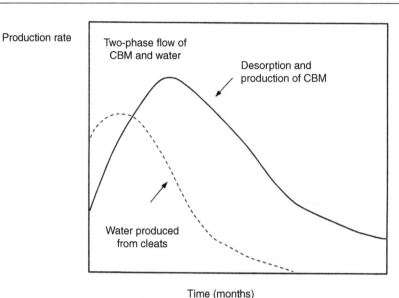

Figure 22.14 **Performance of CBM wells highlighting the dominance of water production at the initial stages.** CBM is produced initially at a low rate; however, the rate increases with time and a peak rate is attained in later stages. Following months or years of production of CBM, eventual decline in rate is observed.

- Availability of log data from other wells in the area
- Regional experience and reservoir characterization, lack of available information may pose obstacles
- Volume and quality of water in coal seams
- Water disposal issues
- Access to pipelines and markets

Other resources of unconventional gas [33]

Deep gas resources are located below 15,000 ft. where the cost of development and technological challenges remain high. In past decades, several thousand wells have been drilled in the United States to produce natural gas from deeply seated reservoirs located in Gulf Coast, Anadarko, and Permian basins. Certain deep gas wells reached impressive depths, around 30,000 ft. or more. Deep gas formations are generally older than shallower deposits of hydrocarbons requiring special technologies to produce economically. Technological challenges included deep drilling issues, high reservoir temperature of 300°C or more, and handling of sour gas. Deep gas development, like any other unconventional gas resources, had been sensitive to market conditions. With the fall of oil prices in the early 1980s, deep gas resources faced significant challenges to develop further. Deep gas development was viewed as high risk and resource intensive ventures. Even in the 1980s, deep gas wells cost several million dollars, requiring a breakeven price of gas of about $20/MCF in extreme cases. According to one study, the probability of hitting a dry hole was about twice that of conventional gas wells located at shallower depths. Nevertheless, deep gas plays a significant role in the supply of natural gas in the United States and other countries. According to United States Geological Survey estimates, potential quantities of deep gas remain undiscovered or unexploited in various regions, including the Rocky Mountains, Gulf Coast, and Alaska. The key to successful development of deep gas reservoirs includes the following:

- Focused drilling leading to high success rates in finding deep resources
- Advancement in deep drilling techniques requiring fewer resources
- Better completion techniques facilitating economic production
- Reduction in sour gas production and more efficient processing in surface facilities

Summing up

Introduction: Unconventional gas resources include reservoirs from where the natural gas is difficult to recover as the technology has not matured or nontraditional methods are required to produce the reservoirs. Moreover, the cost of production is often higher than that of conventional gas reservoirs. Major deposits of unconventional gas that are produced currently include shale gas, tight gas, and CBM. Other types of unconventional gas, such as deep basin gas, have seen limited developments due to relatively high cost, dry hole occurrences, presence of sour gas, and various technological challenges. Methane hydrates, found in deep sea or in arctic regions, have not been extracted at a commercial scale. However, unconventional gas production is substantial in the United States and constitutes 65% of total production as of 2013. The number is expected to increase in coming years. Unconventional gas resources in the world exceed 32,000 TCF (32 × 10^{12} ft.^{3}), with the three major deposits located in North America, the former Soviet Union, and China. Global estimates of shale gas reserves are about 5700 TCF.

Shale gas technology: Currently, the most prominent of all unconventional gas resources is shale gas. Shale is a source rock for the generation of petroleum. However, permeability of shale is quite low, often in the order of tens or hundreds of nanodarcies (10^{-9} D). Even a few years ago, such ultralow permeability was considered to be a seal for conventional oil and gas reservoirs. Commercial development of shale reservoirs became a reality only after the successful implementation of horizontal drilling and multistage fracturing technologies. The latter is sometimes referred to as fracking. Microseismic surveys are conducted to determine the extent, characteristic, and density of the hydraulic and natural fractures.

Target areas for development: Although major shale gas reservoirs are continuous over very large areas, shale is highly heterogeneous in nature; rock properties vary considerably from one location to another. In order to develop shale gas successfully, "sweet spots" are sought for drilling new wells. The sweet spots include rocks having high organic carbon content (known as "black shale"), relatively high permeability, appropriate thermal maturity belonging to the gas window, and naturally occurring swarms of fractures, among others. Furthermore, the geomechanical properties of rock, such as relatively high Young's modulus, low Poisson's ratio, and appropriate fracture stress, are desired for having favorable hydraulic fracturing characteristics.

Production and recovery factor: The first shale gas well dates back to the early nineteenth century and was drilled in the Appalachian Basin. Shale gas production has increased by 1400% in recent years in the United States. As indicated earlier, the astounding development of shale gas reservoirs is made possible by horizontal drilling and multistage hydraulic fracturing. As technology matures, the length of the lateral grew longer and more stages of hydraulic fracturing were implemented. The laterals can be 10,000 ft. long and as many as 30–40 hydraulic fracturings are done in stages in order to optimize production at a commercial scale. Typical shale gas production can be substantial at first, followed by a steep decline over the first few months or a year. The rate of decline slows in subsequent periods, and the trend may follow exponential decline for many years, sometimes for 20 years or more. The average EUR from shale gas wells varies widely, ranging between 1 BCF and 3 BCF per well. Recovery factor is quite low, estimated to be less than 20% in most cases. In contrast, recovery from conventional gas reservoirs can be 80% under favorable conditions. The extent of recovery from shale reservoirs depends in large part on the effectiveness of the fractures in providing conduits for gas flow.

Shale gas data for major deposits: Shale gas reservoirs in the United States are Barnett, Fayetteville, Haynesville-Bossier, Marcellus, Woodford, and Eagle Ford, among others. Marcellus shale extends over 100,000 square miles in several states including New York, Pennsylvania, Ohio, Maryland, and West Virginia. In Table 22.4, geological, geochemical, petrophysical, well, reservoir, and economic data are presented for the vast deposits of shale gas noted previously. Porosity of shale as encountered is generally low, about 4–10%. Permeability values range in hundreds of nanodarcies in the accumulations listed previously. Thousands of wells, both horizontal and vertical, have been drilled. Over 90,000 wells have been drilled in Marcellus shale alone. Currently, the cost of each horizontal well is as high as $6 million or more.

Mechanisms of gas storage and transport: Natural gas is stored in shale in free and adsorbed states. Free gas is found to occupy the pores and fractures, while a portion of gas is adsorbed on the organic matter contained in shale. Hence, determination of gas in place involves the estimation of both gas volumes. The amount of adsorbed gas is estimated by the Langmuir isotherm, which correlates the amount of gas adsorbed per unit weight of rock. The relationship is nonlinear. As reservoir pressure declines desorption occurs at a relatively low rate, followed by acceleration in rate below a certain value of pressure. The Langmuir isotherm is unique for each basin. Transport of free gas may occur as Darcy flow or non-Darcy flow, the latter results when gas velocity is sufficiently high in hydraulic fractures. Flow of gas within rock matrix involves a diffusional process as well as Darcy flow.

Development strategy: Successful development of shale gas depends on the following:

- Identification of "sweet spots" as drilling targets
- Favorable fracturing characteristics of rock

- Presence of natural fractures
- Optimum length of horizontal well lateral
- Optimum number of fracturing stages
- Optimum design of hydraulic fracturing operation
- Ability of fractures to stay open for sustained production
- EUR that leads to favorable economics
- Environmental and other issues

Hydraulic fracturing: Hydraulic fracturing is accomplished by injecting the fracturing fluid under very high pressure and at a predetermined rate, which props open fractures in the ultratight shale and facilitate the flow of gas. The fracturing fluid is mostly water, about 98%. Sand is added to the fluid, which acts as proppant to keep the fractures open. Lubricants are added to facilitate the transport of various additives; hence, the fracturing fluid is referred to as slick water. A small amount of chemical additives are also added. Hydraulic fracturing design is based on sophisticated 3D modeling that takes into account the depth, thickness, lithology, fracture stress, and other properties of formation, among others. Models allow an optimized design of fracturing operation where the height, length, and orientation of the fractures can be most effective as well as economic. Currently, hydraulic fracturing is performed in stages at every few hundred feet of the lateral. For example, a horizontal well 10,000 ft. long may have 30–40 stages of fracturing to create adequate pathways for sustained production.

Shale gas production characteristics: Shale gas is produced initially at a high rate, which usually declines notably after a period of several months. Gas stored in the highly conductive fracture network is produced first, leading to peak production rates. Eventually, gas is produced from the very low conductivity shale matrix, resulting in slow decline in rates over the years. In many cases, the initial gas production follows a hyperbolic decline.

Coalbed methane: CBM, stored in the cleats and micropores of coal, is a major source of unconventional gas. Cleats are natural fractures that develop due to the prevailing stresses on the subsurface geologic formation. Large deposits of coal are found in basins and produced commercially. CBM deposits are found in over 60 countries. According to one estimate, CBM reserves in the top 20 countries are 1800 TCF. Large CBM development projects have been undertaken in China, India, and Australia, among others. It is estimated that 6–7% of the supply of natural gas in the United States is based on the production of CBM.

The distinct features of CBM reservoirs as compared to conventional gas reservoirs are shown in Table 22.6.

Deep basin gas: Deep gas resources are located below 15,000 ft. where the cost of development and technological challenges remain high. In past decades, natural gas has been produced from deeply seated reservoirs located in Gulf Coast, Anadarko, and Permian basins. Issues related to deep gas development include: high risk, substantial investment, handling of sour gas, and technology. However, there are vast potentials in the Rocky Mountains, Gulf Coast, and Alaska, according to a study conducted by the United States Geological Survey.

Table 22.6 Distinct features of CBM reservoirs as compared to conventional gas

Characteristic	Conventional	CBM	Notes
Type of rock	Sandstone, limestone, and dolomite	Coal formation	
Generation and migration of gas	Gas is generated in source rock followed by migration and entrapment in reservoir rock.	Gas is generated as well as produced to the surface from source rock.	
Reservoir depth	Conventional gas is usually found at depths corresponding to a "gas window."	CBM reservoirs are located at shallow depths, 4000 ft. or less.	
Storage mechanism	Gas is stored in the free state in porous network.	Gas is stored in an adsorbed state in coal. The quantity of gas is several times more than what the pore volume of coal would suggest.	
Permeability	May vary widely, from less than 1 mD to a few darcies	Generally low, between 1 mD and 25 mD	CBM reservoir development often requires horizontal drilling and hydraulic fracturing due to the low permeability range.
Mechanism of transport	Gas transport may involve Darcy flow and non-Darcy flow at high velocities.	Desorbed gas is initially transported by diffusion.	Diffusional process can be represented by Fick's law.
Determination of gas in place	Requires the knowledge of reservoir pore volume, fluid saturation, and pressure.	Requires the knowledge of the adsorption characteristic of gas per unit weight of rock, in addition to the estimates of free gas in pores and fractures.	
Production characteristic	Gas is initially produced at high rate followed by eventual decline. If the gas reservoir has a strong aquifer effect, water can be produced in later stages.	Water contained in the cleats or seams of coal are produced initially. Production of CBM increases gradually with the decline in water production and reaches a peak in months or years.	

Questions and assignments

1. What are the major resources of unconventional gas? Discuss the potential of each in terms of current technological development.
2. Describe the characteristics that classify reservoirs as unconventional. Are the characteristics common among all types of unconventional gas reserves?
3. Describe in detail the various technologies that have led to the astounding increase in unconventional gas production in recent times.
4. How does the determination of gas in place differ between conventional and unconventional reserves? How does typical recovery vary between the two reservoirs? Explain with examples.
5. Describe the reservoir properties and production potential of major shale gas basins and plays of the world. Why are shale drilling and production more successful in some reservoirs than others? Explain with examples.
6. What rock properties are important in shale gas development? What are sweet spots in shale gas drilling? How are these spots identified?
7. How can a shale gas well be designed for optimum performance? Provide a case study based on a literature review.
8. Describe the factors that may influence hydraulic fracturing design. Why is hydraulic fracturing s modeled? Describe the role of additives used in fracturing fluids.
9. What are the typical production characteristics of shale gas? What EUR can be expected from a shale gas well to be economically feasible? Describe the factors considered in the economic analysis.
10. Describe the mechanism of shale gas flow in porous medium. How does it differ from conventional reservoirs?
11. Your company is planning to develop a large area in Marcellus shale. Describe a detailed development plan including the expected reservoir characteristics, locations to drill, horizontal well design, multistage fracturing, completion technique, and workover. Make necessary assumptions. Include potential environmental issues in your plan and how these can be addressed.
12. What is CBM and how it is used accumulated and produced?
13. Distinguish between the characteristics of CBM and conventional reservoirs in detail. Describe how the differences may affect reservoir development and well performance.
14. Distinguish between adsorption characteristics of CBM and shale gas.
15. How does the production of CBM differ from conventional gas? What can be done to enhance the production of CBM?

References

[1] Modern shale gas development in the United States: an update. National Energy Technology Laboratory; 2013.
[2] Technically recoverable shale oil and shale gas resources: an assessment of 137 shale formations in 41 countries outside the United States; 2013. Available from: http://www.eia.gov/analysis/studies/worldshalegas/.
[3] Modern shale gas development in the United States: a primer. National Energy Technology Laboratory; 2009.
[4] Baihly J, Altman R, Malpani R, Luo F. Study assesses shale decline rate. AOGR; May 2011.

[5] Murray R. Shale gas reservoirs similar yet so different. Available from: https://www.
     transformsw.com/wp-content/uploads/2013/05/Shale-Gas-Reservoirs-Similar-yet-so-
     different-2010-RMAG-DGS-3D-Symposium-Roth.pdf [accessed 15.01.15].
[6] Navarette M. Unconventional workflow: a holistic approach to shale gas development.
     Available from: http://www.seapex.org/im_images/pdf/Simon/7%20Mike%20Navarette_
     %20Unconventional%20Workflow%20A%20Holistic%20Approach%20to%20
     Shale%20Gas%20Development.pdf [accessed 21.01.15].
[7] Steffen K. Techniques for assessment of shale gas and their applicability to plays from
     early exploration to production. Available from: http://www.uschinaogf.org/Forum10/
     pdfs/11%20-%20ExxonMobil%20-%20Mericle%20-%20EN.pdf [accessed 09.10.14].
[8] Barnett Shale Model-2 (conclusion): Barnett study determines full-field reserves, produc-
     tion forecast. O & GJ; September 2, 2013.
[9] Kaufman P, Penny GS, Paktinat J. Critical evaluations additives used in shale slickwater
     fracs. 2008 SPE Shale Gas Production Conference. SPE 119900: Forth Worth (TX); 2008.
[10] Lee DS, Herman JD, Elsworth D, Kim HT, Lee HS. A critical evaluation of unconven-
     tional gas recovery from the Marcellus shale, northeastern United States. Energy Geo-
     technol. KSCE J. Civil Eng. 2011;15(4):679–87.
[11] Milici R, Swezey C. Assessment of Appalachian basin oil and gas resources: Devonian
     shale – middle and upper Paleozoic total petroleum system. U.S. Geological Survey,
     Open-File Report; 2006. p. 2006–1237.
[12] Harper J. The Marcellus shale: an old "new" gas reservoir in Pennsylvania. Pennsylvania
     Geol. 2008;38(1):2–13.
[13] Harper J. The Marcellus and other shale plays in Pennsylvania: are they really worth all
     the fuss? State College, PA: PSU EarthTalks Series; 2009.
[14] Myers R. Marcellus shale update. Independent Oil & Gas Association of West Virginia;
     2008; Ottaviani W. Gas pains: technical and operational challenges in developing the
     Marcellus shale. PSU EarthTalks Series. State College, PA; 2009.
[15] Soeder DJ. Porosity and permeability of eastern Devonian gas shale. Soc. Petrol. Eng.
     Form. Eval 1988;3(2):116 24.
[16] Engelder T, Lash GG. Marcellus shale play's vast resource potential creating stir in Ap-
     palachia. Am. Oil Gas Report. 2008;51(6):76–87.
[17] Gas pains: technical & operational challenges in developing the Marcellus shale. PSU
     EarthTalks Series. State College, PA 2009.
[18] Agbaji A, Lee B, Kuma H, Belvalkar R, Eslambolchi S, Guiadem S, et al. Sustainable
     development and design of Marcellus shale play in Susquehanna, PA. Report of EME580.
     Penn State University, State College, PA; 2009.
[19] Arthur JD, Bohm B, Coughlin BJ, Layne M. Evaluating the environmental implications
     of hydraulic fracturing in shale gas reservoirs. 2009 SPE Americas E&P Environmental
     & Safety Conference; 2009San Antonio, TX. SPE 121038.
[20] Beauduy T. Development of the Marcellus shale formation in the Susquehanna River
     basin. State College, PA: PSU EarthTalks Series; 2009.
[21] Bell MRG, Hardesty JT, Clark NG. Reactive perforating: conventional and unconven-
     tional applications, learning and opportunities. 2009 SPE European Formation Damage
     Conference. SPE 122174: Scheveningen, The Netherlands; 2009.
[22] Ottaviani W. Gas pains: technical & operational challenges in developing the Marcellus
     shale. State College, PA: PSU EarthTalks Series; 2009.
[23] Kundert D, Mullen M. Proper evaluation of shale gas reservoirs leads to a more effective
     hydraulic-fracture stimulation. 2009 SPE Rocky Mountain Petroleum Technology Con-
     ference; 2009Denver, CO. SPE 123586.

[24] Ozkan E, Brown M, Raghavan R, Kazemi H. Comparison of fractured horizontal-well performance in conventional and unconventional reservoirs. 2009 SPE Western Regional Meeting; 2009San Jose, CA. SPE 121290.

[25] Engelder T. Geology and resource assessment of the Marcellus shale. State College, PA: PSU EarthTalks Series; 2009.

[26] Kaufman P, Penny GS, Paktinat J. Critical evaluations additives used in shale slickwater fracs. 2008 SPE Shale Gas Production Conference; 2008Forth Worth, TX. SPE 119900.

[27] Swistock B. Water quality impacts from natural gas drilling. State College, PA: PSU EarthTalks Series; 2009.

[28] Gaudlip AW, Paugh LO. Marcellus shale water management challenges in Pennsylvania. 2008 SPE Shale Gas Production Conference; 2008Forth Worth, TX. SPE 119898.

[29] Soeder DJ, Kappel WM. Water resources and natural gas production from the Marcellus shale. USGS Fact Sheet. U.S. Geological Survey; 2009. p. 2009–3032.

[30] Course notes: shale gas modeling, reservoir simulation of shale gas & tight oil reservoirs using IMEX, GEM and CMOST. Computer Modelling Group; 2014.

[31] Coalbed methane: principles and practices. Available from: http://www.halliburton.com/public/pe/contents/Books_and_Catalogs/web/CBM/CBM_Book_Intro.pdf [accessed 25.06.15].

[32] Laubach SE, Marrett RA, Olson JE, Scott AR. Characteristics and origins of coal cleat: a review. Int. J. Coal Geol. 1998;35:175–207.

[33] Seidle J. Fundamentals of coalbed methane – reservoir engineering; Tulsa, OK: Pennwell; 2011.

[34] Reeves SR, Kuuskraa JA, Kuuskraa VA. Deep gas poses opportunities, challenges to U.S. operators.

Conventional and unconventional petroleum reserves – definitions and world outlook

Introduction

Reservoir engineers play an important role in assessing and enhancing petroleum reserves. Generally speaking, petroleum reserves relate to the volumes of oil and gas that can be recovered from reservoirs technically and economically. Technical methodology must be available to produce oil and gas from the subsurface reservoir conditions. Economic considerations include profitable capital investment that factors in the cost of discovery, development, facilities, production, and transport, among others.

This chapter highlights the various topics related to petroleum reserves and attempts to address the following queries:

- What are the petroleum reserves? How are the reserves defined?
- How are various types of oil and gas reserves distinguished from each other?
- How do technological innovations, probability of occurrences, and commercial viability influence the classification of reserves?
- How are unconventional petroleum reserves distinguished from conventional reserves?
- What are the petroleum resources? How are the resources categorized?
- How are reserves estimated and reported for oil and gas reservoirs?
- Do petroleum reserves change with time? If yes, what are the factors responsible for change?
- Are there any inherent uncertainties in petroleum reserves estimation?
- What are the potential sources of errors in estimating reserves?

Petroleum reserves and resources

In essence, four components constitute oil and gas in place; these are (i) reserves, (ii) resources, (iii) the amount already produced, and (iv) unrecoverable quantities of petroleum. In a typical petroleum reservoir, reserves are significantly less than oil and gas in place due to technical, economic, geopolitical, and other constraints. Petroleum reserves are basically tied to the ultimate recovery factor estimated from a reservoir.

Petroleum reserves are defined by various organizations and governments in more ways than one. However, several elements in the definitions and reporting of reserves are common, including, but not limited to, the following:

- Petroleum reserves and resources are hydrocarbon deposits predominantly occurring in subsurface geologic formations.
- Reserves are either discovered or likely to be discovered in the future.

Reservoir Engineering. http://dx.doi.org/10.1016/B978-0-12-800219-3.00023-1

- Reasonable estimates of the fraction of accumulated hydrocarbon volume that can be extracted must be made.
- Technological expertise, either current or in the future, must be in place to exploit the deposits of petroleum.
- Economic feasibility, either current or in the future, must exist to bring the oil and gas reserves from reservoirs to markets.
- Reserves can be reported in the context of a reservoir, field, petroleum basin, or country as a whole.
- A probabilistic range of values of oil and gas reserves are often reported rather than a single value due to the various uncertainties involved in exploration and development.

According to the Petroleum Resources Management System (PRMS) [1] published by SPE/WPC/AAPG/SPEE, the criteria for petroleum reserves include:

- Known accumulations of oil and gas that are already discovered
- Recoverable in commercial quantities
- Remaining quantities based on the field development projects

From a project perspective, PRMS defines oil and gas reserves as (i) what is in production, (ii) what is approved for development, and finally (iii) what is justified for development. It is again emphasized that the petroleum reserves and hydrocarbon in place are not the same. Furthermore, reserves are a subset of total petroleum resources.

Conventional versus unconventional reserves

World unconventional resources exceed conventional resources by a wide margin; however, the former often poses challenges to develop and process. There is more than one way the unconventional reserves are distinguished from conventional reserves, as follows:

- Geology: Conventional reserves are accumulated within a well-defined area delineated by oil−water contact, gas−water contact, impermeable cap rock, or geologic discontinuity. Unconventional reserves are often pervasive over a very large area with no discernible oil−water or gas−water contact in most instances. Furthermore, conventional reserves are discovered in rocks where they have migrated in the past following generation in source rock, whereas unconventional reserves such as shale gas and oil are found in source rock itself where they were generated millions of years ago.
- Technology: Traditional technology in well drilling, reservoir development, and production are utilized in extracting conventional reserves. However, novel techniques and innovative methods are necessary in producing unconventional petroleum reserves economically.
- Oil and gas mobility: Conventional reserves are either relatively less viscous or accumulate in rocks having favorable characteristics that facilitate fluid mobility. In contrast, unconventional reserves are either highly viscous or found in rocks having unfavorable characteristics, which cannot be extracted by traditional methods.
- Cost of production: Unconventional reserves are likely to cost more to produce than conventional reserves due to the limitations of technological knowhow, lack of detailed understanding of the process, and associated risk.

Classification of petroleum reserves

PRMS, which is widely recognized by the oil and gas industry in defining petroleum reserves, classifies the reserves in three broad categories, namely, (i) proved, (ii) probable, and (iii) possible. Out of the three categories, proved reserves have the maximum probability of commercial production, usually 90% or better. Proved reserves, also referred to as proven reserves, are further categorized as developed and undeveloped. Developed reserves are based on the currently producing wells and include oil and gas producing from completed and open intervals; also included are shut-in and behind-the-pipe reserves. The latter is referred to as developed non-producing reserves. The undeveloped reserves, on the other hand, would require future investments to drill more wells, for instance.

Probable and possible reserves, sometimes referred to as unproved reserves, are usually associated with relatively less probability of commercial production, 50% and 10%, respectively. Nevertheless, the reserve estimates are based on geologic and engineering data similar to that associated with the proved reserves category; however, certain factors preclude them from being proved reserves. The two most prevalent reasons are the lack of available technology or unsustainable cost of operation even if the appropriate technology is available. Logistics or government regulations could also be a factor in producing oil and gas from a field (Figure 23.1).

It is interesting to note that in the case of a matured reservoir nearing abandonment, very few uncertainties remain about future recovery. Hence, probable or possible reserves approach zero.

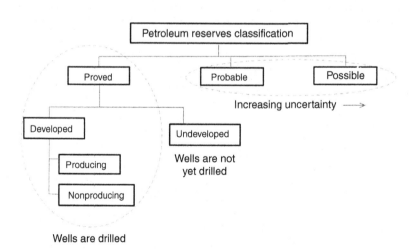

Figure 23.1 Classification of petroleum reserves according to PRMS. The definition of various types of reserves and resources evolved over decades along with technological innovation, industry practices, financial market, and regulatory oversight.

Methods of reporting petroleum reserves

In addition to the probabilistic method of reporting petroleum reserves, PRMS is inclusive of deterministic approaches, which may report reserves in discrete quantities. The reservoir and fluid properties in the deterministic approach represent the "best value" that can be used in analysis. Last but not least, multiscenario methods can also be used, which are a combination of probabilistic and deterministic methods.

Petroleum plays and resources

Petroleum resources are further down the probability tree as far as future production and commerciality is concerned. The quest for petroleum resources begins with play, which can be defined as accumulations of hydrocarbon, either known or postulated, in similar geologic settings with regard to source rock, migration, geologic age, and trapping mechanism, among others. Resources can either be (i) contingent or (ii) prospective. The important distinction between the two subcategories is whether the resource is discovered or not. Contingent resources are discovered, but the development of such resources is either (i) pending, (ii) on hold, or (iii) not viable under the present circumstances. Prospective resources are yet to be discovered. These are based on exploration and related studies in a prospect, lead, or play. Last but not least, certain petroleum resources are deemed unrecoverable due to technological, economic, or various other impediments.

Resources are also categorized on the basis of technical and economic considerations. Technically recoverable resource, referred to as TRR, is the volume of oil and gas that can be extracted by utilizing current technology without any reference to economic feasibility. In contrast, economically recoverable resource, referred to as ERR, refers to the portion of technically recoverable resource that can be extracted economically. With the advancement of technology and favorable economic conditions, the ERR part increases within the TRR.

Many unconventional resources cannot be easily converted to the proved reserves category by utilizing the present day technology. Examples of unconventional oil resources include shale oil, extra heavy oil, tar sands, and bitumen, the commercial extraction of which are difficult and cost intensive.

Reserves estimation methods

The simplest approach to estimate reserves is based on analogy when pertinent data are not available. Performance of a reservoir in a similar geologic setting is taken as a basis for estimating reserves. However, the popular methodologies adopted by the industry to estimate oil and gas reserves of a field are the volumetric method and decline curve analysis where applicable. The material balance technique based on a "tank model" of input and output of reservoir fluids is also used to estimate oil and gas reserves. While the volumetric method solely relies on static data, decline curve analysis and the material balance method are based on production data that are expected to increase the degree of confidence in the estimation process. However, in large and complex reservoirs, robust reservoir models are needed to estimate reserves where the model is matched with production history. Furthermore, it is interesting to note that more than one method of reserves estimation leads to the comparative evaluation of results, which may enhance the degree of confidence in predicting the reserves. Reserve estimation methods are shown in Table 23.1.

Table 23.1 Reserves estimation methods

Method	Type of data required	Applicability	Usability	Degree of confidence
Reservoir analogy	Recovery trend from other developed reservoirs in similar geologic settings	Any reservoir in the absence of "hard data"	Only method available at the earliest stage of the reservoir life cycle when sufficient data are not available	Low to moderate
Volumetric analysis	Static data including reservoir pore volume, rock, and fluid properties; recovery efficiency	Any reservoir, from small to large if reservoir characteristics are known with some degree of confidence	Used in the early stages of reservoir life cycle	Low to high
Decline curve analysis	Reservoir performance history including rate decline	Best suited for small fields where decline trends are clearly identifiable	Used when initial production data are available	Moderate to high when appropriate decline model is used
Material balance	Reservoir performance history	Small to medium sized reservoirs where reservoir complexities are not predominant	Used when production data are available	Moderate to high in relatively simple cases
Reservoir simulation	Detailed reservoir characteristics and production history	Appropriate for complex reservoirs having many producers and injectors	Used when ample reservoir and production data are available	Moderate to high when reservoir model is history matched

Probability distribution of petroleum reserves

There are inherent uncertainties involved in estimating and reporting oil and gas reserves of a newly discovered field as reservoir characteristics are not known with certainty. Reporting of hitherto undiscovered reserves brings in even more uncertainty. Proved reserves are referred to as 1P in the oil and gas industry, which suggests that the probability of commercial production is 90% or higher. Proved and probable reserves combined are referred to as 2P. Proved, probable, and possible reserves added together are referred to as 3P, and represent the total reserves for a field. It is obvious that proved reserves are less than probable reserves; and the latter is less than possible reserves (Figure 23.2). For contingent resources, the equivalent terms are 1C, 2C, and 3C, respectively.

Sources of uncertainty

The following are sources of uncertainty, among others:

- Geologic formations are invariably heterogeneous to a varying degree
- Important rock characteristics are known with certainty at certain well locations only
- Overoptimistic assumption of pore volumes leads to higher estimates of hydrocarbon in place than the actual value
- Overoptimistic assumption of rock permeability may lead to higher recovery than what is likely

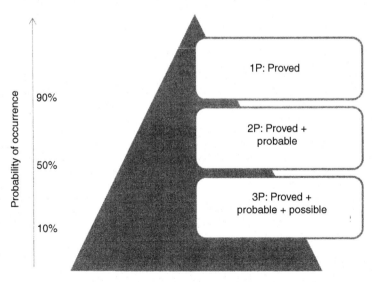

Figure 23.2 Hierarchy of 1P, 2P, and 3P reserves. 1P represents the proved or proven reserves and is included in 2P and 3P estimates. 2P refers to proved and probable reserves, which are added to possible reserves in 3P.

- True reservoir extent cannot be ascertained in many instances due to the limited number of wells drilled
- Ultimate recovery from the reservoir may be less than expected as the recovery mechanisms are not fully understood
- Unexpected decline in reservoir performance due to the presence of unknown hetero-geneities such as faults, fractures, compartments, thief zones, facies change, and water encroachment
- Small changes in oil–water contact or gas–water contact may alter reserves estimates significantly
- Prediction of hitherto undiscovered reserves depends heavily on regional trends, earlier discoveries in the same basin, and the experience of earth scientists in the absence of "hard data"

Monte Carlo simulation

The uncertainties associated with reservoir characteristics, rock and fluid properties, and recovery estimates lead to the determination of oil and gas reserves as a set of values rather than a unique value for a petroleum reservoir. Estimates of oil and gas volumes in a reservoir or field are often accomplished by Monte Carlo simulation, which is based on the probability distribution of rock properties, fluid saturation, and recovery efficiency relevant to the analysis. Common distribution patterns for the above-mentioned parameters include normal, log-normal, and triangular distributions. Consider the volumetric method of estimating oil reserves presented earlier in Chapter 12:

$$OOIP = [7758 \, A \, h\phi (1 - S_{wi})] / B_{oi} \qquad (12.1)$$

Instead of using single average values for A, h, ϕ, S_{wi}, and B_{oi} in the above equation, probability distribution of the parameters can be used leading to more realistic treatment of reservoir, rock, and fluid characteristics. As a result, distribution of values of original oil in place (OOIP) with a high, mean, and low range of probability is estimated. Monte Carlo simulation accomplishes this by conducting thousands of iterations of Equation (12.1). Each iteration computes a value for OOIP based on the values of porosity, thickness, water saturation, etc., which are generated randomly based on individual distribution patterns. Figure 23.3 shows the normal distribution of porosity and triangular distribution of net to gross thickness ratio of the formation as examples. The distribution of porosity is obtained from analyzing a large number of core samples from various wells. Typical results from Monte Carlo simulation are plotted in Figure 23.4, where each value of OOIP reflects the probable outcome. Of note is that some of the parameters used in Monte Carlo simulation such as porosity and permeability are correlated. For example, an increase in porosity in any iteration should reflect a corresponding increase in permeability during simulation.

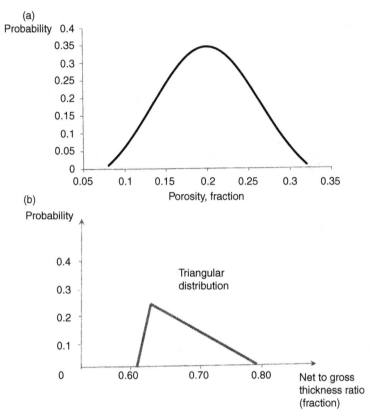

Figure 23.3 (a) Normal distribution of porosity and (b) triangular distribution of net to gross thickness ratio of the formation.

Sources of inaccuracy in reserves estimates

There are large number possible sources of inaccuracies that can be introduced in estimating and reporting of oil and gas reserves. Some of the sources are listed in the following:

- Analysis mostly focused on best case scenario, including highest reservoir quality, hydrocarbon pore volume, and recovery efficiency
- Overoptimistic estimate of reserves based on analogy with a similar field that performed extremely well
- Volumetric calculations with questionable assumptions
- Inaccuracies in structural, isopach, net-to-gross, and isoHCPV maps
- Lack of knowledge of reservoir heterogeneities and how they affect reservoir performance
- Lack of understanding of primary, secondary, and tertiary recovery mechanisms

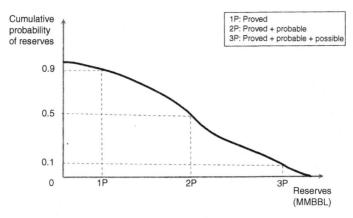

Figure 23.4 Proved, probable, and possible reserves of a typical oil field. 1P, 2P, and 3P reserves are also shown.

- Inaccuracies in assuming oil – water contact or gas – water contact
- Expectation of unrealistic recovery efficiency without a sound technical basis
- Unrealistic assumptions in economic analysis and future technologies

Update of field reserves

As time progresses, it is inevitable that technological innovations occur and economic conditions change, and so does our technical knowledge about a reservoir or petroleum basin based on the development of new wells and reservoir analyses. What is a probable reserve today may very well become a proved reserve in the near future. Consequently, oil and gas reserves of a field or a nation as a whole are usually updated on a regular basis. A case in point is the estimation of unconventional reserves of shale oil and gas in North America and all over the world. Although the hydrocarbon in place remained same since ancient times, estimates of reserves grew by leaps and bounds since the beginning of the present century with the advancement of horizontal drilling and multistage fracturing technology. Current technologies are capable of producing natural gas from extremely tight shale reservoirs where permeability is in nanodarcies. Equally noteworthy, the current market price of natural gas supports the utilization of the shale gas extraction technology.

Studies have shown that reserves of a field can be revised upward in a significant manner during the production phase as more wells are drilled, detailed reservoir data become available, and new technologies are implemented.

In certain other fields, presence of an unknown heterogeneity or unexpected phenomenon may lead to downward revision of reserves. For example, a compartmental reservoir can only be identified after time lapse seismic surveys are conducted during waterflood operation. In another scenario, a number of producers may be beset with severe water cut issues following a short period of water injection, which may lead to unfavorable reservoir performance and early abandonment.

World outlook

Figure 23.5 highlights the growth of petroleum reserves worldwide as reported in *Oil and Gas Journal* [2].

Oil and reserves grow with time as new petroleum horizons are explored in various parts of the world, innovative technologies are adopted in the industry, smart tools are deployed in the field, and reserves of the older fields are updated.

Summing up

Petroleum reserves are defined as the known accumulations of oil and gas that are producible in commercial quantities. Total reserves for a field refer to the volume that has been produced, that is in production, and that is approved and justified for future development. Reserves are firmly anchored to the technical expertise to produce from the reservoirs and economic feasibility to bring it to the market. Based on the probability of occurrence in geologic formations, petroleum reserves are categorized as shown in Table 23.2.

Petroleum resources are broadly categorized as contingent and prospective. The most important distinction between the two is that the contingent resources are discovered while the prospective resources are not. The criteria of categorization are shown in Table 23.3.

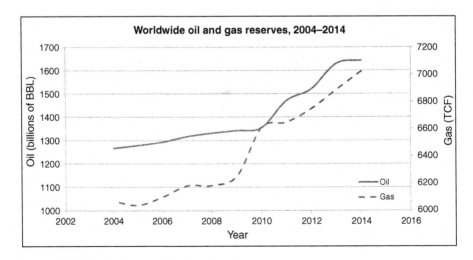

Figure 23.5 Worldwide growth of petroleum reserves.

Table 23.2 Petroleum reserve based on the probability of occurrence in geologic formations

Petroleum reserves type	Reserve highlights	Probability of occurrence of petroleum	Notes
Proven or proved	What is currently producing through the wells; included are shut-in and behind-the-pipe reserves, and reserves to be obtained from future wells	90% or more	Referred to as 1P
Probable	Known accumulations but technical and economic challenges remain to a moderate degree	50% or more	Proved + probable is referred to as 2P
Possible	Known accumulations but technical and economic challenges remain to a high degree	10% or more	Proved + probable + possible is referred to as 3P

Table 23.3 Criteria for categorization of petroleum resources

Resources type	Subcategory	Status	Commerciality
Contingent	1C, 2C, and 3C (analogous to 1P, 2P, and 3P, respectively)	Discovered	Subcommercial. Development is pending, on hold, or not possible for some reason
Prospective	Low, best, and high estimates	Undiscovered	n/a

The volumetric method and decline curve analysis are the two popular methods for estimating oil and gas reserves in the industry. The volumetric method is based on static data only, while decline curve analysis is appropriate where the rate decline trend of individual wells or the field as a whole is identifiable. In the case of large and complex reservoirs having many producers and injectors, reservoir simulation is used to predict the ultimate recovery and reserves.

There are a large number possible sources of uncertainties and inaccuracies that can be introduced in estimating and reporting oil and gas reserves. These include, but are not limited to, inaccuracies in structural, isopach, isoHCPV, and other maps, incorrect oil−water or gas−water contact, lack of knowledge regarding rock heterogeneities, lack of understanding of recovery mechanisms, limited number of wells, incorrect analogy, and overoptimistic assumptions.

Field reserves may be revised upward as new wells are drilled in uncharted portions of a reservoir and new technologies are adopted by the industry to produce additional quantities of oil and gas that were not possible before. Reserves may also be revised downward when unknown heterogeneities such as faults and compartments are discovered, and unexpected events take place, such as premature water breakthrough during waterflood.

Questions and assignments

1. Define petroleum reserves. What criteria must it fulfill to be reported as reserves?
2. How are proved reserves distinguished from probable and possible reserves?
3. How is the probability of the occurrences of petroleum determined?
4. What are 1P, 2P, and 3P reserves? Explain with an example published in the literature.
5. Do all types of reserves have commercial viability? Is the past production volume included in reporting the reserves for an oil field?
6. Distinguish between petroleum reserves, resources, and plays. Have all the resources been discovered?
7. What are the common methods to estimate reserves? Explain the relative merits of each method, including data collection and analysis efforts, inherent strengths and weaknesses, and degree of confidence that can be attached with the analysis.
8. When a new reservoir is discovered, how are its reserves estimated? What degree of confidence can be placed on the estimates?
9. Describe the scenarios where the volumetric method and decline curve analysis are not adequate to estimate reserves.
10. Why are petroleum reserves often reported in the context of probability of occurrences? What tool or technique is frequently used to assign probability values to reserves? Explain.
11. How might estimates of petroleum reserves change over time? Provide a few examples.
12. Develop a workflow to estimate the reserves of a newly discovered oil reservoir. Highlight any distinction between the approaches in estimating reserves of conventional and unconventional reservoirs. Include the potential sources of uncertainties in your analysis.

References

[1] Guidelines for Application of the Petroleum Resources Management System, http://www.spe.org/industry/docs/PRMS_Guidelines_Nov2011.pdf; November 2011.
[2] OGJ Worldwide Production Reports, http://www.ogj.com/articles/print/volume-111/issue-12/special-report-worldwide-report/worldwide-reserves-oil-production-post-modest-rise.html.

Reservoir management economics, risks, and uncertainties

24

Introduction

All petroleum ventures, from basin exploration to reservoir development to matured field revitalization, require capital investment, with an objective of generating revenue. Investments are usually substantial, requiring careful and detailed economic study. For example, the overall cost of developing a large offshore complex (exploration, production, surface handling facilities, transportation, and others) may run into billions of dollars. The goal of reservoir management is to maximize the economic profitability of a project. Once the investment is made, it cannot be recovered if wrong assumptions are made in economic analysis. Making sound business decisions requires that the project will be economically viable, generating profits that meet or exceed the economic goal of the enterprise. In recent decades, a reservoir team is viewed by the management as more of an "asset team," and is expected to add value to the asset (petroleum reserves). The view is that all technical initiatives of a reservoir team must be integrated with the overall asset management goals.

This chapter provides a review of commonly used economic criteria, and a working knowledge of analyzing project economics. Answers are provided for the following:

- What are the objectives of economic analysis in an oil and gas venture?
- What economic decision criteria are considered?
- What is an integrated economic model?
- What are the risks and uncertainties in the petroleum industry?
- How are the economic analyses performed?
- What are the data requirements for performing economic analysis?

The chapter concludes with an economic analysis of future petroleum reservoirs in Alaska in relation to oil price.

Objectives of economic analysis

In the petroleum industry, the major business ventures include, but are not limited to, the following:

- Exploration of petroleum basins
- Oil and gas field development
- Enhancement of reservoir performance by infill drilling and/or enhanced oil recovery (EOR) operations

Economic optimization of petroleum ventures, including competitive oil and gas production costs, is the ultimate goal of best practices in reservoir management. It

Reservoir Engineering. http://dx.doi.org/10.1016/B978-0-12-800219-3.00024-3

involves building multiple what-if scenarios or alternative approaches in order to arrive at the optimum solution. One approach to an optimized solution is producing maximum recovery of oil and gas at least cost within the framework of imposed regulations and constraints. Potential issues in oilfield development include, but are not limited to, the following:

- Exploration strategy: The optimum number of exploratory wells to be drilled in a new area. Decisions to make when an exploration well turns out to be dry or marginally productive.
- Recovery scheme: Natural production methods augmented by waterflooding and EOR methods. Design and timing of implementing improved oil recovery methods.
- Well spacing and design: The number of wells and offshore platforms. Drill single- or multilateral horizontal wells?
- Consider drilling high-density infill wells or initiate an EOR project or both?
- Return on investment/rate of return under best-case and worst-case scenarios.
- Correlation of capital investment with field size as new reserves are discovered.
- Impact of rules, regulations, logistics, and taxes.

The resulting economic analyses and comparative evaluation of what-if scenarios can provide the answers that are sought to make the best business decisions. This may lead to the maximum value added to the petroleum asset given the available technology, size of the reservoir, expected reservoir performance, and market conditions.

Integrated economic model

An integrated approach to develop and manage oil and gas fields requires the evaluation of all relevant technological and economic aspects [1]. These include, but are not limited to, the following:

- Optimum scheduling of wells and fields to be developed, based on reservoir simulation, economic analysis, and contractual obligations
- Optimum scheduling of construction of necessary infrastructure: offshore platforms, surface facilities, and pipelines
- Technical, operational, financial, and other constraints imposed on field development. Examples include the following:
  - Uncertainties in estimating oil and gas reserves
  - Uncertainties in future reservoir performance due to unknown reservoir heterogeneities
  - Number of wells that can be drilled from an offshore platform
  - Logistics and weather. Many oil fields are located in virtually inaccessible regions
  - Project delays due to unavoidable reasons
  - Contractual obligations related to timely delivery of oil and gas
  - Penalties imposed by the government
  - Geopolitical instability and regional conflicts

Sound economic analysis depends on robust reservoir simulation models, among others. The models should be based upon geological, seismic, petrophysical, and other studies that are able to adequately describe structural uncertainties and rock heterogeneities that affect reservoir performance. Real-time data as obtained by downhole gauges and sensors should also be collected and integrated in the model. These include, but are not limited to, flow rates of oil, gas and water, pressure, and temperature. The schematic of an integrated economic model is shown in Figure 24.1.

Model optimization

In optimizing an integrated model, the decisions that are sought can be of multiple types. Decisions related to the construction of offshore platforms, facilities, and pipelines are either true or false. The number of wells to be drilled is based on integers. Yet many other decisions involve continuous numbers such as optimum well rates. Furthermore, the above factors are interrelated in a complex and nonlinear fashion requiring integrated model analysis. The analysis must be performed though the entire life cycle of the reservoir, including the scheduling and drilling of wells, construction of oil and gas handling facilities, design of waterflood and EOR projects, and finally abandonment. This effort leads to the maximization of asset performance under existing contractual obligations or market conditions. Optimization of an integrated economic model requires a multidisciplinary approach, evaluation of a large number of what-if scenarios, and modeling of uncertainties involved in various elements. The latter is discussed in the following sections.

Risk and uncertainty in the petroleum industry

The activities related to exploration and production of petroleum are inherently associated with myriad risks and uncertainties [2]. Some of the constraints, including uncertainties, in the economics of petroleum reservoir development were mentioned earlier. Consider the following scenarios in predicting the success or failure in oil and gas property investments:

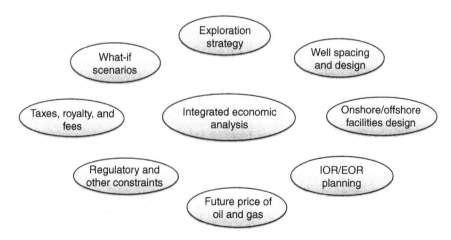

Figure 24.1 Integrated economic model.

- The most significant risks in the petroleum industry are associated with exploration activities. Exploratory wells drilled in a new basin or region may not turn out to be productive.
- An oil or gas field may not generate revenue as expected following initial production. The issue could be rooted in poor reservoir quality and geologic complexities, among other factors.
- Future oil and gas prices could move unpredictably in a rapidly changing world of supply and demand.
- Unforeseen events such as political unrest, regional conflict, or natural calamity may adversely affect the demand, production, and transport of petroleum.
- New governmental policies, regulations, and taxes may significantly influence the way a petroleum company conducts business.
- The inflation factor or other economic indicators in the future cannot be known with certainty.
- As oil and gas prices increase due to ever-increasing world demand, alternate sources of energy may become economically attractive.
- Environmental considerations may play a role in weighing other options of energy in a specific industry or region.

Obviously, economic analysis of a petroleum venture requires the recognition and quantification of risk and uncertainties in wide-ranging areas. In conclusion, the feasibility of a petroleum field may be critically affected by myriad factors. Most of them cannot be controlled by reservoir professionals. Thus, all of the influencing factors are not within the scope of this book.

Workflow for performing integrated economic analysis

As in most facets of petroleum reservoir development and management, performing a sound economic analysis requires integrated team effort. The workflow is outlined in the following:

- Set economic objective: The initial step focuses on setting up economic criteria and standards over a time horizon aligned with a company's goals. The economic criteria may include payout period and rate of return. The common economic yardsticks used in the industry are described later.
- Formulate project development scenarios: The next step includes a general guideline and plan to effectively develop, manage, and produce a reservoir to generate income after all incurred costs.
- Collect data for analysis: The data necessary to perform an economic analysis are collected from various sources as shown in Table 24.1. These include: estimated reserves, oil and gas production rates, projected price of oil and gas, rate of inflation, taxes, royalties, and production sharing, among others.
- Perform economic calculations: These include cash flow (CF) analysis, payout period, rate of return, and other yardsticks of measurements. The analysis can be either probabilistic or deterministic, depending on the requirements. Various software applications are available in the industry to build various scenarios.
- Perform sensitivity analysis: In this step, the effects of various factors that impact the economics of the project are analyzed. For example, the effects of a delay in pipeline construction and the price of oil on the development of the reservoir can be examined. A case study concerning the development of future reservoirs is included in this chapter. Sensitivity

Table 24.1 **Data requirements summary**

Data	Source
Estimated reserves; oil and gas rates versus time	Reservoir simulation and field data
Projected oil and gas prices	Finance and economic professionals
Capital investments including tangible, intangible and operating costs	Finance, engineering, and facilities professionals
Royalty/production sharing	Government and contract professionals
Discount and inflation rates	Finance and economic professionals
State and local taxes (production, severance, ad valorem, etc.)	Finance and economic professionals; strategic planning interpretation
Federal income taxes, depletion, and amortization schedules	Accounting professionals

analysis often includes the number of wells to be drilled, design of wells (single lateral versus multilateral), and timing of waterflood or EOR efforts, among others. Working with management, the integrated reservoir team, including engineers, geologists, and operations staff, decides on the optimum project.

In essence, sound estimates of hydrocarbon in place, reservoir performance forecasts, capital investment, and operating expenses are essential ingredients in any economic analysis. The workflow for performing integrated economic analysis is depicted in Figure 24.2.

Criteria for economic decisions

Various economic standards and metrics are used to evaluate a business venture including investments in oil and gas. Companies have standards based on the following metrics, and others, to accept or reject a project. A brief description of the major yardsticks used in economic analysis is provided in the following:

- Discounted CF return on investment (DCFROI)
- Present worth net profit
- Payout period
- Profit to investment ratio

Cash flow and discounted cash flow

Any economic analysis related to a petroleum reservoir begins with projected CF over the life of the reservoir. CF is defined as follows:

$$CF = \text{Revenue from oil and gas} - \text{Capital investment to drill wells and build facilities} - \text{Operating expenses to manage the reservoir} \qquad (24.1)$$

In Equation (24.1), all values are in dollars or any other currency.

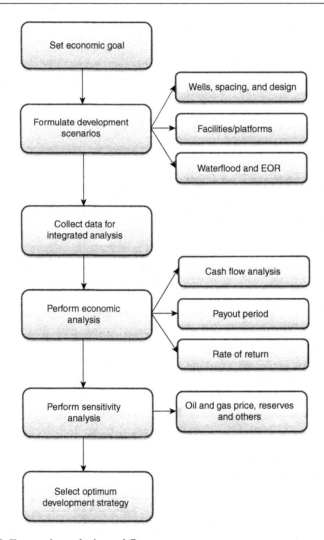

Figure 24.2 Economic analysis workflow.

In the beginning of the project, CF is negative, as more money is spent in exploration and drilling, and building facilities, platforms, and pipelines. As the wells start producing in the next phase, revenue is generated from oil and gas production, and CF turns positive (Figure 24.3). Production may reach a plateau for a period of time, followed by a decline.

However, the amount of CF to be generated in the future is less in actual value than the same amount of money today. The reasons include the following:

- Time value of money
- Inflation
- Uncertainty in petroleum venture

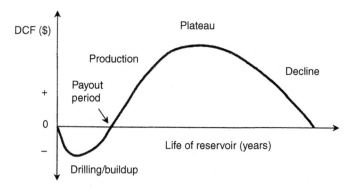

Figure 24.3 Discounted CF over the life of a reservoir. In the early stages, CF is negative due to drilling of wells and development of infrastructure. As wells begin to produce and revenue is generated, CF turns positive. It may reach a plateau during the mid-phase of the life of the reservoir. Finally, well production declines, accompanied by a decline in CF.

Money has a time value. We can deposit $10,000 in a bank that offers 5% interest rate per year, and then withdraw the deposited amount (capital) plus interest after a year. If the rate is compounded annually, the interest would be $500. Mathematically, the future amount including the interest can be calculated as:

$$FV = PV(1+i)^n \qquad\qquad (24.2)$$

where FV = future value, $; PV = present value, $; i = annual interest rate, fraction; n = number of years.

$$FV = 10,000(1.05)^1$$
$$= \$10,500$$

Similarly, the CF to be received in future years can be "discounted" to the present day CF as follows:

$$DCF = CF(1+i)^{-n} \qquad\qquad (24.3)$$

where DCF = discounted CF, $; i = discount factor, fraction.

Equation (24.2) suggests that the amount received far out in the future is discounted further. Consider the following scenario where CF from a business venture is $20,000 for 5 years. Then the total discounted CF over the years can be calculated as:

First year: $20,000(1 + 0.05)^{-1} = \$19,047.63$
Second year: $20,000(1 + 0.05)^{-2} = \$18,140.59$
Third year: $20,000(1 + 0.05)^{-3} = \$17,276.75$

Fourth year: $20,000(1 + 0.05)^{-4} = \$16,454.05$
Fifth year: $20,000(1 + 0.05)^{-5} = \$15,670.52$

Total: $86,589.53.

When the revenues are received mid-year, Equation (24.3) is modified to calculate discounted CF as follows:

$$DCF = CF(1+i)^{-(n-0.5)} \qquad (24.4)$$

In the petroleum industry, discounted CF from oil and gas sales is calculated as follows:

1. The annual revenues are calculated using oil and gas volume and unit sales prices.
2. Costs are calculated accounting for capital investment, drilling, completion, facilities, costs, operating expenses, and production taxes.
3. Annual undiscounted CF is calculated by subtracting total costs from the total revenues as shown in Equation (24.1).
4. Finally the annual discounted CF is calculated by using Equation (24.4).

The above procedure reflects calculation of the amount before federal income tax.

Apart from discounting by interest rate, the amount can be further adjusted down for inflation and risks associated with a petroleum venture. In that case, the total discounted CF would be less than what is computed above.

Present worth net profit

The sum of all the discounted CF over the entire life of the reservoir is the present worth net profit (PWNP) from the business venture. Consider the previous example where the business venture generated a revenue of $20,000 each year for 5 years. If the initial investment for the business was $60,000, then the present net worth is:

$$PWNP = -\$60,000 + \$86,589.53 = \$26,589.53$$

When the value of present worth net profit is 0 or negative, the project becomes unattractive, as we will spend more than the expected revenue.

Payout period

The time needed to recover the invested capital in a venture is referred to as the payout period. The shorter the payout period, the more attractive is the project. Before the payout period, CF is negative. Following the payout period, CF is positive (Figure 24.4). However, the payout period is not the sole criterion for selecting a project as it does not indicate the total CF to be expected over the life of the reservoir.

Discounted cash flow return on investment

DCFROI is the maximum discount rate that leads to the present net worth being zero. Since a more profitable venture generates relatively high present net worth value, a

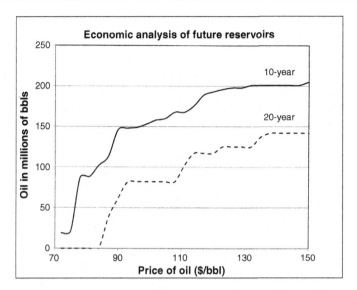

Figure 24.4 Two scenarios evaluate the development of petroleum reservoirs depending on the projected price of oil. The scenarios include 10- and 20-year time horizons to develop pipelines and LNG facilities.

higher discount rate must be applied to bring the present net worth to zero. Hence, the higher the DCFROI, the better it would appear as an option to choose. The DCFROI is referred to as internal rate of return (IRR).

Mathematically, DCFROI can be expressed as follows:

$$0 = -C + CF_1(1+i)^{0.5} + CF_2(1+i)^{1.5} + CF_3(1+i)^{2.5} \ldots\ldots CF_n(1+i)^{(n-0.5)} \quad (24.5)$$

where C = initial investment, \$; CF_n = CF in nth year, \$; i = internal rate of return, fraction.

Consider again the previous example where C = \$60,000 with an annual revenue of \$20,000. Assume that the revenues are expected in the middle of the year. The IRR is then calculated by using Equation (24.5):

$$0 = -60,000 + 20,000(1+i)^{0.5} + 20,000(1+i)^{1.5} + 20,000(1+i)^{2.5}$$
$$+ 20,000(1+i)^{3.5} + 20,000(1+i)^{4.5}$$
IRR = 0.211 or 21.1%, obtained by using trial and error method

Profit to investment ratio

Profit to investment ratio is the total undiscounted CF over the total investment. However, the undiscounted CF does not include the capital investment.

Case Study: Economic Analysis of Conventional Oil and Gas Reserves in Alaska

The United States Geological Survey conducted a detailed economic study of the future oil and gas reservoirs in Alaska over an area over 24 million acres of federal, state, and native land as well as offshore areas [3]. In the first part of the study, the probability distributions of undiscovered and technically recoverable resources were determined based on regional trends, geologic features including depth and thickness of formations, and earlier discoveries of oil and gas reservoirs. Note that the technically recoverable resources can be developed by using available technology; however, the estimates are not based on economic feasibility for development.

The probabilistic distributions of oil and gas were reported in three categories as follows:

Technically recoverable reserves	Probability value		
	95%*	Mean	5%**
Oil, MMBO	336	895	1,707
Associated gas, BCF	348	840	1,327
Nonassociated gas, BCF	43,042	52,821	61,985

* 19 out of 20 chances that the reserves will be higher than the tabulated value.
** 1 out of 20 chances that the reserves will be higher than the tabulated value.

The economic analysis focused on the economically recoverable resources that are the "part of the assessed technically recoverable resources for which the costs of finding, developing, producing, and transporting to market, including a return on capital, can be recovered by production revenues at a particular price." Based on available software, the study took into consideration the following economic factors:

- Exploration costs
- Development costs of oil and gas reservoirs
- Drilling and completion costs of injectors and producers
- Oil and gas production costs
- Transportation costs, including pipeline construction to markets
- Infrastructure development costs
- Price of oil and gas, which may vary over a wide range

The harsh climate of Alaska, access to remote locations, and absence of infrastructure add to the cost and uncertainties in the analysis. In the absence of pipelines, delays are anticipated to market gas. Understandably, the study includes scenarios where pipelines and LNG facilities will be built in the future to conduct the analysis. Two scenarios were evaluated. These included 10- and 20-year time frame to build the above (Figure 24.4). Typical well drainage area was assumed to be 160 acres. Drilling of horizontal wells was also incorporated in the study to augment well productivity leading to better economic scenarios.

Summing up

All petroleum ventures, from basin exploration to matured field revitalization, require capital investment, with an objective of generating profits. Investments are usually substantial. Investment for offshore oil field development may run into billions of dollars. Hence, detailed economic analysis is of paramount importance in oil and gas projects.

Economic optimization, including competitive production costs, is the ultimate goal of sound reservoir management. It involves building multiple scenarios or alternative approaches in order to arrive at the optimum solution. Issues that require detailed economic analysis in reservoir development and management include, but are not limited to, the following:

- Exploration strategy
- Recovery scheme
- Well spacing
- Drill high-density wells or initiate an EOR project
- Return on investment/rate of return under best-case and worst-case scenarios
- Correlation of capital investment with field size as new reserves are discovered

An integrated approach to develop and manage oil and gas fields requires the evaluation of all relevant technological and economic aspects. These include, but are not limited to, the following:

- Optimum scheduling of wells and fields to be developed, based on economic analysis and contractual obligations
- Optimum scheduling of construction of necessary infrastructure, including surface facilities and pipelines
- Technical, operational, and financial constraints imposed on field development
- Design decisions
- Operational decisions
- Nonlinearity in the physical system

The activities related to exploration and production of petroleum are inherently associated with myriad risks and uncertainties. Some of the major risks and uncertainties include:

- Exploratory well or wells turn out to be dry
- Wells drilled in formations where reservoir quality is poor; production is marginal
- Future prices of petroleum move up or down unexpectedly
- High inflation rate
- Regional conflicts
- New laws and regulations adversely affecting reservoir economics
- Innovations in the areas of alternate forms of energy
- Unexpected environmental factors

The tasks in integrated economic analysis require team efforts as follows:

- Setting economic objectives that are aligned with a company's short- and long-term goals
- Collection of production, operation, and economic data. The reservoir asset team, including reservoir engineers, is responsible for economic justification based on all available information, experience, and sound judgment

- Performing economic analysis based on integrated reservoir models
- Formulation of various scenarios related to drilling of wells and building of facilities required to develop the reservoir
- Conducting sensitivity analyses and selecting an optimum project. Optimization is based on available resources and ultimate recovery of oil and gas

Data required for economic analysis include the following:

- Oil and gas production volumes over the life of the reservoir
- Oil and gas price predictions in the future
- Capital investment (tangible and intangible) and operating costs
- Royalty/production sharing
- Discount and inflation rates
- Federal, state, and local taxes

Making a sound business decision requires yardsticks or metrics for measuring the economic value of proposed investments and financial opportunities. Each company has its own economic strategy for conducting business profitably. The major metrics to evaluate a business venture, including petroleum ventures, are as follows:

- DCFROI: The DCFROI, also referred to as IRR, indicates a rate at which the sum of all discounted CF, positive and negative, is zero. Capital investment and operating costs are negative CF. Revenue from sales is positive CF. Any CF received in the future is "discounted" or adjusted down to reflect time value of money, inflation, and uncertainty. Basically DCFROI reflects the overall rate of return the investment will produce over the years. A higher rate of return usually makes the investment attractive.
- Payout period: The time needed to recover the invested capital in a business venture is referred to as the payout period. The shorter the payout period, the more attractive is the project.
- PWNP: The sum of all the discounted CF over the entire life of the reservoir is the PWNP from the business venture.
- Profit to investment ratio: Profit to investment ratio is the total undiscounted CF over the total investment. However, the undiscounted CF does not include the capital investment.

Questions and assignments

1. Why is an integrated economic analysis important in petroleum ventures?
2. What data are required in integrated economic analysis?
3. What are the common criteria used in economic analysis to accept or reject a proposal?
4. Why is CF discounted? How does the discounted CF affect PWNP?
5. Define DCFROI.
6. Describe the uncertainties involved in petroleum economics.
7. What are the critical factors in the economic analysis of a remote oil field?
8. Based on the literature, highlight the differences between economic analyses of onshore and offshore reservoirs.
9. Your company is planning to drill several horizontal wells to produce from a low permeability reservoir. The spacing and design of horizontal wells (single lateral vs. multilateral) are not yet finalized. Develop a workflow for an integrated economic analysis.

10. Develop a spreadsheet program that calculates the payout period and DCFROI given the following data:
    - Capital investment
    - Annual oil and gas sales
    - Unit price of oil and gas
    - Operating expenses
    - Life of reservoir

Factor in the escalation of the price of oil and gas over the years, production decline following peak production, and operating expenses that may change with time.

References

[1] Satter A, Iqbal GM, Buchwalter JA. Practical enhanced reservoir engineering: assisted with simulation software. Tulsa, OK: Pennwell; 2008.
[2] Iqbal G. Course notes, Petrobangla Workshop on Reservoir Economics, Dhaka; 2002.
[3] Attanasi ED, Freeman PA. Economic analysis of the 2010 U.S. Geological Survey Assessment of undiscovered oil and gas in the National Petroleum Reserve in Alaska, US Department of Interior and US Geological Survey, Open File Report 2011–1103.

Subject index

Printed in the United States
By Bookmasters